*Numerical Methods
in Finance*

Numerical Methods in Finance

A MATLAB-Based Introduction

Paolo Brandimarte

A Wiley-Interscience Publication

JOHN WILEY & SONS, INC.

Copyright © 2002 by John Wiley & Sons, Inc., New York. All rights reserved.

Published simultaneously in Canada.

For ordering and customer service, call 1-800-CALL-WILEY.

Library of Congress Cataloging-in-Publication Data:

Brandimarte, Paolo.
    Numerical methods in finance : a MATLAB-based introduction / Paolo Brandimarte.
      p. cm. — (Wiley series in probability and statistics. Financial engineering section)
    "A Wiley-Interscience publication".
    Includes bibliographical references and index.
    ISBN 0-471-39686-9 (cloth : alk. paper)
    1. Finance—Statistical methods.   I. Title.   II. Series.
  HG176.5 .B73 2001
  332'.01'51—dc21   2001026767

Printed in the United States of America

10 9 8 7 6

This book is dedicated to Commander Straker, Lieutenant Ellis, and all SHADO operatives. Thirty years ago they introduced me to the art of using both computers and gut feelings in making decisions.

Contents

Preface

Crossroads are hardly, if ever, points of arrival; but neither are they points of departure. In some sense, crossroads may be disappointing, indeed. You are tired of driving, you are not at home yet, and by Murphy's law there is a far-from-negligible probability of taking the wrong turn. In this book, different paths cross, involving finance, numerical analysis, optimization theory, probability theory, Monte Carlo simulation, and partial differential equations. It is not a point of departure, because although the prerequisites are fairly low, some level of mathematical maturity on the part of the reader is assumed. It is not a point of arrival, as many relevant issues have been omitted, such as hedging exotic options and interest rate derivatives.

The book stems from lectures I give in a Master's course on numerical methods for finance, aimed at graduate students in economics, and in an optimization course aimed at students in industrial engineering. Hence, this is not a research monograph; it is a textbook for students. On the one hand, students in economics usually have little background in numerical methods and lack the ability to translate algorithmic concepts into a working program; on the other hand, students in engineering do not see the potential application of quantitative methods in finance clearly.

Although there is an increasing literature on high-level mathematics applied to financial engineering, and a few books illustrating how cookbook recipes may be applied to a wide variety of problems through use of a spreadsheet, I believe there is some need for an intermediate-level book, both interesting to practitioners and suitable for self-study. I believe that students should:

- Acquire *reasonably* strong foundations in order to appreciate the issues behind the application of numerical methods

- Be able to translate and check ideas quickly in a computational environment

- Gain confidence in their ability to apply methods, even by carrying out the apparently pointless task of using relatively sophisticated tools to pricing a vanilla European option

- Be encouraged to pursue further study by tackling more advanced subjects, from both practical and theoretical perspectives

The material covered in the book has been selected with these aims in mind. Of course, personal tastes are admittedly reflected, and this has something to do with my operations research background. I am afraid the book will not please statisticians, as no econometric model is developed; however, there is a wide and excellent literature on those topics, and I tried to come up with a complementary textbook.

The book is structured basically on three groups of chapters. Some of them are relatively independent, so that the instructor may choose the material to cover based on the time available and her preferences.

- Chapters 1 and 2 deal with finance and numerical analysis, respectively. The first chapter is aimed primarily at students of engineering; the second, at students of economics. The purpose of this first group of chapters is to lay down a homogeneous background for the heterogeneous groups of students at which the book is targeted.

- Chapters 3, 4, and 5 deal with the three main topics of the book: optimization theory, Monte Carlo simulation (including low-discrepancy sequences), and finite difference methods for partial differential equations.

- Finally, chapters 6, 7, and 8 give a few illustrative examples of the application of each of the three methodologies covered in the book.

The text is interspersed with MATLAB snapshots and pieces of code, to make the material as lively as possible and of immediate use. MATLAB is a flexible high-level computing environment which allows us to implement nontrivial algorithms with a few lines of code. It has also been chosen because of its increasing potential for specific financial applications.

It may be argued that the book is more successful at raising questions than at giving answers. This is a necessary evil given the space available to cover such a wide array of topics. Yet if after reading this book, students will want to read others, my job will have been accomplished. This was meant to be a crossroads, after all.

PS1. Despite all of my effort, the book is likely to contain some errors and typos. I will maintain a list of errata, which will be updated based on reader feedback. Any comment or suggestion on the book will also be appreciated. Please contact me at the following e-mail address: `brandimarte@polito.it`.

PS2. The list of errata will be posted on a Web page which will also include additional material and MATLAB programs. The current URL is

- `http://www.polito.it/~brandimarte`

An up-to-date link will be maintained on Wiley Web page:

- `http://www.wiley.com/mathematics`

PS3. And if (what a shame ...) you are wondering who Commander Straker is, take a look at the following Web sites:

- `http://www.ufoseries.com`

- `http://www.isoshado.org`

P. Brandimarte

Part I

Background

1

Financial problems and numerical methods

The purpose of this chapter is to outline some basic financial problems that can be tackled by numerical methods, to provide the reader with some background and motivation for the rest of the book. Some of these problems are rather easy to solve, at least in their basic form, by standard numerical methods; indeed, there are different software packages, including (but not limited to) MATLAB,[1] that offer built-in functionalities to cope with the basic versions of portfolio optimization problems and option pricing. However, when the limitations of such basic problem formulations are recognized, the need for the more sophisticated methods described in the rest of the book is apparent. Besides motivating the use of numerical methods such as mathematical programming (including mixed-integer and stochastic programming), Monte Carlo and quasi-Monte Carlo simulation methods, and finite differences to solve partial differential equations, we have to get familiar with MATLAB. The basic use of the MATLAB interactive environment and its syntax as a high-level programming language are described in appendix A, to which the reader is referred. Some background on probability and statistics is given in appendix B. This chapter is also meant as a reference to those unfamiliar with financial problems. In fact, there are (at least) two different potential users of this book:

- Students in finance wishing an introduction to numerical methods

---

[1]MATLAB is a registered trademark of The MathWorks, Inc.

- Students in a quantitative discipline (such as mathematics, engineering, or computer science) who would like to see how their skills may come handy when dealing with finance

It is a bit difficult to write a book suitable to both these groups without boring one and/or losing the other one in the first chapter. Here we have tried to come up with a compromise by structuring the chapter around the core functionalities of the Financial toolbox of MATLAB, i.e., the analysis of simple fixed-income securities (section 1.2), mean-variance portfolio theory (section 1.3), and the pricing of vanilla options (section 1.4). We also introduce the value-at-risk (VaR) concept (section 1.5).

The treatment will be rather brief, and it is certainly not meant as a substitute for a good book on finance (see the references at the end of the chapter), but it will be complemented by short MATLAB snapshots to make it immediately useful. A last caveat is that we deal here with concepts such as portfolio immunization, mean-variance efficiency, and VaR, which do have many well-known limitations and have been the subject of quite a bit of controversy. The point here is not to suggest that they should be used as they are stated, but just to pave the way for further developments.

1.1 MATLAB ENVIRONMENT

MATLAB is an interactive computing environment, providing both basic and sophisticated functions to cope with numerical problems. You may use built-in functions to solve possibly complex but standard problems, or devise your own programs by writing them as M-files, i.e., as text files including sequences of instructions written in a high-level matrix-oriented language. MATLAB also has a rich set of graphical capabilities, which we will use in a very limited fashion. Refer to appendix A for a quick tour of MATLAB programming.

Some classical numerical problems are readily solved by MATLAB built-in functions. They include:

- Solving systems of linear equations

- Solving nonlinear equations in a single unknown variable (including polynomial equations as a special case)

- Finding minima and maxima of functions of a single variable

- Approximating and interpolating functions

- Computing integrals

- Solving ordinary differential equations

Some background of the underlying methods is given in chapter 2. Although there is no need for a deep knowledge of numerical analysis in order to use

MATLAB, it is useful to have at least a basic and intuitive grasp of it in order to choose the most appropriate algorithm when alternatives are given, and to understand what is happening when the results do not seem reasonable.

More complex versions of these problems may be solved by other MATLAB ready-to-use functions, but you have to get the appropriate toolbox. A toolbox is simply a set of functions written in the MATLAB language, and it is usually provided in source form, so that the user may customize or use the code as a starting point for further work. For instance, the Optimization toolbox is needed to solve complex optimization problems, involving several decision variables and possibly complex constrains, as well as to solve systems of nonlinear equations. Some background on optimization methods is given in chapter 3. Another relevant toolbox for finance is the Statistics toolbox, which includes many more functions than we will use. In particular, it offers functions to generate pseudorandom numbers that are needed to carry out Monte Carlo simulations, which are treated in chapter 4. Another numerical problem to which a significant part of the book is devoted is the solution of partial differential equations (PDEs; see chapter 5). We implement methods to solve PDEs from scratch, but it is worth noting that a toolbox for PDEs is also available.

Based on the Statistics and Optimization toolboxes, a Financial toolbox has been devised, which includes different groups of functionalities; some are low-level functions such as date and calendar manipulation or finance-oriented charting, which are building blocks for real-life applications; others deal with simple fixed-income assets, portfolio optimization, and derivatives pricing. It should be noted that these functions are just aimed at basic problems, but they are a good learning tool in view of further developments.

Other financial toolboxes have been written to analyze financial time series and to deal with GARCH models. We will not deal with such econometrics-oriented toolboxes here, but some information can be obtained by browsing The MathWorks' Web site (http://www.mathworks.com). Another recent addition to the financial set of toolboxes is the Financial Derivatives toolbox, which is oriented mainly to interest rate derivatives.

1.1.1 Why MATLAB?

Why bother using MATLAB when user-friendly spreadsheets are available? Sure, when you have to carry out simple computations, there's little point in resorting to a full-fledged computing environment. Indeed, there are spreadsheet-based books showing how optimization and simulation methods may be applied to financial problems. Spreadsheets are equipped with solvers able to cope with small-scale mathematical programming problems, and extensions are available to run Monte Carlo simulations or optimization by genetic algorithms. The point is that the extra effort in learning a programming language pays off when you have to program a complex numerical method which goes beyond what is standard and readily available. Indeed, there is no way to really learn numerical methods without some knowledge of a programming language. With MATLAB (and similar products) you have a much more powerful and versatile tool than a spreadsheet, without the need to go down to the level of a general-purpose language such as C/C++ or Java. So MATLAB can be thought of as a suitable compromise between conflicting requirements.

Needless to say, this does not imply that MATLAB has no definite limitations. When one has to deal with large-scale optimization problems, it is necessary to resort to specialized packages such as CPLEX[2] or OSL,[3] with which MATLAB is unlikely to be competitive (it should be noted that the Optimization toolbox is aimed at general nonlinear programming, whereas many optimization packages deal only with linear and quadratic programming). Furthermore, mixed-integer programming problems simply cannot be solved, at present, by MATLAB. Even worse, when you have a large optimization model, loading the data in a form suitable to a numerical routine is a difficult and error-prone task without the support of algebraic modeling languages such as AMPL[4] or GAMS[5] (whose use is illustrated in section 6.2). This remark applies to any optimization library; however, there are many commercially available (and reliable) links between well-established solvers and algebraic modeling languages, whereas only experimental links are available to MATLAB.

However, we should also note that MATLAB toolboxes are improving continuously, and that writing a program to call a numerical procedure requires programming skills which are certainly not acquired by using a spreadsheet; furthermore, some programming is needed to extend spreadsheet capabilities. So we may argue that a tool like MATLAB is the best *single* tool to lay down good foundations in numerical methods. Cheap MATLAB student editions are available, and its use in finance is spreading. So we believe that learning MATLAB is definitely an asset for students in finance.

1.2   FIXED-INCOME SECURITIES: ANALYSIS AND PORTFOLIO IMMUNIZATION

Fixed-income securities are one of the instruments that firms and public administrations use to fund their activity; they are debt instruments which, unlike stocks, do not imply any ownership of a firm on the part of the buyer. The prototype fixed-income security is the simple fixed-coupon bond. Such a bond is characterized by a *face value*, also called the *par value*, and a maturity date; the bond is purchased at a price that is not necessarily the face value;

---

[2]CPLEX is a registered trademark of ILOG.
[3]The Optimization Solutions and Library (OSL) is a product of IBM Corporation.
[4]AMPL was originally developed at Bell Laboratories. At present it is available in many versions through different sellers.
[5]GAMS is a product of GAMS Development, Washington, DC.

bond prices are quoted as a percentage of the face value, so the actual face value is not relevant for certain types of analysis. The bond entitles the owner to a stream of periodic payments (the coupons, which may be expressed as a percentage of the par value) and to the final refund of the face value at maturity. For instance, given a face value of $100 and a maturity of five years, a coupon rate of 8% implies that we will receive four yearly payments of $8 and a final payment of $108 (the face value is refunded together with the last coupon). Most often, coupons are paid twice a year; in the previous case, this means that we would receive a payment of $4 every six months and a final payment of $104. Another class of fixed-income securities, called *zero-coupon securities*, just promise the payment of the face value at maturity. We see that when we buy a bond, we actually buy a cash flow. The term *fixed income* stems from the fact that in simple bonds the cash flow is fixed and known from the beginning (at least if we assume that the issuer will not default). Since there is a well-developed secondary market for bonds, there is no need to buy a bond right when it is issued, nor to keep it to maturity.

There are two basic problems one has to deal with: *valuing* bonds, i.e., determining a fair price for them, and *managing a portfolio* of bonds, shaped according to one's particular needs. It would seem that fixed-income instruments are easy to deal with, since everything is certain, but the reality is quite different. To begin with, *fixed income* is a somewhat misleading term, since in fact there are many bonds whose coupon rates depend on some other financial quantity. If the coupon rate is not certain, analyzing a bond may be difficult. Even if the coupon rate is fixed, bond prices may differ depending on the probability of default; bonds affected by credit risk must sell at lower prices, or promise higher coupon rates. Furthermore, some bonds have embedded options which complicate the analysis. For instance, a callable bond may be redeemed by the issuer before maturity at a certain price; again, since the issuer may redeem the bond when she finds this advantageous, this must be somehow reflected in the bond price and/or the coupon rate. In this chapter we deal only with simple bonds.

1.2.1   Basic valuation of fixed-income securities

Consider a zero-coupon bond, with a face value F, a maturity of one year, and a price P on the issue date. We purchase the security at P, and we get F on the maturity date; hence, we may define a return

$$R = \frac{F}{P}$$

and a rate of return

$$r = R - 1 = \frac{F}{P} - 1.$$

Then, an obvious relationship between r, F, and P is

$$P = \frac{F}{1+r}. \tag{1.1}$$

We may see this relationship the other way around. If we fix F and r, this may be interpreted as a pricing relationship, which is actually nothing more than a discounting formula. It is well known that if we have a cash flow, i.e., a stream of periodic payments C_t at discrete-time instants $t = 0, 1, \ldots, n$, the present value of the cash flow may be computed as

$$PV = \sum_{t=0}^{n} \frac{C_t}{(1+r)^t},$$

where r is some interest rate for a single time period (note that the cash flows need not be positive; typically, in investment analysis we have $C_0 < 0$, corresponding to an initial cash outlay). When the nominal interest rate is quoted yearly but the payments occur more frequently, the formula may easily be adapted. For instance, if there are m payments per year at regular time intervals, we have

$$PV = \sum_{k=0}^{n} \frac{C_t}{(1+r/m)^k}, \tag{1.2}$$

where k indexes the periods and n is in this case the number of years times the number of periods within one year. Another type of adjustment is needed when a bond is purchased after the issue date. For instance, suppose that the one-year zero-coupon bond of equation (1.1) is purchased three months after the issue date. Since the bond matures in nine months, a plausible pricing formula could be

$$P = \frac{F}{(1+r)^{9/12}},$$

assuming that all months consist of the same number of days.

What rate r should we use in pricing? If the bond is default-free, as in the case of government bonds, this should be the prevailing risk-free interest rate: no more, no less. To see why, we may use a common principle in finance, i.e., the *no-arbitrage* principle. Assume that the bond is underpriced, i.e., it sells for a price P_1 such that

$$P_1 < P = \frac{F}{1+r},$$

and that we may take out a loan at the risk-free interest rate r (we are assuming that the borrowing and lending rates are equal). Then we can borrow an amount L and use it to purchase L/P_1 bonds. Note that the net cash flow at the beginning is zero. Then, at maturity, we must pay $L(1+r)$ to our money lender, and we get an amount FL/P_1 when the face value is refunded for each bond. But since, by hypothesis,

$$\frac{F}{P_1} > 1+r,$$

the net cash flow at maturity will be

$$L\frac{F}{P_1} - L(1+r) = L\left(\frac{F}{P_1} - 1 - r\right) > 0.$$

Hence, we pay nothing at the beginning and receive a positive amount in the future; since the bargain is an interesting one, we might well exploit it, in the limit, to ensure an unbounded profit for increasing L. This is what is called *arbitrage*. Of course, limitless borrowing is not available; more important, purchasing a huge amount of those bonds would raise their prices, and the arbitrage opportunity would soon disappear. Indeed, a common assumption in many financial problems is that arbitrage opportunities do not exist. Note that this does not imply that they actually do not exist; on the contrary, it is the very fact that many people are there to exploit those opportunities which tends to eliminate them quickly. The argument may be repeated similarly if the inequality is reversed and the bond is overpriced:

$$P_1 > P = \frac{F}{1+r}.$$

In this case we should borrow the bond itself rather than the cash needed to buy it. Then we may sell the overpriced bond and invest the proceeds at the risk-free rate; let us assume that we borrow bonds for a total value L, we sell them at price P_1, and we invest the money we obtain. The immediate net cash flow is again zero. At maturity, we get $L(1+r)$ from our investment, and we have to pay the face value F to the owner for each bond that we have borrowed. Hence the net cash flow at maturity is again positive:

$$-L\frac{F}{P_1} + L(1+r) = L\left(-\frac{F}{P_1} + 1 + r\right) > 0.$$

The practice of borrowing an asset to sell it immediately is known as *short-selling*. There are many limitations to short-selling in practice, but for pricing models it is often (not always) reasonable to assume that it is possible. We have also assumed implicitly that transaction costs are negligible and that we may lend or borrow money at the same rate. Again, these assumptions are violated in practice, but they may be close enough to reality, at least for some large investors, to warrant their use. The reader may have the impression that the arbitrage argument is, at least in this case, an unnecessary complication to obtain an almost obvious result. However, the no-arbitrage principle is used, with some modification, to price quite complex securities where intuition does not help (as in the case of options; see section 1.4.2).

Now let us consider the case of a bond with an associated stream of coupons. If we assume that the bond is default-free and that the same riskless rate can be applied for any period length (provided that we account for compounding),[6]

---

[6] Actually, this is not the case, as different period lengths are associated with different interest rates; see section 1.2.4.

the fair bond price may be obtained simply by computing the present value of the cash flow

$$\text{PV} = \sum_{i=1}^{n} \frac{C}{(1+r)^i} + \frac{F}{(1+r)^n},$$

where C is the coupon per period and F is the face value. Actually, if r is quoted yearly and there is more than one coupon payment per year, the formula could be adjusted in the same vein as equation (1.2). If this simple formula were generally applicable, any bond with the same coupon rate and maturity date should have the same price, which is actually not the case. A first point is that not all bonds are issued by institutions with the same credit rating. Although a bond issued by some governments may be default-free, a corporate bond may not be of the same quality; hence, all other things being equal, you would require a lower price for it. Indeed, a bond with a price P is characterized by the *yield*, which is the solution λ of the following equation:

$$P = \sum_{i=1}^{n} \frac{C}{(1+\lambda)^i} + \frac{F}{(1+\lambda)^n}.$$

The yield is actually the internal rate of return of the cash flow stream (see example 1.1).[7] If more than one coupon payment is made during a year, the formula is

$$\sum_{i=1}^{n} \frac{C/m}{(1+\lambda/m)^i} + \frac{F}{(1+\lambda/m)^n}.$$

Example 1.1 Given a stream of cash flows C_t $(t = 0, 1, 2, \ldots, n)$, the internal rate of return is defined as a value ρ such that the present value of the stream is zero. In other words, it is a solution of the nonlinear equation

$$\sum_{t=0}^{n} \frac{C_t}{(1+\rho)^t} = 0.$$

With the change of variable $h = 1/(1+\rho)$, we may transform this equation into a polynomial equation:

$$\sum_{t=0}^{n} C_t h^t = 0,$$

which is readily solved by the MATLAB function `roots` (see section 2.4 for some information about basic methods to solve nonlinear and polynomial equations). All we have to do is to represent a cash flow stream as a vector, as done in the following MATLAB interaction snapshot (>> is simply the

---

[7]Actually, there are different concepts of yield (see, e.g., [4] or [5]), but we will stick to this one.

MATLAB prompt you see at the beginning of any session; you will see it for any line we type in):

```
>> cf=[-100 8 8 8 8 108]
cf =
  -100     8     8     8     8    108
>> h=roots(fliplr(cf))
h =
    0.3090 + 0.9511i
    0.3090 - 0.9511i
   -0.8090 + 0.5878i
   -0.8090 - 0.5878i
    0.9259
>> rho=1./h -1
rho =
   -0.6910 - 0.9511i
   -0.6910 + 0.9511i
   -1.8090 - 0.5878i
   -1.8090 + 0.5878i
    0.0800
```

A few comments are in order. First, we define a variable cf and we associate with it a cash flow. Then, in a single command line, we flip the cash flow from left to right with the function fliplr and we invoke the roots function to assign the roots of the resulting polynomial to the variable h. Flipping the cash flow vector is necessary since roots assumes that a polynomial is represented by a vector in which the first components correspond to the highest power terms in the polynomial, whereas when we represent cash flows we put such terms at the end. After obtaining the solution in terms of h, we go back to the original variable ρ (note that the dot in ./ is necessary since h is a vector of solutions). Since in this example $n = 5$, we have a vector of five roots: four are complex conjugates, and the one we are interested in is the real one, i.e., $\rho = 0.08$. Indeed, it can be shown that for a cash flow stream with $C_0 < 0$ and $C_t \geq 0$ $(t = 1, \ldots, n)$ and $\sum_{t=1}^{n} C_t > 0$, we have a unique real and positive solution of the nonlinear equation (see, e.g., [13, chapter 2]). All the work above (including filtering the complex roots out) is done by the irr function available in the Financial toolbox:

```
>> irr(cf)
ans =
    0.0800
```
☐

Example 1.2 The Financial toolbox includes different functions to analyze cash flow streams, including pvvar, which computes the present value of a stream given a discount rate. We could use it to find the price of a five-year bond with face value 100 and a 8% coupon rate, for different required yields:

```
>> cf=[0 8 8 8 8 108]
cf =
     0    8    8    8    8   108
>> pvvar(cf,0.08)
ans =
  100.0000
>> pvvar(cf,0.09)
ans =
   96.1103
```

Note that the cash flow vector has a zero in the first position, because we will receive the first coupon at the end of the first year. We see that when the yield and the coupon rate are equal, the bond is sold *at par*, i.e., its price is equal to the face value. If we raise the yield from 0.08 to 0.09, the bond price is decreased.   ⬜

Intuitively, the higher the required yield, the lower the price. Higher yields must be offered for risky bonds. If the credit rating of the bond issuer changes, the bond price will change accordingly to reflect the new situation. But is credit risk the only source of risk for bonds? Unfortunately, the answer is no. To begin with, coupon rates may depend on some other economic or financial variable, resulting in some uncertainty in the cash flow, so we have a form of financial risk. Another point is that some bonds have embedded options which may be unfavorable for the holder; for instance, the issuer may *call* the bond, that is, redeem it before maturity, which results in reinvestment risk since we would have to reinvest the cash we receive from the bond issuer (bonds with embedded options may be analyzed using techniques we discuss later when we deal with options). But even if all of these risks are ruled out, there may still be a form of risk, depending on the intended use of the security. If a default-free, strictly fixed-income security is purchased and held to maturity, perhaps to fund a fixed stream of liabilities, we may say that this security is risk-free from our point of view. However, if you purchase the bond, hold it for awhile, and then decide to sell it on the secondary market (see section 1.2.4 to see why bond trading is actually needed to fund liabilities), you face interest rate risk. The problem is that the risk-free rate is not constant over time; it may change, depending, e.g., on inflation. If the risk-free rate moves up, a higher yield will be required for new bonds of the same characteristics. For the old ones, an increase in the yield results in a decrease in the price at which they may be sold. On the contrary, if interest rates drop, we may gain something from the decrease in the required yield, which results in an increase in the price. Depending on the maturity and the coupon rate, a bond may be more or less sensitive to yield changes.

Example 1.3 Consider a five-year zero-coupon bond, with face value 100, sold with required yield $r_1 = 0.08$. Which is the percentage change in its price if the yield is increased immediately to $r_2 = 0.09$?

```
>> r1=0.08;
>> r2=0.09;
>> P1=100/(1+r1)^5
P1 =
    68.0583
>> P2=100/(1+r2)^5
P2 =
    64.9931
>> (P2-P1)/P1
ans =
    -0.0450
```

We see that we have a 4.5% decrease in its value. Now what if the maturity is 20 rather than five years?

```
>> P1=100/(1+r1)^20
P1 =
    21.4548
>> P2=100/(1+r2)^20
P2 =
    17.8431
>> (P2-P1)/P1
ans =
    -0.1683
```

We see that the loss is now much larger, almost 17%. Although zero-coupon bonds with long maturities may not be available easily, it is a general rule that the longer the maturity, the more sensitive to yield changes the bond price is. Coupon rates play some role, too. We may compare two bonds with coupon rates of 4% and 8%, respectively.

```
>> cf1=[0 8 8 8 8 8 8 8 8 8 108];
>> cf2=[0 4 4 4 4 4 4 4 4 4 104];
>> P1=pvvar(cf1,0.08)
P1 =
   100.0000
>> P2=pvvar(cf1,0.09)
P2 =
    93.5823
>> (P2-P1)/P1
ans =
    -0.0642
>> P1=pvvar(cf2,0.08)
P1 =
    73.1597
>> P2=pvvar(cf2,0.09)
```

```
P2 =
   67.9117
>> (P2-P1)/P1
ans =
   -0.0717
```

We see that a lower coupon rate implies a larger sensitivity.   ⬜

These examples show that bond characteristics determine the degree of sensitivity of the bond price to changes in the required yield. We need a formal way to measure the interest risk associated with bonds, in order to figure out a way to shape a fixed-income portfolio. A relatively simple answer is represented by the duration and convexity concepts discussed in the next section.

1.2.2   Interest rate sensitivity and bond portfolio immunization

Imagine that you are an investor facing a stream of known liabilities in the future and you want to hold a portfolio of bonds such that you may meet the liabilities. On the one hand, you would like to do it at minimum cost, but you would also like to hold a portfolio that is not likely to get you in trouble in case of changes in the interest rates. As a simple example, imagine that you have one liability L to be paid in five years. If you may find a safe zero-coupon bond maturing in five years, with face value F, you may just buy L/F of these bonds. However, if the bond maturity is less than five years, you will face reinvestment risk; if the bond maturity is more than five years, you will face interest rate risk, as we have seen in example 1.3. Ideally, you would like to find a zero-coupon bond with maturity corresponding exactly to the date of each liability. Unfortunately, it is practically impossible to do so, and we must find another way to protect the bond portfolio against interest rate uncertainty. Immunization is a possible, and simple, solution.

Formally, we have a function $P(\lambda)$ that gives the relationship between the yield and the price of a bond. We may draw this curve (how this may be done in MATLAB is explained in example 1.5), obtaining something like the curve illustrated in figure 1.1. We see that the curve is convex,[8] which is actually the case for usual bonds. Now consider small movements in the required yield; we would like to find out a way to approximate the change in price with respect to a change in yield. Indeed, there are two concepts, duration and convexity, which can be used to this aim.

---

[8] Formally, a function f is convex if $f(\lambda x + (1 - \lambda)y) \le \lambda f(x) + (1 - \lambda)f(y)$ holds for $0 \le \lambda \le 1$; more on this in supplement S3.1.

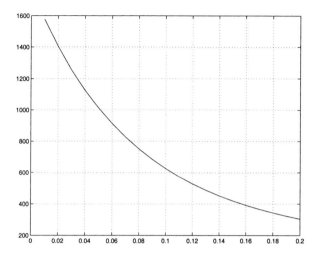

Fig. 1.1   Price-yield curve.

Given a stream of cash flows occurring at times $t_0, t_1, \ldots, t_n$, the *duration* of the stream is defined as

$$D = \frac{PV(t_0)t_0 + PV(t_1)t_1 + PV(t_2)t_2 + \cdots + PV(t_n)t_n}{PV},$$

where PV is the present value of the stream and $PV(t_i)$ is the present value of the cash flow c_i occurring at time t_i, $i = 0, 1, \ldots, n$. In some sense, the duration looks like a weighted average of cash flow times, where the weights are the present values of the cash flows. Note that for a zero-coupon bond, which has a single cash flow, the duration is simply the time to maturity. When we consider a generic bond and use the yield as the discount rate in computing the present values, we get the Macaulay duration:

$$D = \frac{\displaystyle\sum_{k=1}^{n} \frac{k}{m} \frac{c_k}{(1 + \lambda/m)^k}}{\displaystyle\sum_{k=1}^{n} \frac{c_k}{(1 + \lambda/m)^k}},$$

where it is assumed that there are m coupon payments per year. In order to see why duration is useful, let us compute the derivative of the price with respect to yield:

$$
\begin{aligned}
\frac{dP}{d\lambda} &= \frac{d}{d\lambda}\left(\sum_{k=1}^{n} \frac{c_k}{(1 + \lambda/m)^k}\right) \\
&= \sum_{k=1}^{n} c_k \frac{d}{d\lambda}\left[\frac{1}{(1 + \lambda/m)^k}\right] = -\sum_{k=1}^{n} \frac{k}{m} \frac{c_k}{(1 + \lambda/m)^{k+1}}.
\end{aligned}
\qquad (1.3)
$$

If we define the modified duration $D_M \equiv D/(1 + \lambda/m)$, we get

$$\frac{dP}{d\lambda} = -D_M P.$$

Thus, we see that the modified duration is related to the slope of the price-yield curve at a given point; technically speaking, it is the price elasticity of the bond with respect to changes in the yield. This suggests the opportunity of using a first-order approximation:

$$\delta P \approx -D_M P \, \delta\lambda.$$

An even better approximation may be obtained by using a second-order approximation. This may be done by defining the *convexity*:

$$C = \frac{1}{P} \frac{d^2 P}{d\lambda^2}.$$

It turns out that, for a bond with m coupons per year,

$$C = \frac{1}{P(1 + \lambda/m)^2} \sum_{k=1}^{n} \frac{k(k+1)}{m^2} \frac{c_k}{(1 + \lambda/m)^k}.$$

Note that the unit of measure of convexity is time squared. Convexity is actually a desirable property of a bond, since a large convexity implies a slower decrease in value when the required yield increases, and a faster increase in value if the required yield decreases. Using both convexity and duration, we have the second-order approximation

$$\delta P \approx -D_M P \, \delta\lambda + \frac{PC}{2}(\delta\lambda)^2.$$

Example 1.4 We may check the quality of the price change approximation based on duration and convexity with a simple example. Let us consider a stream of four cash flows $(10, 10, 10, 10)$ occurring at times $t = 1, 2, 3, 4$. We may compute the present values of this stream under different yield values using the MATLAB function **pvvar**:

```
>> cf = [10 10 10 10]
cf =
    10    10    10    10
>> p1=pvvar([0, cf], 0.05)
p1 =
   35.4595
>> p2=pvvar([0, cf], 0.055)
p2 =
   35.0515
>> p2-p1
```

```
ans =
   -0.4080
```

Note that we have to add a 0 in front of the cash flow vector cf since pvvar assumes that the first cash flow occurs at time 0. We see that increasing the yield by 0.005 results in a price drop of 0.4080. Now we may compute the modified duration and the convexity using the functions cfdur and cfconv. The function cfdur returns both Macauley and modified duration; for our purposes, we must pick up the second output value.

```
>> [d1 dm] = cfdur(cf,0.05)
d1 =
   2.4391
dm =
   2.3229
>> cv = cfconv(cf,0.05)
cv =
   8.7397
>> -dm*p1*0.005
ans =
   -0.4118
>> -dm*p1*0.005+0.5*cv*p1*(0.005)^2
ans =
   -0.4080
```

We see that at least for a small change in the yield, the first-order approximation is satisfactory and the second-order approximation is practically exact.

□

We have defined duration and convexity for a single bond; what about a bond portfolio? If the yield is the same for all the bonds, it can be shown that the duration of the portfolio is simply a weighted average of all the durations (the weight is given simply by the weight of each bond within the portfolio). This is not exactly true if yields are not the same; however, the weighted average of the durations may be used as an approximation. How can we take advantage of this? In the case of asset liability management, one possible approach is to match the duration (and possibly the convexity) of the portfolio of bonds and the portfolio of liabilities. This process is called *immunization*. To carry out the necessary calculations, we may use the functions available in the Financial toolbox.

1.2.3 MATLAB functions to deal with fixed-income securities

When turning our attention from simple cash flows streams to real-life bonds, various complications arise. The first one is that in order to represent the settlement date and the maturity date of a bond correctly, we must be able to cope with a calendar, taking leap years into account. MATLAB has an internal way of dealing with dates, which is based on converting a date to an integer number. For instance, if we type today, MATLAB replies with a number corresponding to the current date; this number may be converted to a more meaningful string by using datestr:

```
>> today
ans =
     730701
>> datestr(today)
ans =
03-Aug-2000
```

You may wish to check which date corresponds to day 1. The inverse of datestr is datenum:

```
>> datenum('03-aug-2000')
ans =
     730701
```

There is a wide variety of string formats that you may use to input a date in MATLAB; the one you see above is only one of them (note that it is necessary to enclose the string between quotes). Dates must be taken into account for different reasons. Consider buying a bond after it is issued; if you buy a bond at a date between two coupon payments, the time elapsed from the last coupon payment date must be taken into account. If not, you would receive a coupon benefit to which the previous owner is partially entitled. Actually, by computing the present value of the cash flow stream you would take it into account; however, the market convention is to quote a bond price without considering this issue. What you read is the *clean price*, to which the accrued interest must be added in order to obtain the correct price. Accrued interest may be computed by prorating the coupon payment over the period between two payments. Roughly speaking, if coupons are paid every six months and you buy a bond two months before the next coupon payment, you owe something like two-thirds of the coupon to the previous owner. However, there are different day count conventions to make the necessary calculations. These issues are considered in the bndprice function, which is used to price a bond, for a given yield value. To understand the input arguments required, we may use the online help (we have included only the first few lines appearing on the screen):

```
>> help bndprice

BNDPRICE Price a fixed income security from yield to maturity.
   Given NBONDS with SIA date parameters and semi-annual yields to
   maturity, return the clean prices and the accrued interest due.
```

```
[Price, AccruedInt] = bndprice(Yield, CouponRate, Settle, Maturity)

[Price, AccruedInt] = bndprice(Yield, CouponRate, Settle, Maturity, ...
     Period, Basis, EndMonthRule, IssueDate, FirstCouponDate, ...
     LastCouponDate, StartDate, Face)
```

We see that as usual in MATLAB, this function may be called with a minimal set of input arguments, which are the required yield, the coupon rate, the settlement date (i.e., when the bond is purchased), and the maturity date. The two output values are the clean price and the accrued interest, which must be summed in order to get the real (dirty) price.

```
>> [clPr accrInt] = bndprice(0.08, 0.1, '10-aug-2007', '31-dec-2020')
clPr =
   116.2366
accrInt =
     1.1141
>> clPr+accrInt
ans =
   117.3507
```

When calling the function this way, all the other arguments take a default value. For instance, the `Period` parameter, which is the number of coupon payments per year, is assumed to be two, and the face value (`Face`) is assumed to be 100. Another possibly important parameter is `Basis`, which controls the day count convention in computing the accrued interest; the default value is 0, which corresponds to the actual/actual convention; if the parameter is set to 1, the convention is 30/360 (i.e., it is assumed that all months consist of 30 days). To appreciate the difference between the day count conventions, we may compute the number of days between two dates by the 30/360 convention and the actual number of days:

```
>> days360('27-feb-1999', '4-apr-1999')
ans =
    37
>> daysact('27-feb-1999', '4-apr-1999')
ans =
    36
```

Other day count conventions are possible and used for different securities (see, e.g., [5]). The remaining parameters are related to the coupon structure and are described in the Financial toolbox manual.

Example 1.5 To obtain the price-yield curve of figure 1.1, we may use the following code fragment:

```
settle    = '19-Mar-2000';
maturity  = '15-Jun-2015';
```

```
face       = 1000;
couponRate = 0.05;
yields = 0.01:0.01:0.20;
[cleanPrices , accrInts] = bndprice(yields, couponRate, settle, ...
   maturity, 2, 0, [] , [] , [] , [], [] , face);
plot(yields, cleanPrices+accrInts);
grid on
```

Note that when we have to provide a function with an optional argument, such as the face value, but we do not want to use optional arguments which should occur before that one, we have to pass empty vectors represented by [] so that the arguments are properly matched.   ▯

For now, we have computed a price given a required yield. We may also go the other way around; we may compute the yield given the price, using another predefined function:

```
>> settledates=[datenum('10-jan-2001'), datenum('15-mar-2001'), ...
datenum('20-jun-2001')]
settledates =
      730861
      730925
      731022
>> bndyield(100, 0.08, settledates, '31-dec-2025')
ans =
    0.0800
    0.0800
    0.0800
```

The minimal set of parameters for the bndyield function are: the clean price, 100 in this case, with no accrued interest; the coupon rate; the settlement date; and the maturity date. In this case we have used a common feature of MAT-LAB functions. If a scalar input argument is passed as a vector, the output is, in most cases, the vector of the results obtained by applying the function to each component of the input vector. Here we have used different settlement dates, and we see that a bond selling at par has a yield corresponding to the coupon rate; since we provide the clean price, the settlement date does not influence the yield. Optional parameters may be passed to bndyield which are similar to the parameters of bndprice.

Other useful functions may be used to compute duration and convexity given the price or the yield of a bond. They are best illustrated by a simple immunization example.

Example 1.6 A common problem in bond portfolio management is to shape a portfolio with a given (modified) duration D and convexity C. Suppose that we have a set of three bonds; we would like to find a set of portfolio weights w_1, w_2, w_3 for each bond, such that

```
% SET BOND FEATURES
settle     = '28-Aug-2000';
maturities  = ['15-Jun-2005' ; '31-Oct-2010' ; '01-Mar-2020'];
couponRates = [0.07 ;  0.06 ; 0.08];
yields = [0.06 ; 0.07 ; 0.075];

% COMPUTE DURATIONS AND CONVEXITIES
durations = bnddury(yields, couponRates, settle, maturities);
convexities = bndconvy(yields, couponRates, settle, maturities);

% COMPUTE PORTFOLIO WEIGHTS
A = [durations'
     convexities'
     1 1 1];
b = [ 10
      160
        1];
weights = A\b
```

Fig. 1.2 Simple code for bond portfolio immunization.

$$\sum_{i=1}^{3} D_i w_i = D$$

$$\sum_{i=1}^{3} C_i w_i = C$$

$$\sum_{i=1}^{3} w_i = 1,$$

where C_i and D_i are the bond durations and convexities, respectively ($i = 1, 2, 3$). Note that we have assumed that both the duration and the convexity of the portfolio can be computed as weighted combinations of the bond characteristics; actually, this is not true in general, but for the moment we will consider this as a simple approximation. All we have to do is to compute the coefficients C_i and D_i and to solve a system of three equations and three unknowns. This is easily accomplished by the code in figure 1.2.

Note that we have assumed a given yield, and we have used the functions bnddury and bndconvy to compute durations and convexities. It is possible to carry out a similar computation starting from the clean bond prices; we have just to use functions bnddurp and bndconvp. By running this piece of code we obtain the following solution:

```
weights =
    0.1209
   -0.4169
    1.2960
```

Note that we have to sell bond 2 short, which may not be feasible.   ▯

In the immunization example above, we have solved a linear system to find a bond portfolio with specific immunization properties. The problem was set up in such a way that there was a unique solution. However, when many bonds are available, more than one solution can be found. In such a case, it might make sense to look for the "best" solution among the feasible ones. Defining *best* is not so easy, but we may try using a simple approach.[9] One possible idea is maximizing the average yield of the portfolio, given that the portfolio must have duration D and convexity C; we add the requirement that short sales are not allowed. This results in the following linear programming (LP) model:

$$\max \quad \sum_{i=1}^{N} Y_i w_i$$

$$\text{s.t.} \quad \sum_{i=1}^{N} d_i w_i = D$$

$$\sum_{i=1}^{N} c_i w_i = C$$

$$\sum_{i=1}^{N} w_i = 1$$

$$w_i \geq 0 \quad \forall i.$$

Note that without the nonnegativity constraints on the weights w_i, we may easily end up with an unbounded solution (see [16] for conditions ensuring the finiteness of the solution and for generalizations of the model). It is easy to write a MATLAB function solving this problem.

Example 1.7 We consider first the selection of an "optimal" portfolio based on five bonds. The code is illustrated in figure 1.3. Since all the functions dealing with bonds are able to cope with vector arguments, provided that they are of compatible size, we group the bond characteristics in vectors. Here we assume to know the clean price for each bond, and we use **bndyield**

---

[9]It may well be argued that it is a *simplistic*, rather than a *simple* approach, as the choice of the objective function is debatable. Nevertheless, let us accept the model for now and explore the type of solution we get.

```
% BOND CHARACTERISTICS FOR SET 1
settle     = '19-Mar-2000';
maturity1  = ...
    ['15-Jun-2015' ; '02-Oct-2010' ; '01-Mar-2025' ; ...
     '01-Mar-2020' ; '01-Mar-2005'];
Face1      = [500  ;  1000  ;   250 ; 100 ; 100];
couponRate1 = [0.07 ;  0.066 ; 0.08 ; 0.06 ; 0.05];
cleanPrice1 = [ 549.42 ; 970.49 ; 264.00 ; 112.53 ; 87.93 ];

% COMPUTE YIELDS AND SENSITIVITIES
yields1 = bndyield(cleanPrice1, couponRate1, settle, maturity1, ...
           2, 0, [] , [] , [] , [], [] , Face1);
durations1 = bnddury(yields1, couponRate1, settle, maturity1, ...
           2, 0, [] , [] , [] , [], [] , Face1);
convexities1 = bndconvy(yields1, couponRate1, settle, maturity1, ...
           2, 0, [] , [] , [] , [], [] , Face1);

% SET UP AND SOLVE LP PROBLEM
A1 = [durations1'
      convexities1'
      ones(1,5)];
b = [ 10.3181 ; 157.6346 ; 1];
weights1 = LINPROG(-yields1,[],[],A1,b,zeros(1,5))
```

Fig. 1.3 Code to set up and solve a linear programming model for bond portfolio optimization.

to compute the corresponding yield and **bnddury** and **bndconvy** to obtain the sensitivities. The function `linprog` is used to solve the corresponding linear program. We describe this function in section 3.4.5. For now, it is enough to note the following. The first argument is the vector of cost coefficients, which we must change in sign because we want to maximize the objective function rather than minimizing it; the next four arguments contain the coefficient matrix and the right-hand side of inequality and equality constraints (since we have only equality constraints in this model, the first two arguments are empty); finally, we have a vector of zeros representing the lower bound on the decision variables. Running this piece of code, we get the following output:

```
Optimization terminated successfully.
weights1 =
      0.4954
      0.0000
      0.4358
      0.0688
```

```
% BOND CHARACTERISTICS FOR SET 2
maturity2  = [maturity1 ; ...
    '15-Jan-2013' ; '10-Sep-2004' ; '01-Aug-2017' ; ...
    '01-Mar-2010' ; '01-May-2007'];
Face2       = [Face1 ; 100   ;  500   ;  200 ; 1000 ; 100];
couponRate2 = [couponRate1 ; 0.08 ;  0.07 ; 0.075 ; 0.07 ; 0.06];
cleanPrice2 = [ cleanPrice1 ; ...
        108.36 ; 519.36 ; 232.07 ; 1155.26 ; 89.29 ];

% COMPUTE YIELDS AND SENSITIVITIES
yields2 = bndyield(cleanPrice2, couponRate2, settle, maturity2, ...
        2, 0, [] , [] , [] , [], [] , Face2);
durations2 = bnddury(yields2, couponRate2, settle, maturity2, ...
        2, 0, [] , [] , [] , [], [] , Face2);
convexities2 = bndconvy(yields2, couponRate2, settle, maturity2, ...
        2, 0, [] , [] , [] , [], [] , Face2);

% SET UP AND SOLVE LP PROBLEM
A2 = [durations2'
     convexities2'
     ones(1,10)];
weights2 = LINPROG(-yields2,[],[],A2,b,zeros(1,10))
```

Fig. 1.4 Expanding the set of available bonds in the linear programming model for bond portfolio optimization.

```
    0.0000
```

We may also enlarge the universe of bonds we are considering by adding five more bonds, as illustrated in figure 1.4. Solving this new model, we get the following output:

```
Optimization terminated successfully.
weights2 =
    0.0000
    0.0000
    0.3810
    0.0000
    0.0000
    0.0793
    0.0000
    0.5397
    0.0000
    0.0000
```

You may notice that in both cases only three bonds are included in the portfolio. This might appear a bit odd, since one would assume that considering more bonds leaves more space for diversification. Actually, this does not happen by chance. We will see in chapter 3 that if we have only M equality constraints in a linear program, there is an optimal solution (provided that the problem is bounded and feasible) with at most M decision variables which take a nonzero value at optimality. Since here $M = 3$, the optimal portfolio will always include just three bonds, even if many more are available. If we considered only duration constraints, we would include just two bonds, whose durations would bracket the target duration. $\square$

From the results of this example we start seeing that maybe our LP model is too simple. In the next section we discuss a few problems with this naive immunization approach. Now, one might wonder why we don't just purchase bonds in such a way as to meet all the liabilities. In fact, this is possible, at least in principle. Consider a set of N bonds, each with a price P_i ($i = 1, \ldots, N$). Assume further that the planning horizon is discretized in T periods and that the liability in period $t = 1, \ldots, T$ is L_t. If the cash flow from a unit of security i at time t is represented by F_{it}, we may consider the following cash flow matching model:

$$\min \quad \sum_{i=1}^{N} P_i x_i$$

$$\text{s.t.} \quad \sum_{i=1}^{N} F_{it} x_i \geq L_t \qquad \forall t$$

$$x_i \geq 0.$$

Here the decision variables x_i represent the amount of bond i purchased rather than the weight in the portfolio. If we neglect the possibility of default and assume that the liabilities are fixed, the resulting portfolio would certainly meet the obligations; unfortunately, it is likely to be quite expensive. Unless bond maturities are matched to the liabilities, we will have to meet the obligations with coupon payments, requiring a possibly large number of bonds. Note also that we must meet the liabilities with an inequality constraint, since it is unlikely that a perfect match of cash flows and liabilities may be obtained with a given set of bonds. In the case of a long planning horizon, the lack of suitable long-term bonds may compound these difficulties. In practice we would also sell bonds along the way, and we would also use short-term lending and borrowing. What we need is a dynamic model, taking all bond trading activities fully into account. But this once again opens the door to interest rate risk. If we also consider that liabilities may be uncertain, with respect both to their amount and the timing, we understand that naive linear programming models are not able to solve real-life problems.

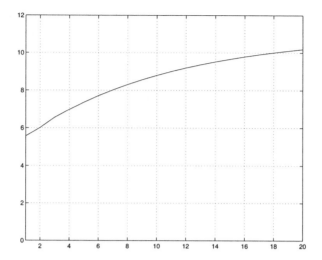

Fig. 1.5 Term structure of the interest rate; years are reported on the horizontal axis, and the corresponding (percentage) spot interest rates are plotted.

1.2.4   Critique

The naive immunization and cash flow matching models we have discussed leave room for many criticisms. The first one is that we have assumed that bonds are priced using only one required yield. In practice, different discount rates should be used for the cash flows occurring in different periods:

$$P = \sum_{t=1}^{T} \frac{C_t}{(1 + \lambda_t)^t} + \frac{F_T}{(1 + \lambda_T)^T}.$$

The set of discount factors λ_t is related to the *term structure of interest rates.* The idea is depicted in figure 1.5, where we see an upward-sloping structure; this corresponds to the intuitive notion that longer interest rates are usually associated with longer terms. Actually, other shapes are possible in general. We see that the simple duration immunization approach is actually based on a flat term structure and on parallel shifts of it. In practice, shape changes are possible, calling for more sophisticated sensitivity measures. A full account of the theory of the term structure is beyond the scope of this book, and we refer the reader to the references at the end of the chapter. We just note here that the Financial toolbox is equipped with functions to analyze term structures and to price bonds in this context.

The last observation in example 1.7 suggests that there could be something wrong with the linear programming immunization model. To begin with, computing the portfolio yield as a weighted average of the individual bond yields is quite debatable, since the maturities of the various bonds included in the portfolio will differ in general (see, e.g., [5, pp. 85-87]). Another point is that by maximizing the yield we are likely to include quite risky bonds; indeed, this naive model does not consider credit risk issues, only interest rate risk.[10] And even the interest rate risk is dealt with in a limited way, since we may expect some immunization for small changes in the required yield. But after such a change, the duration and convexity are changed and the portfolio is no longer immunized. In fact, we are not paying due attention to the dynamic character of portfolio management. In the limit, consider a portfolio consisting of two bonds, one with a short and the other with a long duration, bracketing the target duration. It may be the case that the first bond has a short maturity; when maturity is reached, we are left with only one bond and a portfolio that is far from immunized. More generally, since linear programming yields "extreme" solutions, changes in the immunization requirements will call for drastic portfolio rebalancing, resulting in high transaction costs.

An alternative is to use dynamic optimization models, accounting for uncertainty in the interest rates and for dynamic trading. This leads to stochastic programming models. Basically, stochastic programming models are large-scale linear programming models which must be tackled by advanced numerical methods like those described in chapter 3. Some models in this vein are discussed in chapter 6. It must also be mentioned that many financial instruments have been devised to manage interest rate risk. Such interest rate derivatives may be analyzed by the same numerical methods as those employed for analyzing stock derivatives, which will be introduced in section 1.4.

1.3 PORTFOLIO OPTIMIZATION

In this section we deal with asset allocation decisions with risky securities. Actually, we have seen in previous sections that *fixed income* is a somewhat misleading term, since bond portfolios face different sources of risk. Here we consider stocks, which are assets to which a random cash flow is associated. Consider a set of n stocks. We may purchase stock i, $i = 1, \ldots, n$, at a known price P_i; after holding it for a period, we may sell it for a price Q_i, which is, however, a random variable. Thus, both the return

$$R_i = \frac{Q_i}{P_i}$$

and the rate of return

$$r_i = \frac{Q_i - P_i}{P_i} = R_i - 1$$

are random variables, too. Stocks may pay dividends during the holding period, which would contribute to the return. The rate of return may well be

---

[10]See, e.g., [6] for an outline of LP models dealing with risk issues in fixed-income portfolio management.

negative, but it is bounded below by -1. The worst that can happen is that you lose your investment completely. This is not a trivial statement, as it may seem, since unlike corporate bonds, stocks imply some degree of ownership of a firm; however, stocks are limited liability assets, so if you own a stock, you do not bear responsibilities linked to the firm's debts and liabilities.

Basically, managing a stock portfolio requires a way to to trade off risks and potential returns. In this section we limit our treatment to the single-period problem. We do not consider here the possibility of rebalancing our portfolio after awhile; multiperiod problems are considered in chapter 6. We are given an initial wealth W_0, and we should allocate it among a set of n risky stocks. We would like to maximize our wealth W at the end of the investment period. However, since returns are random, W is a random variable and maximizing it does not make sense. We have to cope with a stochastic optimization problem, and some thought is needed to define a sensible objective function. One possibility would be to maximize the expected value $E[W]$, but in so doing we would ignore risks completely. Indeed, if we want to trade off risks and returns, we must find out a way to quantify risk. Assume that the rate of return of a stock is normally distributed. In this case, the rate of return is fully characterized by two parameters, the expected return $\bar{r}_i$ and the standard deviation σ_i. The standard deviation may be considered as a measure of risk; the larger the standard deviation, the larger the possibility of having both large positive and large negative returns. Actually, using standard deviations or variances to measure risk would be debatable in the case of asymmetrically distributed returns, but let's go on with the assumption that returns have a symmetrical distribution. Now, suppose that we have two stocks and that $\bar{r}_1 > \bar{r}_2$ and $\sigma_1 > \sigma_2$; in other words, stock 1 has better potential than stock 2, but it is riskier. Which one is better? Of course, there is no easy answer, in that it depends on the subjective attitude toward risk.

The customary way to model investors' preferences is by adopting a utility function $U(W)$, depending on the wealth. It is reasonable to assume that U is an increasing function, since we prefer more wealth than less. Another reasonable assumption is that U is concave, i.e.,

$$U(\lambda x + (1 - \lambda)y) \geq \lambda U(x) + (1 - \lambda)U(y) \qquad 0 \leq \lambda \leq 1. \qquad (1.4)$$

That a concave utility function implies some degree of risk aversion may be seen from figure 1.6. Consider two wealth levels, W_1 and W_2, with $W_1 < W_2$. You are offered two choices. You may get a "lottery," whereby you may win one of the two wealth levels, each with probability $1/2$; alternatively, you may simply get the wealth $W^* = 0.5W_1 + 0.5W_2$. Note that with this "safe" offer you would get just the expected value of the wealth you win with the lottery. In the utility function framework, decision makers act by maximizing their expected utility. The expected utility of the lottery is then

$$0.5\, U(W_1) + 0.5\, U(W_2),$$

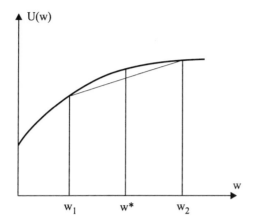

Fig. 1.6 How concave utility functions imply risk aversion.

whereas the expected utility of the alternative choice is

$$U(0.5W_1 + 0.5W_2).$$

From condition (1.4), with $\lambda = 0.5$, we see that the expected utility of the safe offer will be larger than the expected utility of the lottery for any concave utility function, even if the expected values of the wins are equal. Typical utility functions are the logarithmic utility,

$$U(W) = \log(W), \tag{1.5}$$

and the quadratic utility,

$$U(W) = W - bW^2. \tag{1.6}$$

Note that the quadratic function only makes sense in the domain over which it is increasing [i.e., $W \leq 1/(2b)$]. It is also worth noting that utility functions are used just to rank random alternatives and that their absolute value is irrelevant. Hence, they may be scaled and translated, since the linearity of the expected value operator implies that this will not affect the ranking of alternatives. Now, the portfolio optimization can be stated as

$$\max \quad E\left[U\left(\sum_{i=1}^{n} \theta_i Q_i\right)\right]$$

$$\text{s.t.} \quad \sum_{i=1}^{n} \theta_i P_i = W_0,$$

where the decision variables θ_i are the number of shares of stock i purchased (or sold short if we do not require that $\theta_i \geq 0$). Actually, some assumptions are required to ensure that this problem has a solution (which need not exist if arbitrage is possible) and that we invest all of our initial wealth W_0 (see, e.g., [9]).

Apart from the qualitative implication of the concavity, we may also measure the degree of risk aversion associated with a utility function. One such measure is the *Arrow-Pratt absolute risk aversion coefficient*, defined as

$$A(W) = -\frac{U''(W)}{U'(W)}.$$

The utility function is assumed increasing, hence we have $U'(W) \geq 0$; furthermore, concavity implies that $U''(W) \leq 0$. These facts imply that the coefficient $A(W)$ is nonnegative. Note that according to this definition, risk aversion may change, depending on the level of wealth. Indeed, some investors might tend to be less risk averse when wealth increases; whether or not this is really the case depends on the utility function. If we assume the logarithmic utility function (1.5), we have

$$A(W) = -\frac{U''(W)}{U'(W)} = -\frac{-1/W^2}{1/W} = 1/W,$$

and risk aversion decreases with increasing wealth. For the quadratic utility (1.6) we have

$$A(W) = \frac{2b}{1 - 2bW},$$

showing that in this case risk aversion increases with increasing wealth.

Specifying a utility function may be a difficult task, since assessing the trade-off between risk and return is far from trivial. A relatively simple approach is based on the idea of restricting the choice to "reasonable" portfolios. If you fix the expected return you want to get from the investment, you would like to find the portfolio achieving that expected return with minimal risk. By the same token, if you fix the level of risk you are willing to take, you would like to select a portfolio maximizing the expected return. This approach leads to mean-variance portfolio theory, which, despite considerable criticism, underlies quite a significant part of financial theory.

1.3.1 Basics of mean-variance portfolio optimization

Let us return to the asset allocation problem when only two risky assets are available. Apparently, the problem is trivially solved when $\bar{r}_1 > \bar{r}_2$ and $\sigma_1 < \sigma_2$. In this case, stock 1 has a larger expected return than stock 2, and it is also less risky; hence, a naive argument would lead to the conclusion that asset 2 should not be considered at all. Actually, this may not be the case, since we have neglected the possible correlation between the two assets. The inclusion of asset 2 may, in fact, be beneficial in reducing risk. So we see that there is some need for formalization in order to solve the problem.

Assume that we are interested in defining the portfolio weights, w_1 and w_2 in our case. A natural constraint is

$$w_1 + w_2 = 1.$$

Note that we are not considering the initial wealth level W_0, since we deal with the allocation of fractions of wealth. If we want to rule out short selling, we must also require that $w_i \geq 0$. Elementary probability theory tells us that the portfolio rate of return will be

$$r = w_1 r_1 + w_2 r_2,$$

and the expected return will be

$$\bar{r} = w_1 \bar{r}_1 + w_2 \bar{r}_2.$$

More generally, when we must devise a portfolio of n risky assets, the expected return is given by

$$\bar{r} = \sum_{i=1}^{n} w_i \bar{r}_i = \mathbf{w}^T \bar{\mathbf{r}}.$$

The variance of r is given, for the two-asset case, by

$$\sigma^2 = \text{Var}(w_1 r_1 + w_2 r_2) = w_1^2 \sigma_1^2 + 2 w_1 w_2 \sigma_{12} + w_2^2 \sigma_2^2,$$

where σ_{12} is the covariance between r_1 and r_2. For n assets we have

$$\sigma^2 = \sum_{i,j=1}^{n} w_i w_j \sigma_{ij} = \mathbf{w}^T \Sigma \mathbf{w},$$

where Σ is the covariance matrix.

By choosing the weights w_i, we will get different portfolios characterized by the expected value of the return and by its variance or standard deviation, which we assume as a risk measure. Any investor would like both to maximize the expected return and to minimize variance. Since these two objectives are, in general, conflicting, we must find a trade-off. The exact trade-off will depend on the degree of risk aversion, which is hard to assess, but it is reasonable to assume that for a given target value $\bar{r}_T$ of the expected return, one would like to minimize variance. This is obtained by solving the following optimization problem:

$$
\begin{aligned}
\min \quad & \mathbf{w}^T \Sigma \mathbf{w} \\
\text{s.t.} \quad & \mathbf{w}^T \bar{\mathbf{r}} = \bar{r}_T \\
& \sum_{i=1}^{n} w_i = 1 \\
& w_i \geq 0.
\end{aligned}
\tag{1.7}
$$

This is a quadratic programming problem, which is easily solved by MATLAB (see section 1.3.2). Changing the target expected return, one may obtain a set of *efficient* portfolios. A portfolio is efficient if it is not possible to obtain a higher expected return without increasing risk. There are infinite efficient portfolios in general, and it is reasonable to assume that the preferred portfolio will be one of them.

It is instructive to go back to the case of two assets. Assume the following data:

$$\bar{r}_1 = 0.2 \qquad \bar{r}_2 = 0.1$$
$$\sigma_1^2 = 0.2 \qquad \sigma_2^2 = 0.4$$
$$\sigma_{12} = -0.1.$$

Note that asset 2 is apparently useless, but it is negatively correlated with asset 1; hence, when asset 1 performs poorly, we may hope that asset 2 will perform well (and vice versa). Including asset 2 may result in some diversification. The Financial toolbox provides the user with a ready-to-use function, frontcon, to analyze mean-variance efficient portfolios. Using this function, it is easy to trace the efficient set:

```
>> r = [0.2 0.1];
>> s = [0.2 -0.1 ; -0.1 0.4];
>> frontcon(r,s,10)
```

The function frontcon receives the vector r of expected returns and the covariance matrix s and plots the efficient frontier depicted in figure 1.7. The third argument is simply the number of efficient portfolios used in plotting the frontier (more about this in the next section). It is interesting to note that it is possible to obtain portfolios whose standard deviation of return is lower than the standard deviation of both assets. It is possible to trace the efficient set, adding further constraints on the portfolio composition, and it is possible to spot an "optimal" portfolio by selecting a quadratic utility function. The MATLAB functions to perform such tasks are analyzed below.

1.3.2 MATLAB functions to deal with mean-variance portfolio optimization

MATLAB includes a set of functions based on mean-variance portfolio theory. They rely on the Optimization toolbox to solve quadratic and linear programming problems. We have already met linprog to solve LP problems; the function quadprog is used to cope with quadratic programs. To get acquainted with these functions, we may devise a function to trace the efficient frontier, given the vector of expected returns and the covariance matrix. The MATLAB code is illustrated in figure 1.8. The function naiveMV is a simplified version of frontcon and takes three arguments: the vector of the

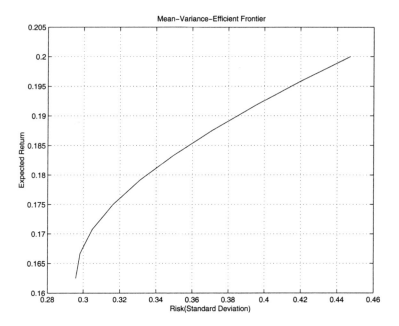

Fig. 1.7 Efficient frontier for a risky portfolio.

expected rates of return, ERet, the covariance matrix, ECov, and the number of efficient portfolios desired, NPts. For each efficient portfolio we get the risk (standard deviation), the expected rate of return, and the weights for each asset.

The approach is based on solving problem (1.7) with different values of the target expected return $\bar{r}_T$. To select proper target values we have to spot both the maximum return achievable and the return associated with the minimum variance portfolio. The first return is obtained by solving

$$\text{max} \quad \mathbf{w}^T \bar{\mathbf{r}}$$
$$\sum_{i=1}^{n} w_i = 1$$
$$w_i \geq 0.$$

The second return is obtained by finding the minimum risk portfolio:

$$\text{min} \quad \mathbf{w}^T \mathbf{\Sigma} \mathbf{w}$$
$$\sum_{i=1}^{n} w_i = 1$$
$$w_i \geq 0$$

and by computing its return (we take for granted that the solution of this problem is unique). These are the two extreme efficient portfolios. If they

```
% naiveMV.m
function [PRisk, PRoR, PWts] = naiveMV(ERet, ECov, NPts)
ERet = ERet(:);        % makes sure it is a column vector
NASSETS = length(ERet);  % get number of assets
V0 = zeros(1, NASSETS);
V1 = ones(1, NASSETS);

% Find the maximum expected return
MaxReturnWeights = linprog(-ERet, [], [], V1, 1, V0);
MaxReturn = MaxReturnWeights'*ERet;

% Find the minimum variance return
MinVarWeights = quadprog(ECov, V0, [], [], V1, 1, V0);
MinVarReturn = MinVarWeights'*ERet;
MinVarStd = sqrt(MinVarWeights' * ECov * MinVarWeights);

if MaxReturn > MinVarReturn
   RTarget = linspace(MinVarReturn, MaxReturn, NPts);
   NumFrontPoints = NPts;
else
     RTarget = MaxReturn;
     NumFrontPoints = 1;
end

PRoR = zeros(NumFrontPoints, 1);
PRisk = zeros(NumFrontPoints, 1);
PWts = zeros(NumFrontPoints, NASSETS);
PRoR(1) = MinVarReturn;
PRisk(1) = MinVarStd;
PWts(1,:) = MinVarWeights(:)';

VConstr = ERet';
A = [V1 ; VConstr ];
B = [1 ; 0];
for point = 2:NumFrontPoints
B(2) = RTarget(point);
Weights = quadprog(ECov, V0, [], [], A, B, V0);
PRoR(point) = dot(Weights, ERet);
PRisk(point) = sqrt(Weights'*ECov*Weights);
PWts(point, :) = Weights(:)';
end
```

Fig. 1.8 Simple MATLAB code to trace the mean-variance efficient frontier.

```
% CallNaiveMv.m
ExpRet = [ 0.15 0.2 0.08];
CovMat = [ 0.2 0.05 -0.01 ; 0.05 0.3 0.015 ; ...
        -0.01 0.015 0.1];
[PRisk, PRoR, PWts] = naiveMV(ExpRet, CovMat, 10);
[PRoR , PRisk]


ans =
    0.1080    0.1414
    0.1182    0.1751
    0.1284    0.2320
    0.1387    0.2795
    0.1489    0.3218
    0.1591    0.3607
    0.1693    0.3972
    0.1796    0.4320
    0.1898    0.4804
    0.2000    0.5477
```

Fig. 1.9   Generating the efficient frontier.

are equal, there is a unique portfolio maximizing return and minimizing risk: an unlikely event in practice, which is taken into account by the function (in this case the number NumFrontPoints of efficient points in the frontier is 1). To find other efficient portfolios, we use the function linspace to specify the vector of NPts target returns between the two extremes. Then we solve a sequence of risk minimization problems, obtaining the risk/return characteristics and the composition of each portfolio. For instance, an example of calling our naiveMV function and the results obtained are illustrated in figure 1.9. The same results are obtained by calling frontcon:

```
>> frontcon(ExpRet, CovMat, 10);
```

Unlike naiveMV, frontcon is safe in the sense that some consistency checks are carried out on the input arguments. For instance, a covariance matrix must be positive semidefinite, but we did not check this in our function. The reader is urged to try frontcon with the following covariance matrix:

```
CovMat = [0.2 0.1 -0.1 ; 0.1 0.2 0.15 ; -0.1 0.15 0.2]
```

To obtain a plot like the one in figure 1.7, you have just to call frontcon without output arguments. Finally, optional arguments can be used to enforce lower and upper bounds on the allocation to single assets or groups of assets. This may make sense if you want to limit the exposure to certain risky stocks

```
NAssets = 5;
AssetMin = 0;
AssetMax = [0.35 0.3 0.3 0.4 0.5];
Groups = [1 1 0 0 0 ; 0 0 1 1 1];
GroupMin = [ 0.2 0.3 ];
GroupMax = [ 0.6 0.7 ];

ConstrMatrix = portcons(...
   'AssetLims', AssetMin, AssetMax, NAssets, ...
   'GroupLims', Groups, GroupMin, GroupMax)
```

Fig. 1.10 How to use `portcons` to build the constraint matrix.

or to market sectors (e.g., telecommunications or energy). A richer function, from this point of view, is `portopt`. To illustrate, consider a problem involving five assets. Suppose that you do not want to consider short-selling and that the following upper bounds are given on each asset weight in the portfolio:

$$0.35 \quad 0.3 \quad 0.3 \quad 0.4 \quad 0.5.$$

Furthermore, the assets can be partitioned into two groups, consisting of assets 1 and 2 and of assets 3, 4, and 5, respectively. You might wish to enforce both lower and upper bounds on asset allocation to each group; say the lower bounds are 0.2 and 0.3 and the upper bounds are 0.6 and 0.7. Formally, this would result in a constraint set like the following, which should be added to our quadratic programming problems:

$$0 \leq w_1 \leq 0.35 \quad 0 \leq w_2 \leq 0.3 \quad 0 \leq w_3 \leq 0.3$$
$$0 \leq w_4 \leq 0.4 \quad 0 \leq w_5 \leq 0.35$$
$$0.2 \leq w_1 + w_2 \leq 0.6$$
$$0.3 \leq w_3 + w_4 + w_5 \leq 0.7.$$

The optimization functions available in MATLAB can easily cope with such constraints, but it is up to the user to put them into a matrix form like $\mathbf{Aw} \leq \mathbf{b}$ or $\mathbf{A}_{eq}\mathbf{w} = \mathbf{b}_{eq}$. To help the user, the `portcons` function is provided, which helps in building the matrix. For our small example, we would call this function as illustrated in figure 1.10, obtaining the constraint matrix in figure 1.11. Then the matrix may be used by calling `portopt` as follows (some optional arguments are omitted; see the manual):

```
>> [PRisk,PRoR,PWts] = portopt(ExpRet,CovMat,[],[],ConstrMatrix);
```

Functions such as `portcons`, which may be used to build more complex matrices than the one in our example, are certainly helpful. However, it should

```
ConstrMatrix =
    1.0000         0         0         0         0    0.3500
         0    1.0000         0         0         0    0.3000
         0         0    1.0000         0         0    0.3000
         0         0         0    1.0000         0    0.4000
         0         0         0         0    1.0000    0.5000
   -1.0000         0         0         0         0         0
         0   -1.0000         0         0         0         0
         0         0   -1.0000         0         0         0
         0         0         0   -1.0000         0         0
         0         0         0         0   -1.0000         0
   -1.0000   -1.0000         0         0         0   -0.2000
         0         0   -1.0000   -1.0000   -1.0000   -0.3000
    1.0000    1.0000         0         0         0    0.6000
         0         0    1.0000    1.0000    1.0000    0.7000
```

Fig. 1.11 Sample constraint matrix built by `portcons`.

be noted that they can be used only to build matrices corresponding to predefined constraints. Algebraic modeling languages such as AMPL and GAMS,
which are illustrated in chapter 6, do the same job for arbitrary constraints.[11]

A last function we describe here may be used to find an optimal portfolio.
So far, we have dealt with efficient portfolios, leaving the risk/return trade-off
unresolved. We may resolve this trade-off by linking mean-variance portfolio
theory to the more general utility theory. In fact, mean-variance theory is
not necessarily compatible with an arbitrary utility function: i.e., an optimal
portfolio for some utility function need not be on the mean-variance efficient
frontier. It can be shown that this inconsistency does not arise if the returns
are normally distributed or if the utility function is quadratic (see, e.g., [9] or
[13]). The last point implies that if may specify a quadratic utility function
such as (1.6), the optimal solution will be a mean-variance efficient portfolio.
All we have to do is to specify the b parameter according to our degree of risk
aversion. In the Financial toolbox the function `portalloc` is provided, which
yields the optimal portfolio assuming the quadratic utility function

$$U = \bar{r} - 0.005 \cdot A \cdot \sigma^2,$$

where the parameter A is linked to risk aversion; its default value is 3, and
suggested alternative values range between 2 and 4. There is still another

---

[11]Another consideration is that `portcons` generates a full matrix with many zero entries.
Good optimization solvers deal with sparse matrices, which avoid storing zero entries in
order to save memory space. Algebraic languages exploit this possibility, which is essential
to deal with large-scale problems with special structure.

```
% CallPortAlloc.m
ExpRet = [ 0.18 0.25 0.2];
CovMat = [ 0.2 0.05 -0.01 ; 0.05 0.3 0.015 ; ...
      -0.01 0.015 0.1];
RisklessRate = 0.05;
BorrowRate = 0.06;
RiskAversion = 3;

[PRisk, PRoR, PWts] = frontcon(ExpRet, CovMat, 10);

[RiskyRisk , RiskyReturn, RiskyWts, RiskyFraction, ...
   PortRisk, PortReturn] = portalloc(PRisk, PRoR, PWts, ...
      RisklessRate, BorrowRate, RiskAversion)
```

Fig. 1.12 Calling portalloc.

fundamental issue that we have neglected so far. We have considered mean-variance efficient portfolios, assuming that only risky assets were available. However, we may obtain a known return by investing in a bank account with a fixed interest rate or in a safe zero-coupon bond (with maturity equal to our investment horizon, to avoid interest rate risk issues). What is the effect of the inclusion of such a risk-free asset in our portfolio? A detailed analysis of this issue is rich in implications in financial theory, but it would lead us too far. For our purposes it is sufficient to say that the optimal portfolio will be a combination of the risk-free asset and one particular efficient portfolio. The amounts invested in the risk-free asset and in the risky portfolio depend on our risk aversion, but the risky portfolio involved does not. An important implication of this, if we believe the theory, is that investors could live with just one "mutual" fund, mixing it with the risk-free asset. The Financial toolbox includes a function called portalloc to obtain the optimal combination of the risky portfolio and the risk-free asset; it assumes further that cash may be borrowed at some rate. Figure 1.12 illustrates a script to call this function.

Some explanation is in order. First, we give the vector of the expected rates of return and the covariance matrix, which are used by frontcon to generate an approximation of the efficient frontier with a given number of points. We also give a riskless rate (for investing), a borrowing rate (which is larger than the riskless rate), and a risk aversion coefficient. Calling portalloc with these parameters will produce the output reported in figure 1.13. RiskyRisk, RiskyReturn, and RiskyWts are the risk, the expected return, and the composition of the ideal fund. RiskyFraction is the fraction we should invest in the risky portfolio; PortRisk and PortReturn are the risk and return of the portfolio consisting of the risky portfolio and the risk-free asset. A quick check is instructive. Let us denote the fractions invested in the risk-free asset and

```
RiskyRisk =
    0.2450
RiskyReturn =
    0.2048
RiskyWts =
    0.2334    0.1903    0.5762
RiskyFraction =
    0.8601
PortRisk =
    0.2107
PortReturn =
    0.1832
```

Fig. 1.13 Sample output of `portalloc`.

in the mutual fund by w_f and w_m; similarly, let r_f and r_m be the respective rates of return. Then we should have that the expected return of the portfolio is

$$\bar{r} = w_f r_f + w_m \mathrm{E}[r_m] = (1 - 0.8601) \cdot 0.05 + 0.8601 \cdot 0.2048 = 0.1831,$$

which (apart from rounding errors) agrees with what we get. Similarly, when considering risks,

$$\sigma = w_m \sigma_m = 0.8601 \cdot 0.2450 = 0.2107,$$

where σ_m is the risk of the mutual fund (by definition, for the risk-free asset $\sigma_f = 0$). The reader is urged to try changing the risk aversion coefficient from 3 to 1. The fraction allocated to the risky portfolio will rise to 2.4082. So the fraction is larger than 1, but this is not an error; a risk seeking investor would simply borrow at the borrowing rate in order to invest more in the risky portfolio, resulting in a higher overall expected rate of return.

1.3.3 Critique

Mean-variance portfolio theory leads to relatively simple numerical problems. However, despite its prominent role in financial theory, the approach has been the subject of widespread criticism. We have pointed out that mean-variance portfolio theory is consistent with the utility function approach in the case of normally distributed returns and in the case of a quadratic utility function. Both conditions may be debated.[12]

---

[12]See, e.g., [11] for a discussion of alternative utility functions in portfolio optimization.

One important feature of the normal distribution is its symmetry. If the return distribution is symmetric, then using variance or standard deviation as a measure of risk may make sense; in fact, variance takes into account returns that are both higher and lower than the average. The former are actually desirable, but in the case of normal distribution a potential for good performance is exactly counterbalanced by the risk of underperformance. However, if the distribution is not symmetric, we must distinguish the upside potential from the downside risk. While symmetric returns may be assumed for stocks, derivative assets such as those we shall describe shortly may lead to more complex distributions. As for the quadratic utility function, we have seen that it implies increasing absolute risk aversion, which is itself a counterintuitive behavior for the usual investor. A solution to both issues would be the use of alternative utility functions. We could even enforce some constraints on the probability of large losses; if L is the random variable modeling the portfolio loss, we could require something like

$$P\{L > w\} \leq \alpha,$$

where α is a small probability and w is a threshold parameter. All of these approaches lead to more complex optimization problems, namely stochastic programming problems, which are dealt with in chapters 3 and 6.

A further reason for using stochastic programming models is another difficulty in mean-variance theory. The covariance matrix is assumed constant over time. Unfortunately, it is likely that correlation may rise in stock crashes, just when diversification should help. So we should use more complex models in describing the uncertainty. Stochastic programming does so by building a set of multiperiod scenarios. This also enables us to consider another feature that is disregarded by mean-variance models: the dynamic nature of portfolio management, which is not considered in single-period models. Portfolios are revised in time, and the impact of transaction costs should not be neglected.

Modeling transaction costs exactly may be rather difficult. They depend in a nontrivial way on the amounts traded. For instance, it may be preferable to buy and sell stocks in precise lots, since trading in odd lots will increase transaction costs in anticipation of liquidity problems. It might also be advisable to avoid a portfolio with a small weight on some assets; the benefit of diversification will probably be lost because of increasing transaction costs. So we could require that if a stock enters the portfolio, it does so with a minimal weight. Such constraints require the introduction of integer programming models. We give some examples in chapter 6. For now we just give a small investment example illustrating the use of integer variables in optimization models.

Example 1.8 We consider a trivial model for a capital budgeting problem. We must allocate a given amount W to a set of N potential investments. For each investment opportunity we know:

- The initial capital outlay C_i, $i = 1, \ldots, N$

- The revenue R_i we will get from the investment (assumed certain)

This looks like a portfolio model, the key difference being that our decision must be "all-or-nothing." For each opportunity, we may invest or not, but we cannot buy a share of the investment. Indeed, in portfolio models, stocks are assumed infinitely divisible, which may often be a reasonable approximation, but not in this case.

The decision variables must reflect the logical nature of our decision. This is obtained by restricting the decision variables as follows:

$$x_i = \begin{cases} 1 & \text{if we invest in project } i \\ 0 & \text{otherwise.} \end{cases}$$

Now it is easy to build an optimization model:

$$\max \quad \sum_{i=1}^{n} R_i x_i$$

$$\text{s.t.} \quad \sum_{i=1}^{N} C_i x_i \leq W$$

$$x_i \in \{0, 1\}.$$

This model is grossly simplified, but it is a first example of an integer programming model. It is also well known as the *knapsack problem*, as each investment may be interpreted as an object of given value and volume, and we want to determine the maximum value subset of objects that may fit the knapsack capacity W. In general, integer programming models are relatively difficult to solve (see chapter 3). ⬚

1.4 DERIVATIVES

Derivatives are a broad family of financial contracts, owing their name to the dependency of their payoff on the value of some underlying variable, which may be a stock price, a set of stock prices, an interest rate, an index, or a generic nonfinancial asset. For instance, a *forward contract* binds two parties to, respectively, buy and sell a certain asset, in a certain quantity, at a certain date, and at a fixed price. The party agreeing to buy is said to hold the *long position*, whereas the seller holds the *short position*. By entering a forward contract you basically lock in a fixed price for the underlying asset. You may have two quite different reasons. You might wish to eliminate, or reduce, risk; in fact, by locking the price for an asset you have to buy or sell, you eliminate the effect of price uncertainty. This does not mean that the final outcome will necessarily be more favorable. If you hold the long position in a forward specifying a price F, and eventually the price of the asset when the delivery takes place is $S_T < F$, in a sense you have lost an amount $F - S_T$; if, on the contrary, $S_T > F$, you have gained something. The point is that if you really need to buy or sell that asset, it may be wise to lock in a certain price than to take chances. However, you could also be a speculator with a very precise idea of where the price S_T will be, and you may enter a forward contract as a bet. Since there is a payoff depending on the difference $F - S_T$, the forward contract is a simple example of a derivative.

Derivatives may be private contracts issued by two parties for possibly very peculiar and specific reasons. Alternatively, they may be traded actively on exchanges and quoted on newspapers. Two such examples are futures and options. A *future contract* is similar to a forward contract; the main difference is that there is an intermediation process such that the detailed working is different. A common feature of forward and future contracts is that the two parties are compelled to buy and sell the asset (unless you sell the contract to someone else, as is actually often the case with futures). With an *option*, you get the *right*, but not the obligation, to buy or sell a certain asset for a specified price. The two simplest option contracts are the European call and put.

When you buy a call option, you get the right to buy the underlying asset for a price X, called the *exercise price* (or *strike price*), at a certain date T, called expiration date or maturity. Suppose that the underlying asset is a stock whose price is a random variable $S(t)$, depending on time t. If at the maturity date the actual price $S(T)$ is larger than the exercise price X, you would exercise the option and buy the stock, since you may sell the stock immediately and gain $S(T) - X$. If the contrary holds, you would not exercise the option, which expires worthless. Thus, the payoff of this option is

$$\max\{S(T) - X, 0\}.$$

If at time t we have $S(t) > X$, we say that the call option is *in-the-money*; this means that we would get an immediate profit by exercising the option. If $S(t) < X$, the call option is said to be *out-the-money*. If $S(t) = X$, the option is said to be *at-the-money*.[13] With a put option, you have the right to sell the stock. In this case, you would exercise the option only if the exercise price is larger than the actual price. So the payoff is

$$\max\{X - S(T), 0\}.$$

With a European option you may exercise your right only at maturity; an *American option* may be exercised whenever you wish within a prescribed time. These are *vanilla options*, owing their name to their simplicity. A *Bermudan option* is halfway between an American and a European option. It may be exercised at a set of prescribed dates within the horizon. *Asian*

---

[13] A simplistic consideration would suggest that an at-the-money option is not worth exercising; however, when considering the transaction costs involved in purchasing a stock, we see that there are circumstances where exercising an at-the-money option may be interesting.

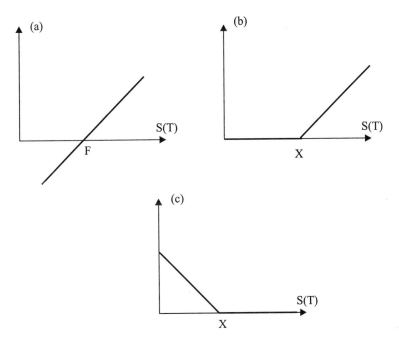

Fig. 1.14 Payoff diagrams for the long position in a forward contract (a), a call option (b), and a put option (c).

options have a payoff depending on the average price of a stock; thus they depend on a set of stock prices. Indeed, quite complex *exotic* options are actually designed and traded.

The payoff diagrams for vanilla European call and puts are depicted in figure 1.14, together with the payoff of a forward contract. The payoff for an option cannot be negative. Does this imply that you cannot lose money? Well, as you can imagine, the option comes with a price. With a forward contract, you pay nothing when you enter the contract, whereas the option has a price. So figures 1.14b and 1.14c are not quite correct, as the payoffs should be shifted down to account for the option price. Indeed, finding this price is the major concern with options, and this is why numerical methods are so important.

Why are options traded? As with futures and forwards, there are two basic reasons. On the one hand, they can be used to control risks. If you hold a stock in your portfolio and you are worried about the possibility of a large drop in its price, you may reduce the risk by buying a protective put. This comes with a price, but in this way you avoid the potential of a large loss. By the same token, you may reduce the interest rate risk of a fixed-income portfolio by buying interest rate derivatives. On the other hand, options may also be used for speculation.

Example 1.9 Suppose that a stock price is $50, and you believe that it will rise in the near future. You could then buy the stock anticipating a large return. Let's say that you are right and the price rises to $55. Then your rate of return will be

$$\frac{55 - 50}{50} = 10\%.$$

But now imagine that a call option is available with a strike price $50, and that this option costs $5 (this may or may not be a reasonable price). In this case you will exercise the option, and the return will be much larger:

$$\frac{55 - 50}{5} = 100\%.$$

This effect is called *leverage* or *gearing*. As you may expect, there is another side to the coin. If you are wrong and the stock price drops to $45, then by buying the stock you will lose $10; with the call option you will lose everything. You are also exposed to other sources of risk if you are interested in selling the option before maturity, as unfavorable movements in the factors determining the option value may have an adverse impact on the value of your portfolio.

▯

There are two basic issues in dealing with derivatives. The first issue is pricing. What is the fair price of a forward or an option contract? The second issue is hedging. Suppose that you are the writer of an option rather than the holder. In some sense the holder is at an advantage, since she is not forced to exercise the option if the circumstances are unfavorable (although example 1.9 shows that careless management of an option portfolio may lead to a disaster). If you are the writer of an option and this is exercised, you have to meet your obligation, and in principle there is no limit to your loss. Thus, you are interested in trading policies to reduce the risk to which you are exposed. We will not pursue hedging in any detail in this book (see, e.g., [19]), but it is worth noting that hedging is related to pricing.

A key role in pricing is played by the no-arbitrage argument we have already used, in a trivial situation, for bond pricing. This is best illustrated by a couple of examples. In the first we derive the price of a forward contract. In the second we derive a fundamental relationship between the price of a call and the price of a put, called *put-call parity*.

Example 1.10 Consider a forward contract for delivery at time T of an asset whose spot price now is $S(0)$. The spot price $S(T)$ at delivery is a random variable; hence, it would seem that randomness is involved in finding the fair forward price F. Actually, a simple arbitrage argument shows that this is not the case.

Suppose that we hold the short position in the contract and consider the following portfolio. We may borrow an amount $S(0)$ at the risk-free interest rate r, assuming continuous compounding, to buy the asset. The net cash flow now is zero. Then at time T we may deliver the asset at price F, and we must pay back $S(0)e^{rT}$. Despite the randomness in the spot price, the value of our portfolio at T is deterministic and given by $F - S(0)e^{rT}$. But since the portfolio value at time $t = 0$ is zero, the same must hold at time $t = T$. Hence,

$$F = S(0)e^{rT}.$$

Any different forward price would lead to an arbitrage opportunity. If $F > S(0)e^{rT}$, the portfolio above will lead to a safe gain $F - S(0)e^{rT}$, with no initial commitment. If $F < S(0)e^{rT}$, we may reverse the portfolio by short-selling the asset and investing the proceeds. The reasoning assumes that short-selling the asset is possible and that no storage charge is paid for keeping the asset. See [8] for a full account of forward pricing.   □

Example 1.11 Consider an asset with spot price $S(0)$, and both call and put European options on the asset, with the same exercise price X and maturity T. For now, we are not able to figure out the fair prices C and P for the two options, but it is easy to see that a precise relationship must hold between the two. Consider two portfolios:

1. Portfolio P_1 consists of one European call option and an amount of cash equal to Xe^{-rT}, where r is the risk-free interest rate

2. Portfolio P_2 consists of one European put option and one share of the stock

The value of portfolio P_1 at time $t = 0$ is $C + Xe^{-rT}$; the value of portfolio P_2 at time $t = 0$ is $P + S(0)$. At time T, we may have two cases, depending on the price $S(T)$. If $S(T) > X$, the call option will be exercised and the put option will not. Hence, under this hypothesis, portfolio P_1 at time $t = T$ will be worth

$$[S(T) - X] + X = S(T),$$

and portfolio P_2 will be worth

$$0 + S(T) = S(T).$$

If $S(T) < X$, the put option will be exercised and the call option will not. In this case, portfolio P_1 is worth

$$0 + X = X$$

and portfolio P_2

$$[X - S(T)] + S(T) = X.$$

In both cases the two portfolios have the same value at time T. Hence, their values at time $t = 0$ must be equal; otherwise, there will be an arbitrage opportunity. Hence, the following put-call parity relationship must hold:

$$C + Xe^{-rT} = P + S(0).$$

This implies that if we are able to find the fair price for one of the options, the other one is obtained automatically.   ▯

We will see that the use of arbitrage arguments leads to pricing equations in the form of partial differential equations. These may sometimes be solved analytically to yield a pricing formula in closed form, as in the case of Black and Scholes. In other cases, an analytical approach to option pricing may lead to useful approximated pricing formulas. In general, we need to resort to numerical procedures. There are basically three numerical approaches to price a derivative:

- Solving a partial differential equation, e.g., by finite difference approximations

- Monte Carlo simulation

- Binomial or trinomial lattices

Numerical methods for partial differential equations and Monte Carlo simulation are discussed in later chapters; binomial lattices are illustrated later in this chapter. In any case, the starting point in complex derivative pricing is finding a suitable model of the dynamics of the underlying asset.

1.4.1   Modeling the dynamics of asset prices

A model of the dynamics of asset prices must reflect the random nature of price movements. The asset price $S(t)$ must be described as a stochastic process, and the most common model is based on random walks. It is a good idea to introduce the idea by starting with a discrete-time model and then derive a continuous-time model. Consider a time interval $[0, T]$, and imagine that we discretize the interval with a time step δt such that $T = N \cdot \delta t$; then we may index the discrete-time instants by $t = 0, 1, 2, \ldots, N$. Let S_t be the stock price at time t. One possible and reasonable model is the multiplicative form:

$$S_{t+1} = u_t S_t, \tag{1.8}$$

where u_t is a nonnegative random variable and the initial price S_0 is known. The random variables u_t are identically distributed and independent; independency is an assumption linked to market efficiency. The model is reasonable since it ensures that prices will stay nonnegative, which is an obvious requirement for stock prices. If we used an additive model such as $S_{t+1} = u_t + S_t$, we should admit negative values for the random variables u_t to model price drops, and we would not have the guarantee $S_t \geq 0$. With the multiplicative form, a price drops when $u_t < 1$, but it stays positive. Furthermore, the actual price change depends on the present stock price (a \$1 increase is different if the present price is \$100 rather than \$5), and this is easily accounted for by the multiplicative form.

In order to determine a plausible probability distribution for the random variables u_t, it is helpful to consider the natural logarithm of the stock price:

$$\ln S_{t+1} = \ln S_t + \ln u_t = \ln S_t + w_t.$$

A common assumption is that w_t is normally distributed, which means that u_t is lognormal. Starting from the initial price S_0 and unfolding (1.8), we get

$$S_t = \prod_{k=0}^{t-1} u_k S_0,$$

which implies that

$$\ln S_t = \ln S_0 + \sum_{k=0}^{t-1} w_k.$$

Since the sum of normal variables is still a normal variable (see appendix B), we have that $\ln S_t$ is normally distributed, which in turn implies that stock prices are lognormally distributed according to this model. Using the notation

$$E[w_t] = \nu \qquad \text{Var}(w_t) = \sigma^2,$$

we see that due to the independence among the w_t,

$$E[\ln S_t] \quad = \quad \ln S_0 + \nu t \qquad\qquad (1.9)$$
$$\text{Var}(\ln S_t) \quad = \quad t\sigma^2. \qquad\qquad (1.10)$$

So the expected value and the variance of the logarithm of the stock price vary linearly with time; this implies that the standard deviation grows with the square root of time.

The next step is to obtain a model in continuous time. In the deterministic case, when you take the limit of a difference equation, you get a differential equation. In our case you get a trickier object, called a *stochastic differential equation*; a solution of a stochastic differential equation is a stochastic process rather than a deterministic function of time. This topic is quite difficult to deal with rigorously, as it requires some background in measure theory and stochastic calculus (see the references at the end of the chapter). Yet we may at least argue that if we take the limit of

$$\ln S_{t+1} - \ln S_t = w_t$$

for $\delta t \rightarrow 0$, we get the stochastic differential equation

$$d\ln S(t) = \nu \, dt + \sigma \, dz. \qquad\qquad (1.11)$$

The intuitive argument is based on the fact that this equation leads to a continuous-time stochastic process $S(t)$ whose behavior is consistent with equations (1.9) and (1.10). Let us start with a simple case, assuming that $\sigma = 0$. Then we have a familiar differential equation

$$d \ln S(t) = \nu \, dt,$$

which is easily integrated:

$$\ln S(t) - \ln S(0) = \nu t \qquad \Rightarrow \qquad S(t) = S(0) e^{\nu t}.$$

Thus, the value $S(t)$ grows exponentially in time, and the growth rate is linked to ν; this parameter plays the role of a drift. Investing at a risk-free interest rate r with continuous compounding would be described by this equation with $\nu = r$. The parameter σ plays the role of a volatility, as it multiplies a term dz which is linked to a particular stochastic process called a *Wiener process* or *Brownian motion*. Informally, we may think of dz as a normal random variable with zero mean and variance dt. The larger σ, the larger the effect of the random shocks on the stock price, whose evolution looks like a random walk. The Wiener process may also be regarded as the continuous limit of the discrete-time process:

$$z_{t+1} = z_t + \epsilon_t \sqrt{\delta t},$$

where ϵ_t is a sequence of independent standard normal variables. Then we see that for $k > j$,

$$z_k - z_j = \sum_{i=j}^{k-1} \epsilon_i \sqrt{\delta t},$$

which implies that

$$
\begin{aligned}
E[z_k - z_j] &= 0 \\
\operatorname{Var}(z_k - z_j) &= (k - j)\, \delta t.
\end{aligned}
$$

Hence, we may write

$$dz = \epsilon(t)\sqrt{dt}.$$

This supports the intuitive view of dz as a normal random variable with distribution $N(0, dt)$. The Wiener process may be generalized by defining processes which are the solution of more complex stochastic differential equations. The process $x(t)$ satisfying

$$dx = a \, dt + b \, dz,$$

where a and b are constant parameters, is called the *generalized Wiener process*. Direct integration yields

$$x(t) = x(0) + at + bz(t).$$

Since equation (1.11) describes a generalized Wiener process for $x(t) = \ln S(t)$, we see from the last equation that

$$\ln S(t) = \ln S(0) + \nu t + \sigma z(t).$$

This in turn implies that $\ln S(t)$ has the following distribution:

$$\ln S(t) \sim N(\ln S(0) + \nu t, \sigma^2 t),$$

which is consistent with (1.9) and (1.10). We see that the drift ν is linked to the stock price growth and that the volatility σ is linked to the uncertainty in the future prices.

A further generalization of the Wiener process is the *Ito process*, which is the solution of the stochastic differential equation:

$$dx = a(x,t)\ dt + b(x,t)\ dz, \tag{1.12}$$

where now the drift and the volatility parameters both depend on the state x and time t. In general, solution of this equation calls for numerical methods. The simplest way of dealing with stochastic differential equations numerically is to discretize it and use Monte Carlo simulation. More sophisticated solution strategies involve a deeper understanding of stochastic differential equations, which is beyond the scope of this book. We would just like to point out that the notation we have used is only shorthand for a stochastic integral representation; some hints on a more rigorous way to interpret stochastic differential equations are described in supplement S1.1 but are not needed for most applications.

Before going on with our model for the stock prices, it is worth noting that this is not the only plausible model. For instance, it does not account for abrupt jumps in the stock price which are actually observed in the markets. Even if we do not want to take these jumps into account, we could at least model the fact that nervousness periods are observed in which volatility increases, whereas volatility is constant in the basic model. Another point is that the expected value of S grows without bound, which would not be the case if dividends are paid,[14] and it is certainly not acceptable if we are modeling other quantities, such as the interest rates. Interest rates are characterized by mean reversion; i.e., they tend to swing around a long-term average but cannot grow without bound. Modeling interest rates is needed when dealing with interest rate derivatives which are used to control risk in fixed-income portfolios. Since the numerical methods for interest rate derivatives are not much different from those needed to deal with options on stocks, we do not deal with them in this book. We just note that we should write a model looking like

$$dr = a(\hat{r} - r) + \sigma\ dz,$$

where $a > 0$. There is much to say about a model like this, since we should investigate consistency with the entire term structure of interest rates and with no-arbitrage properties. Yet it is easy to see that the process $r(t)$ tends to swing around the value $\hat{r}$. If $r > \hat{r}$, the drift term is negative, and $r(t)$ tends

---

[14]No-arbitrage arguments show that the stock price should drop when dividends are paid.

to drop; if $r < \hat{r}$, the drift term is positive and $r(t)$ tends to increase. Similar considerations hold when modeling a stochastic and time-varying volatility $\sigma(t)$.

Now that we have a more or less plausible model for $\ln S(t)$, we would perhaps like to have a model for $S(t)$ itself. Furthermore, since we started all of this to analyze option values, which can be thought of as functions $f(S, t)$ of the underlying asset price S and time t, it would be helpful to derive an equation to characterize the option value directly. More precisely, for a vanilla option we know the value $f(S, T)$, since at expiration we have an expression for the payoff. From this final condition and the model of the asset price we would like to obtain the value $f(S_0, 0)$. The answer to both questions may be obtained by a theorem in stochastic calculus known as *Ito's lemma*. This theorem is a generalization of the chain rule for normal derivatives; for instance, we know that in ordinary calculus we may write

$$d \ln S(t) = \frac{dS}{S},$$

and it would be tempting just to use this formula to derive a stochastic differential equation for $S(t)$. However, the answer is a little more complicated. Given an Ito process such as (1.12) and a function $F(x, t)$, Ito's lemma states that the process F satisfies

$$dF = \left(a \frac{\partial F}{\partial x} + \frac{\partial F}{\partial t} + \frac{1}{2} b^2 \frac{\partial^2 F}{\partial x^2} \right) dt + b \frac{\partial F}{\partial x} dz.$$

If we set $b = 0$, i.e., there is no random term due to the Wiener process in the differential equation, this equation boils down the chain rule for derivatives. If we have a deterministic function $F(x, t)$, we know that

$$\frac{dF}{dt} = \frac{\partial F}{\partial x} \frac{dx}{dt} + \frac{\partial F}{\partial t},$$

and thus, given the differential equation (1.12) for x,

$$dF = a \frac{\partial F}{\partial x} dt + \frac{\partial F}{\partial t} dt.$$

Here we have a term in dz which is expected given the input stochastic process, and an unexpected term

$$\frac{1}{2} b^2 \frac{\partial^2 F}{\partial x^2}.$$

An informal proof of Ito's lemma is given in supplement S1.1, and it explains the occurrence of this term.

Example 1.12 We may derive an equation for $S(t)$ by introducing the random variable $Y(t) = \ln S(t)$ and applying Ito's lemma to the function

$$S = e^Y = F(Y),$$

with $a = \nu$ and $b = \sigma$. We calculate the partial derivatives

$$\frac{\partial F}{\partial Y} = e^Y = S$$

$$\frac{\partial^2 F}{\partial Y^2} = e^Y = S$$

$$\frac{\partial F}{\partial t} = 0.$$

So, applying Ito's lemma, we get

$$dS = S\left(\nu + \frac{1}{2}\sigma^2\right) dt + S\sigma \ dz,$$

whereas careless application of ordinary differentiation would miss the term $\sigma^2/2$. It is customary to set $\mu = \nu + \sigma^2/2$ and to write the equation as

$$dS = \mu S \ dt + \sigma S \ dz. \tag{1.13}$$

This equation defines the *geometric Brownian motion*. The reader is urged to try reversing the process. Start from (1.13) and obtain the equation for $F(S) = \ln S$. It is also useful to see the relationship between μ and ν in terms of the expected values of the normal and lognormal distributions:

$$E[S(t)] = S(0)e^{(\nu+\sigma^2/2)t} = S(0)e^{\mu t}. \tag{1.14}$$

This is consistent with the relationships given in appendix B on page 386.

☐

Example 1.13 To derive an equation for the value $f(S, t)$ of an option, we may start from equation (1.13). Applying Ito's lemma with $a = \mu S$ and $b = \sigma S$ yields

$$\begin{aligned} df &= \left(\mu S \frac{\partial f}{\partial S} + \frac{\partial f}{\partial t} + \frac{1}{2}\sigma^2 S^2 \frac{\partial^2 f}{\partial S^2}\right) + \sigma S \frac{\partial f}{\partial S} dz \\ &= \frac{\partial f}{\partial t} dt + \frac{\partial f}{\partial S} dS + \frac{1}{2}\sigma^2 S^2 \frac{\partial^2 f}{\partial S^2} dt. \end{aligned} \tag{1.15}$$

This looks like an intractable equation, since it is a stochastic partial differential equation. Actually, by exploiting the no-arbitrage principle, it can be simplified and transformed to a deterministic partial differential equation, which is amenable to solution by numerical methods. In some cases it may be even solved analytically. The most notable case leads to the Black-Scholes pricing formula for vanilla European options, which is given in the next section. ☐

1.4.2   Black-Scholes model

Equation (1.15) would look a little bit nicer without the random term dS. Remember that by using no-arbitrage arguments, we have obtained deterministic relationships in examples 1.10 and 1.11 despite the randomness involved. To get rid of randomness, we may try to use options and stock shares to build a portfolio whose value is deterministic. Consider a portfolio consisting of an option and a short position in a certain number, say Δ, of stock shares. The value of this portfolio is

$$\Pi = f(S,t) - \Delta\, S.$$

Differentiating Π and using equation (1.15), we get

$$d\Pi = df - \Delta\, dS = \left(\frac{\partial f}{\partial t} + \frac{1}{2}\sigma^2 S^2 \frac{\partial^2 f}{\partial S^2}\right) dt + \left(\frac{\partial f}{\partial S} - \Delta\right) dS. \qquad (1.16)$$

We may eliminate the term in dS by choosing

$$\Delta = \frac{\partial f}{\partial S}.$$

With this choice of Δ, our portfolio is riskless; hence, by no-arbitrage arguments, it must earn the risk-free interest rate r:

$$d\Pi = r\Pi dt. \qquad (1.17)$$

Eliminating $d\Pi$ between equations (1.16) and (1.17), we obtain

$$\left(\frac{\partial f}{\partial t} + \frac{1}{2}\sigma^2 S^2 \frac{\partial^2 f}{\partial S^2}\right) dt = r\left(f - S\frac{\partial f}{\partial S}\right) dt$$

and finally,

$$\frac{\partial f}{\partial t} + \frac{1}{2}\sigma^2 S^2 \frac{\partial^2 f}{\partial S^2} + rS\frac{\partial f}{\partial S} - rf = 0. \qquad (1.18)$$

Now we have a deterministic partial differential equation describing an option value $f(S,t)$; but which option exactly? That depends on the boundary conditions we enforce. In the case of a vanilla European call we have a final condition at time T:

$$f(S,T) = \max\{S - X, 0\}.$$

By the same token, the boundary condition for a put is

$$f(S,T) = \max\{S - X, 0\}.$$

Other boundary conditions are used to find the value of different options; in some cases, the same line of reasoning must be applied to derive slightly different equations for more complex exotic options.

A remarkable and counterintuitive feature of equation (1.18) is that the drift μ does not play any role. Only the risk-free interest rate r is involved.

This an example of a general and far-reaching principle called *risk-neutral pricing*. This principle is deeply linked with no-arbitrage conditions, and (very) roughly speaking, it states that complex securities and investments may be evaluated in a risk-neutral world where all the asset earn the same, risk-free, interest rate. We will see some justification for this in section 1.4.4.

In general, a partial differential equation is too difficult to use to get a solution in closed form, and it must be solved by numerical approaches; the difficulty stems partly from the equation itself and partly from the boundary conditions. We illustrate rather simple methods in chapter 5, and their application to option pricing is described in chapter 8. However, there are a few cases where equation (1.18) can be solved analytically. The most celebrated case is due to Black and Scholes, who were able to show that the solution for a European call is

$$C = S_0 N(d_1) - X e^{-rT} N(d_2), \tag{1.19}$$

where

$$d_1 = \frac{\ln(S_0/X) + (r + \sigma^2/2)T}{\sigma\sqrt{T}}$$

$$d_2 = \frac{\ln(S_0/X) + (r - \sigma^2/2)T}{\sigma\sqrt{T}} = d1 - \sigma\sqrt{T},$$

and N is the distribution function for the standard normal distribution:

$$N(x) = \frac{1}{\sqrt{2\pi}} \int_{-\infty}^{x} e^{-y^2/2} \, dy.$$

By using put-call parity, it can be shown that the value of a vanilla European put is

$$P = X e^{-rT} N(-d_2) - S_0 N(-d_1). \tag{1.20}$$

It is also possible to give a value to the number Δ of shares we should sell short to build the riskless portfolio Π:

$$\Delta = \left. \frac{\partial C}{\partial S} \right|_{S=S_0} = N(d_1).$$

For a generic option of value $f(S,t)$,

$$\Delta = \frac{\partial f(S,t)}{\partial S}$$

measures the sensitivity of the option price to small variations in the stock price. Other sensitivities may be obtained, such as

$$\Gamma = \frac{\partial^2 f(S,t)}{\partial S^2}, \qquad \Theta = \frac{\partial f(S,t)}{\partial t}, \qquad \rho = \frac{\partial f(S,t)}{\partial r}, \qquad \mathcal{V} = \frac{\partial f(S,t)}{\partial \sigma}.$$

These sensitivities, collectively nicknamed *the Greeks*, may be used to evaluate the risk involved in holding a portfolio of options. They are known in closed form for some options and must be estimated numerically in general. Δ and Γ play a somewhat similar role to that of duration and convexity in bond portfolios. Θ measures the change in option value as the expiration date is approached, whereas ρ and $\mathcal{V}$ (vega) measure the sensitivity to changes in the riskless rate and in volatility. Δ is particularly significant due to its role in the riskless portfolio we have used to derive the Black-Scholes equation. In fact, the writer of an option might use that portfolio to hedge the option. In principle, this requires a continuous portfolio rebalancing since Δ will change in time; since practical considerations and transaction costs make continuous rebalancing impossible, some hedging error will result.

1.4.3   Black-Scholes model in MATLAB

Implementing the Black-Scholes formula in MATLAB is quite easy. We may take advantage of the `normcdf` function provided by the Statistics toolbox to compute the cumulative distribution function for the standard normal distribution. Straightforward translation of equation (1.19) gives

```
d1 = (log(S0/X)+(r+sigma^2/2)*T) / (sigma * sqrt(T));
d2 = d1 - (sigma*sqrt(T));
C = S0 * normcdf(d1) - X * (exp(-r*T)*normcdf(d2));
P = X*exp(-r*T) * normcdf(-d2) - S0 * normcdf(-d1);
```

where the variables `S0, X, R, T, sigma` are self-explanatory. The Financial toolbox function `blsprice` implements these formulas with a couple of extensions. First, it may take vector arguments to compute a set of option prices at once; second, it may take into account a continuous dividend rate q (whose default value is zero). A dividend rate q means that a stock share yields in a small time interval dt a dividend proportional to the asset price, $qS(t)\,dt$. This may seem an unreasonable assumption, since stocks pay dividends at discrete-time instants. Actually, it may be an acceptable approximation if the underlying asset is a stock index; in this case we may think that the stocks on which the index is based yield an aggregate continuous yield. It is easy to adjust the Black-Scholes model and pricing formula to cope with a continuous dividend rate (see [21, chapter 5]). The following is an example of calling `blsprice`:

```
>> S0 = 50;
>> X = 52;
>> r = 0.1;
>> T = 5/12;
>> sigma = 0.4;
>> q = 0;
>> [C, P] = blsprice(S0, X, r, T, sigma, q)
C =
    5.1911
```

```
% PlotBLS.m
S0 = 30:1:70;
X = 50;
r = 0.08;
sigma = 0.4;
for T=2:-0.25:0
    plot(S0,blsprice(S0,X,r,T,sigma));
    hold on;
end
axis([30 70 -5 35]);
grid on
```

Fig. 1.15 Valuing a European call for different current prices of the underlying stock while approaching the expiration date.

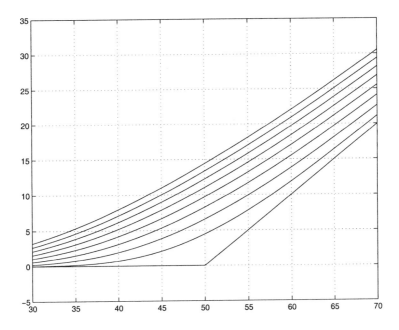

Fig. 1.16 Option value approaching the expiration date.

```
P =
     5.0689
```

It is interesting to plot the value of an option, say a vanilla European call, for different values of the current stock price while approaching the maturity. Running the code illustrated in figure 1.15, we get the plot of figure 1.16. We see that as time progresses, the plot approaches the kinky payoff diagram. An important point is that we have to be consistent in specifying the risk-free interest rate, the volatility, and the expiration date. In the snapshot above everything is expressed in a yearly base; hence, the expiration date is in five months. Similar functions are available to compute the Greeks, too; they are best illustrated through a simple example.

Example 1.14 The Greeks may be used to approximate the change in an option portfolio with respect to risk factors, just like duration and convexity for a bond portfolio, where the main risk factor is interest rate uncertainty. For instance, consider the change in the price of a call option due to an increase in the price of the underlying asset. Using a second-order Taylor expansion, we get the following approximation of this change:

$$C(S_0 + \delta S) \approx C(S_0) + \Delta \cdot \delta S + \frac{1}{2}\Gamma \cdot (\delta S)^2. \tag{1.21}$$

In MATLAB we may use such an approximation by exploiting the functions `blsdelta` and `blsgamma`. It is important to note that unlike the other two functions, `blsgamma` returns only one argument, as it can be shown that Γ is the same for a call and a put. A simple MATLAB snapshot shows that the approximation is fairly good:

```
>> C0 = blsprice(50, 50, 0.1, 5/12, 0.3)
C0 =
     4.8851
>> dS = 2;
>> C1 = blsprice(50+dS, 50, 0.1, 5/12, 0.3)
C1 =
     6.2057
>> delta = blsdelta(50, 50, 0.1, 5/12, 0.3)
delta =
     0.6225
>> gamma = blsgamma(50, 50, 0.1, 5/12, 0.3)
gamma =
     0.0392
>> C0 + delta*dS + 0.5*gamma*dS^2
ans =
     6.2086
```

By using the other Greeks, we could build an approximation of the change in the option price due to a change in risk factors such as the risk-free interest rate and the volatility. The Greeks may be exploited to measure the risk involved in an option portfolio. This point is developed in section 1.5, where we introduce the value-at-risk concept.   ▯

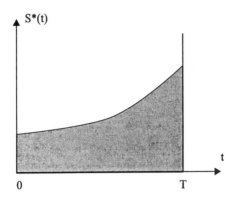

Fig. 1.17 Qualitative sketch of the early exercise boundary for a vanilla American put. The option is exercised within the shaded area.

1.4.4   Pricing American options by binomial lattices

Unlike their European counterparts, American options can be exercised at any date prior to expiration. This seemingly innocent variation makes the analysis of American options much more complex. One easy conclusion is that an American option has a larger value than the corresponding European option, as it gives more opportunity for exercise. From a theoretical point of view, valuing an American option entails the solution of a dynamic stochastic optimization problem. If you hold such an option, you must decide, for each time instant, if it is optimal or not to exercise the option. You should compare the *intrinsic value* of the option, i.e., the immediate payoff you would get from exercising the option early, and the *continuation value*, which is linked to the possibility of waiting for better opportunities. Within the partial differential equation framework, this translates to a *free boundary problem*. Consider a vanilla American put option.[15] What happens is that there is an early exercise boundary, such as the one illustrated in figure 1.17. This boundary specifies a stock price $S^*(t)$ such that if $S(t) < S^*(t)$, i.e., the option is sufficiently deep in-the-money, it is optimal to exercise it.[16] However, finding this boundary is part of the problem.

There is an easy way out of the difficulty, by resorting to a simplified model of the asset price dynamics. We may discretize the time horizon $[0, T]$ with steps of width δt. Consider a single time step. We know the asset price at the beginning of the time step; the price at the end of the period is a random variable. In section 1.4.1 we built a model where the probability distribution

---

[15]We consider a put option, since it can be shown that a vanilla American call option on a non-dividend-paying stock will never be exercised early, and its value is the same as the European call option (see, e.g., [8, p. 175]).

[16]For a detailed treatment of the exercise boundary for American options, see, e.g., [12, chapter 4].

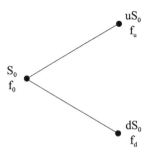

Fig. 1.18   Simple single-period binomial lattice.

is defined on a continuum of values, but we could also build a model based on discrete prices. The simplest model we may think of specifies only two possible values, accounting for the possibility of an increase and a decrease in the stock price. To be specific, let us consider figure 1.18. We start with a price S_0; at the next time instant we assume that the price may have gone up, so $S_1 = S_0u$, where $u > 1$, or it may have gone down, so $S_1 = S_0d$, where $d < 1$. Note the similarity with the multiplicative model of equation (1.8). What we are building is a binomial lattice, since at each step two outcomes are possible, with respective probabilities p_u and p_d. Now, imagine an option whose unknown value is denoted by f_0. If the option can only be exercised after δt, it is easy to find its values f_u and f_d corresponding to the two outcomes. How can we find f_0? We may again exploit the no-arbitrage principle. Let us set up a portfolio like that in equation (1.16). The initial portfolio value is

$$\Pi_0 = f_0 - \Delta S_0,$$

whereas the two possible outcomes after δt are

$$\Pi_u = f_u - \Delta u S_0$$
$$\Pi_d = f_d - \Delta d S_0.$$

We may choose Δ in such a way as to make the portfolio riskless, in which case

$$\Pi_u = \Pi_d \Rightarrow \Delta = \frac{f_u - f_d}{S_0(u - d)}$$

must hold. Note how Δ is a discretized derivative of the option value with respect to changes in the underlying price, which is consistent with our previous findings. But due to the no-arbitrage principle, if the portfolio is riskless, it must earn the risk-free interest rate r. Assuming continuous compounding, we must have

$$f_0 - S_0\Delta = (f_u - \Delta u S_0)e^{-r \cdot \delta t},$$

or

$$f_0 e^{r \cdot \delta t} = e^{r \cdot \delta t} S_0 \Delta + f_u - \Delta u S_0.$$

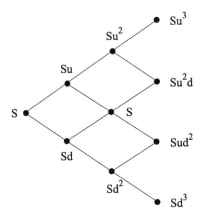

Fig. 1.19   Recombining binomial lattice.

Substituting the expression for Δ and rearranging, we obtain the expression

$$f_0 = e^{-r\cdot\delta t}[pf_u + (1-p)f_d],\qquad(1.22)$$

where

$$p = \frac{e^{-r\cdot\delta t} - d}{u - d}.\qquad(1.23)$$

It is interesting to note that the probabilities p_u and p_d do not play any role. This is again consistent with the Black-Scholes equation, which involves the risk-free rate but not the drifts μ and ν. However, equation (1.23) makes a fundamental interpretation clear; p may be regarded as a probability, since $0 \le p \le 1$. Furthermore, it may be considered a *risk-neutral probability*, in the sense that with this probability measure, the expected stock S return is the risk-free rate:

$$E[S] = pS_0u + (1-p)S_0d = S_0e^{r\cdot\delta t}.$$

We may interpret equation (1.22) as saying that the option value is the expected value of the payoff, according to a risk-neutral measure, discounted at the risk-free rate. The risk-neutral valuation principle has far-reaching consequences, and it is linked to the absence of arbitrage; we refer the reader to a book like [17] for a deeper, yet readable analysis.

Now, to make this discrete model useful, we should try to allow for a wider range of prices. The easiest way to do it is to increase the number of time steps. A convenient requirement is that $u = 1/d$. In this way, an up-step followed by a down-step yields the same price as a down-step followed by an up-step:

$$S_0ud = S_0du = S_0.$$

We obtain the recombining lattice for stock prices illustrated in figure 1.19. The advantage of building a recombining lattice rather than a general tree is evident in terms of computational effort, as we avoid a combinatorial explosion in the number of nodes. How can we select sensible values for u and d? One choice is to calculate them in order to match the stock price volatility σ. Assume for simplicity that $S_0 = 1$, so that the stock price S_1 is simply linked to the rate of return. We know from section 1.4.1 that the variance of the stock return, for small δt, is $\sigma^2 \cdot \delta t$. On the binomial lattice we have

$$
\begin{aligned}
\text{Var}(S_1) &= E[S_1^2] - E^2[S_1] \\
&= pu^2 + (1-p)d^2 - [pu + (1-p)d]^2 \\
&= e^{r \cdot \delta t}(u+d) - 1 - e^{2r \cdot \delta t},
\end{aligned}
$$

where we have used the expression for the risk-neutral probabilities and we have used the condition $ud = 1$. So we should solve the nonlinear equation

$$
e^{r \cdot \delta t}(u+d) - 1 - e^{2r \cdot \delta t} = \sigma^2 \cdot \delta t.
$$

By neglecting terms in $(\delta t)^2$ and higher, an approximate solution is

$$
u = e^{\sigma \sqrt{\delta t}} \qquad d = e^{-\sigma \sqrt{\delta t}}.
$$

It should be stressed that this is not the only plausible approach, and that alternative parameters are proposed in the literature. For pricing purposes, the probabilities for each up and down movement in the lattice are given by the risk-neutral probabilities given by equation (1.23). Assuming that the risk-free interest rate is constant in time, these parameters apply to the entire lattice. For each terminal node in the lattice, it is easy to calculate the option value, as it is simply given by the payoff. Now we should apply equation (1.22) recursively, going backward one step at a time, until we reach the initial node. The binomial lattice approach is best illustrated by its application to a vanilla European call.

Example 1.15 Suppose that we want to find the price of a vanilla European call with $S_0 = X = 50$, $r = 0.1$, $\sigma = 0.4$, and maturity in five months. From the Black-Scholes model, we know the solution:

```
>> call=blsprice(50,50,0.1,5/12,0.4)
call =
     6.1165
```

If we want to obtain the result by a binomial lattice, we must first set up the lattice parameters. Suppose that each time step is one month. Then

$$
\delta t = 1/12 = 0.0833
$$
$$
u = e^{\sigma \sqrt{\delta t}} = 1.1224
$$
$$
d = 1/u = 0.8909
$$
$$
p = \frac{e^{r \, \delta t} - d}{u - d} = 0.5073.
$$

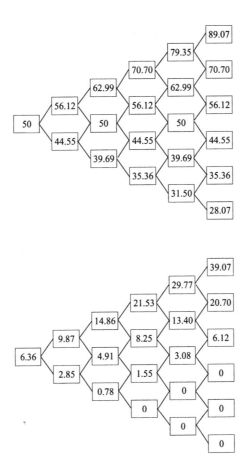

Fig. 1.20   Binomial lattices for the European call of example 1.15.

The resulting lattices for the stock price and the option value are shown in figure 1.20. The rightmost layer of values in the call price lattice is simply obtained by computing the option payoff. To clarify the calculations, let us consider how the first node in the next time layer is obtained:

$$e^{-r \cdot \delta t} \left[p \cdot 39.07 + (1-p) \cdot 20.77 \right]$$
$$= e^{-0.1 \cdot 0.0833} \left[0.5073 \cdot 39.07 + 0.4927 \cdot 20.77 \right] \approx 29.77.$$

Going on recursively, we see that the resulting option price is about 6.36, which is not too close to the exact price; yet we may refine the time step and solve the problem with MATLAB. To implement the approach in MATLAB we require an algebraic expression of the backward evaluation process. Let f_{ij} be the option value in node (i, j), where i refers to period i ($i = 0, \ldots, N$) and j is the jth node in period i (node numbers increase going up in the lattice, $j = 0, \ldots, i$). With these conventions, the underlying price in node (i, j) is $Su^j d^{i-j}$. At expiration we have

$$f_{N,j} = \max\{0, Su^j d^{N-j} - X\} \qquad j = 0, 1, \ldots, N.$$

Going backward in time (decreasing i), we get

$$f_{ij} = e^{-r \, \delta t} [p f_{i+1,j+1} + (1-p) f_{i+1,j}].$$

The implementation in MATLAB is straightforward, and the resulting code is shown in figure 1.21. The only point worth noting is that matrix indexes start from 1 in MATLAB, which requires a little adjustment. The function LatticeEurCall receives the usual arguments, with the addition of the number of time steps N. By increasing the last parameter, we see that we get a more accurate price (with an increase in the computing time):

```
>> call=LatticeEurCall(50,50,0.1,5/12,0.4,5)
call =
    6.3595
>>call=LatticeEurCall(50,50,0.1,5/12,0.4,500)
call =
    6.1140
```

It is interesting to investigate how the price computed by the binomial lattice converges to the correct price. This may be accomplished by the script in figure 1.22, which produces the output shown in figure 1.23. In this case, the error exhibits an oscillatory behavior for an increasing number of time steps.

□

The code we have developed is easily adjusted to cope with a vanilla American put on a non-dividend-paying stock. In each node we have to check for the opportunity of early exercise. With the same notation as above,

$$f_{N,j} = \max\{X - Su^j d^{N-j}, 0\} \qquad j = 0, 1, \ldots, N$$
$$f_{ij} = \max\{X - Su^j d^{i-j}, e^{-r \, \Delta t} [p f_{i+1,j+1} + (1-p) f_{i+1,j}]\}$$
$$i = N-1, N-2, \ldots, 0; \ j = 0, 1, \ldots, i.$$

```
function [price, lattice] = LatticeEurCall(S0,X,r,T,sigma,N)
deltaT = T/N;
u=exp(sigma * sqrt(deltaT));
d=1/u;
p=(exp(r*deltaT) - d)/(u-d);
lattice = zeros(N+1,N+1);
for j=0:N
   lattice(N+1,j+1)=max(0 , S0*(u^j)*(d^(N-j)) - X);
end
for i=N-1:-1:0
   for j=0:i
      lattice(i+1,j+1) = exp(-r*deltaT) * ...
         (p * lattice(i+2,j+2) + (1-p) * lattice(i+2,j+1));
   end
end
price = lattice(1,1);
```

Fig. 1.21   MATLAB code for pricing a European call by a binomial lattice.

```
% CompLatticeBLS.m
S0 = 50;
X = 50;
r = 0.1;
sigma = 0.4;
T = 5/12;
N=50;
BlsC = blsprice(S0,X,r,T,sigma);
LatticeC = zeros(1,N);
for i=(1:N)
   LatticeC(i) = LatticeEurCall(S0,X,r,T,sigma,i);
end
plot(1:N, ones(1,N)*BlsC);
hold on;
plot(1:N, LatticeC);
```

Fig. 1.22   Script to check the accuracy of the binomial lattice for decreasing δt.

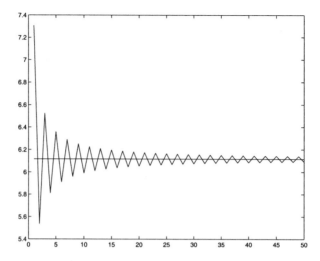

Fig. 1.23 Exact and approximate prices for increasing number of steps in a binomial lattice.

```
function [price, lattice] = LatticeAmPut(S0,X,r,T,sigma,N)
deltaT = T/N;
u=exp(sigma * sqrt(deltaT));
d=1/u;
p=(exp(r*deltaT) - d)/(u-d);
lattice = zeros(N+1,N+1);
for j=0:N
   lattice(N+1,j+1)=max(0 , X-S0*(u^j)*(d^(N-j)));
end
for i=N-1:-1:0
   for j=0:i
      lattice(i+1,j+1) = max( X-S0*u^j*d^(i-j) , ...
         exp(-r*deltaT)*(p*lattice(i+2,j+2) + (1-p)*lattice(i+2,j+1)));
   end
end
price = lattice(1,1);
```

Fig. 1.24 MATLAB code for pricing an American put by a binomial lattice.

The resulting code is shown in figure 1.24.

The Financial toolbox provides the user with a function, `binprice`, which prices vanilla American puts and calls, allowing for the possibility of continuous and lumpy dividends. We may compare `binprice` with `LatticeAmPut` on an American put with the same parameters as in example 1.15:

```
>>[stockpr, optpr] = binprice(50, 50, 0.1, 5/12, 1/12, 0.4, 0);
>> optpr(1,1)
ans =
    4.4885
>>p=LatticeAmPut(50,50,0.1,5/12,0.4,5)
p =
    4.4885
```

The function binprice requires a flag indicating if the option is a put (flag set to 0) or a call (flag set to 1). This parameter is the last one in the snapshot above; we have omitted the optional parameters that may be used to account for dividends.

Binomial lattices are a suitable tool to price many options and many extensions are possible, most notably trinomial lattices, to improve the accuracy. However, there are some disadvantages in the approach, as convergence may be slow and it may be difficult to account for path dependency.

1.4.5 Option pricing by Monte Carlo simulation

The binomial lattice approach is based on risk-neutral pricing, which is in turn linked to the no-arbitrage condition.[17] Equation (1.22) simply says that the option price is obtained by discounting the expected payoff under a risk-neutral probability measure. In more general terms, we may write the price of the derivative asset as

$$P = e^{-rT}\hat{\mathrm{E}}\{f(S_T)\},$$

where $f(S_T)$ is the option payoff and we have assumed that the risk-free rate is constant over time. For instance, in the case of a call option, we have

$$P = e^{-rT}\hat{\mathrm{E}}\{\max[S_T - X, 0]\}.$$

Here we have used the notation $\hat{\mathrm{E}}$ instead of E to emphasize the change in probability measure. Since the binomial lattice is built by discretizing the time horizon, one could wonder if an alternative approach can be devised relying on a continuous-time model. This way is interesting since the payoff could depend not only on the final price S_T of the underlying asset, but could be path dependent; furthermore, we could try to overcome some limitations of the Black-Scholes model by accounting for a stochastic volatility. The basic issue is how to make the stochastic process described by equation (1.13) risk-neutral. An informal argument is the following. From equation (1.14) we know that the expected asset price at time T is, under the "true" probability measure,

$$\mathrm{E}[S_T] = S_0 e^{\mu T}.$$

---

[17]See, e.g., [17] for a thorough treatment of this point.

By risk-neutral pricing we should get something like

$$S_0 = e^{-rT} \hat{\mathrm{E}} \{S_T\},$$

which is obtained if we substitute the risk-free rate r for the drift μ in the stochastic differential equation for the asset price:

$$dS = rS\, dt + \sigma S\, dz.$$

So all we have to do is to generate sample paths of the underlying asset price under the risk-neutral probability measure and to obtain an estimate of the discounted expected payoff. This can be done easily with Monte Carlo simulation, which is described in chapter 4. On the one hand, this approach is quite flexible and relatively easy to apply. On the other hand, it may be time consuming unless care is taken in sampling the price paths; furthermore, it is easily applied to European options but not to American options, since the simulation goes forward in time and it is difficult to account for optimal early exercise.[18]

1.5 VALUE-AT-RISK

Monte Carlo simulation, which we have just introduced for pricing derivatives, plays a role in other financial problems. One of them is the evaluation of value-at-risk (VaR). The VaR concept was introduced as an easy-to-understand measure of portfolio risk. In fact, measuring, monitoring, and managing risk are fundamental activities for any portfolio manager. Bonds and stocks involve different forms of risk, and derivatives, if used for speculation, may be even riskier.

Basically, VaR aims at measuring the maximum portfolio value one could lose, over a given time horizon, within a given confidence level. Suppose that our initial wealth is W_0 and the future (random) wealth is at the end of the time horizon

$$W = W_0(1 + R),$$

where R is the random rate of return. We are interested in characterizing the potential loss

$$\delta W = W - W_0 = W_0 R.$$

The VaR at confidence level α is implicitly defined by the following condition:

$$\mathrm{P}\{\delta W \leq -\mathrm{VaR}\} = 1 - \alpha. \tag{1.24}$$

---

[18]We should note that research is under way on the application of Monte Carlo simulation to American options.

Typical values for the confidence level could be $\alpha = 0.95$ or $\alpha = 0.99$. Let $f(r)$ be the density of the rate of return R. Then we should look for a critical rate of return r^* such that

$$P\{R \le r^*\} = \int_{-\infty}^{r^*} f(r)\, dr = 1 - \alpha.$$

The critical return r^* is obviously linked to a critical wealth w^*, since from equation (1.24) we may deduce

$$w^* - W_0 = -\text{VaR},$$

which in turn implies

$$\text{VaR} = W_0 - w^* = -W_0 r^*.$$

Note that the critical return is usually negative and VaR is positive. Sometimes VaR is defined with respect to the expected wealth:

$$\text{VaR} = E[W] - w^* = -W_0(r^* - E[R]).$$

The two definitions may give approximately the same value for a short time horizon, say a few days. In this case the volatility effect dominates the drift effect and $E[W] \approx W_0$. This assumption is not unreasonable, as regulations suggest using a risk measure in order to set aside enough cash to be able to cover short-term losses.

Computing VaR is easy if one assumes that returns are normally distributed. For simplicity, assume that we hold N shares of an asset whose current price is S. Let σ be the daily volatility for that asset; hence, for a period of length δt days, the volatility is $\sigma\sqrt{\delta t}$. Given a confidence level α, we have to obtain the corresponding critical number $z_{1-\alpha}$ by inverting the cumulative distribution function for the standard normal. For instance, if α is 99% and 95%:

```
>> z = norminv([0.01 0.05], 0, 1)
z =
  -2.3263   -1.6449
```

For the VaR over the time period δt, with a confidence level α, we have

$$\text{VaR} = -z_{1-\alpha}\sigma\sqrt{\delta t}\, NS, \tag{1.25}$$

where the term NS is the current wealth W_0. If the time horizon is longer, we should not neglect the drift due to the expected return. In such a case, we should modify (1.25) as follows:

$$\text{VaR} = NS(\mu\,\delta t - z_{1-\alpha}\sigma\sqrt{\delta t}),$$

where μ is the expected daily return. For a portfolio of assets, computing VaR is again easy. We have just to evaluate the portfolio risk as in mean-variance theory.

Example 1.16 Suppose that we hold a portfolio of two assets. The portfolio weights are $w_1 = 2/3$ and $w_2 = 1/3$, respectively; the two daily volatilities are $\sigma_1 = 2\%$ and $\sigma_2 = 1\%$, and the correlation is $\rho = 0.7$. Let the time horizon δt be 10 days. To obtain the portfolio risk, we compute the variance:

$$\sigma^2 = [w_1 \ w_2] \begin{bmatrix} \sigma_1^2 \, \delta t & \rho \sigma_1 \sigma_2 \, \delta t \\ \rho \sigma_1 \sigma_2 \, \delta t & \sigma_2^2 \, \delta t \end{bmatrix} \begin{bmatrix} w_1 \\ w_2 \end{bmatrix} = 0.0025111;$$

hence $\sigma = 0.05011$. Assuming that the overall portfolio value is \$10 million, and that the confidence level is 99%,

$$\text{Var} = 10^7 \cdot 2.3263 \cdot 0.05011 = \$1,165,709.$$

The same result can be obtained by using the MATLAB functions `portstats` and `portvrisk`. The first one, given the expected return vector for each asset, the covariance matrix, and the portfolio weights, computes the portfolio risk and the expected return:

`[PRisk, PReturn] = portstats (ExpReturn, CovMat, Wts).`

The second one computes the VaR, given the expected portfolio return, its risk, the risk threshold $1 - \alpha$, and the portfolio current value:

`VaR = portvrisk(PReturn, PRisk, RiskThreshold, PValue)`

Using these functions, we get
```
>> format bank
>> s1 = 0.02 * sqrt(10);
>> s2 = 0.01 * sqrt(10);
>> rho = 0.7;
>> CovMat = [ s1^2 rho*s1*s2 ; rho*s1*s2 s2^2];
>> s = PortStats([0 0], CovMat, [2/3 1/3]);
>> var = portvrisk(0,s,0.01,10000000)
var =
      1165755.90
```

The general formula for a portfolio of n assets of current price S_i, $i = 1, \ldots, n$, daily volatility σ_i, correlation ρ_{ij} between assets i and j, where we hold a number N_i of shares for each asset is

$$\text{VaR} = -z_{1-\alpha} \sqrt{\delta t \sum_{i=1}^{n} \sum_{j=1}^{n} N_i N_j \rho_{ij} \sigma_i \sigma_j S_i S_j}.$$

Computing the VaR for a stock portfolio, assuming normality, is easy, but what about a portfolio including nonlinear securities such as options? One possibility is to consider the underlying asset price as a risk factor, and to evaluate the sensitivities of the option value to risk factors; these sensitivities are the Greeks we introduced in section 1.4.2. For instance, given the Δ of an option, an approximation of the standard deviation for the option position is

$$\sigma\sqrt{\delta t}\, S\Delta.$$

This linear approximation may be improved by the delta-gamma approximation, which involves both the Δ and the Γ of an option (see example 1.14). By the same token, if we hold a bond portfolio and the interest rate is the risk factor, we may use duration and convexity to build simple approximations.

Since both option and bond sensitivities are valid only for small changes in the underlying risk factors, one could consider alternative approaches in order to improve our estimates. Monte Carlo simulation may be used to this purpose. We should select a probability distribution for the risk factors, and by proper sampling and evaluation of the portfolio asset prices for each sample we may obtain an accurate estimate of VaR. It should be stressed that for a large portfolio this procedure may be quite time consuming and that the variance reduction methods explained in chapter 4 should be used to ease the computational burden.

Both the VaR concept and the ways it is estimated have been criticized. On the one hand, it has been argued that VaR is not a good risk of measure; for instance, it does not really tell how much one is expected to lose *if* one falls in the unlucky tail of the distribution. Furthermore, it does not meet all of the requirements one would expect in a risk measure. We refer the reader to [1] for a discussion of these points. As to the way VaR is computed, an obvious critical point is the normality assumption. In practice, asset returns are not normal; they exhibit "fat tails," which means that extreme returns are more likely than one would expect in a normal world. This has serious implications for a measure like VaR, which is concerned with one tail of the distribution. An entire theory has been developed for dealing with extreme events and values, and it is particularly relevant for VaR (see [3]). Another point that is critical is the correlation among asset prices. In extreme cases, the correlation is likely to change with respect to what happens under usual circumstances. Finally, we mention an alternative approach for computing VaR, which does not rely on any assumed probability distribution; in *bootstrapping* one uses the historical returns basically by "shuffling" them to generate sample paths that are compatible with both fat tails and correlations. The idea is justified if one assumes that returns on different days are independent (which is basically linked to market efficiency). Bootstrapping has attractive features, but the results one obtains are clearly influenced by the historical data that are used, in particular if they include shocks and if they may be considered representative of the current situation. Despite all these critical points, VaR is a widely used measure, and numerical methods such as Monte Carlo simulation may play a significant role in its estimation.

S1.1   STOCHASTIC DIFFERENTIAL EQUATIONS AND ITO'S LEMMA

We now give an informal argument (following [8, chapter 10]) to justify Ito's lemma. Recall that an Ito process $x(t)$ satisfies a stochastic differential equation such as

$$dx = a(x,t)\, dt + b(x,t)\, dz, \qquad (1.26)$$

which is in some sense the continuous limit of

$$\delta x = a(x,t)\delta t + b(x,t)\epsilon(t)\sqrt{\delta t}, \qquad (1.27)$$

where $\epsilon \sim N(0,1)$ has a standard normal distribution. Our aim is to derive a stochastic differential equation for a function $F(x,t)$ of $x(t)$. One key ingredient is the formula for the differential of a function $G(x,y)$ of two variables:

$$dG = \frac{\partial G}{\partial x}dx + \frac{\partial G}{\partial y}dy,$$

which may be obtained from the Taylor expansion,

$$\delta G = \frac{\partial G}{\partial x}\delta x + \frac{\partial G}{\partial y}\delta y + \frac{1}{2}\frac{\partial^2 G}{\partial x^2}(\delta x)^2 + \frac{1}{2}\frac{\partial^2 G}{\partial y^2}(\delta y)^2 + \frac{\partial^2 G}{\partial x\, \partial y}\delta x\, \delta y + \cdots$$

when $\delta t \to 0$. Now we may apply this Taylor expansion to $F(x,t)$, limiting it to the leading terms. In doing so it is important to notice that the term $\sqrt{\delta t}$ in equation (1.27) needs careful treatment when squared. In fact, we have something like

$$(\delta x)^2 = b^2\epsilon^2\delta t + \cdots,$$

which implies that the term in $(\delta x)^2$ cannot be neglected in the approximation. Since ϵ is a standard normal variable, we have $E[\epsilon^2] = 1$ and $E[\epsilon^2\, \delta t] = \delta t$. A delicate point is the following. It can be shown that as δt tends to zero, the term $\epsilon^2\, \delta t$ can be treated as nonstochastic, and it is equal to its expected value. Hence, when δt tends to zero, in the Taylor expansion we have

$$(\delta x)^2 \to b^2\, dt.$$

Neglecting higher-order terms and taking the limit as both δx and δt tend to zero, we end up with

$$dF = \frac{\partial F}{\partial x}dx + \frac{\partial F}{\partial t}dt + \frac{1}{2}\frac{\partial^2 F}{\partial x^2}b^2\, dt,$$

which, substituting for dx, becomes

$$dF = \left(a\frac{\partial F}{\partial x} + \frac{\partial F}{\partial t} + \frac{1}{2}b^2\frac{\partial^2 F}{\partial x^2} \right) + b\frac{\partial F}{\partial x}dz.$$

Although this proof is far from rigorous, we see that all the trouble is due to the term of order $\sqrt{\delta t}$ linked to the Wiener process. This term plays a fundamental role, since it is the only reasonable modeling choice in order to avoid having a zero or infinite variance. Actually, it turns out the Wiener process is not differentiable, and this raises a thorny question. If we are dealing with a nondifferentiable object, how can something like a stochastic differential equation make sense? In fact, the notation (1.26) should be interpreted in the integral form:

$$x(t) = x(0) + \int_0^t a(x,s)\,ds + \int_0^t b(x,s)\,dz,$$

where the first integral is the usual (Riemann) integral but the second is an Ito stochastic integral (see, e.g., [15]). We will not pursue this any further, but the reader should appreciate why modern integration theory and measure theory play a prominent role in mathematical finance.

For further reading

In the literature

- A book dealing with investments in general and their mathematical modeling is [13]. It is comprehensive and quite readable. A higher-level treatment can be found in [9]. Another general reference is [21], which has a sharper focus on derivatives.

- If you are interested in how a stock exchange actually works, see [20].

- More specific references for bond markets and fixed-income-related assets are [4], [5], [6], and [18].

- Portfolio theory is covered in [2]; you might wish to have a look at chapter 10 to gain a deeper understanding of utility theory.

- The classical reference for options and derivatives in general is [8]. For a more formal treatment, see, e.g., [12].

- A good reference on value-at-risk is [10].

- A book dealing extensively with the intricacies of option hedging is [19]; it is not very readable for the uninitiated, but it gives a precise idea of practical option trading.

- A related topic is interest rate risk management, which is the subject of [7].

- There is a growing literature on continuous-time stochastic calculus in finance. Many books in this vein are quite hard to read; but if you want to find a good compromise between intuition and mathematical rigor, take a look at [15] or [14].

- Discrete-time models are dealt with in [17], which is an excellent reference for an understanding of the relationship between risk-neutral probability measures and the no-arbitrage hypothesis.

On the Web

- To consult a full and updated listing of MATLAB toolboxes, see http://www.mathworks.com.

- The Web page for AMPL is http://www.ampl.com/ampl, where you will find a list of vendors and compatible solvers.

- The GAMS Web page is http://www.gams.com.

- CPLEX and related software are described at http://www.cplex.com or http://www.ilog.com.

- The IBM Optimization Solutions and Library is described at http://www6.software.ibm.com/sos/osl/optimization.htm.

- A site where you may find a list many interesting resources for finance is http://fisher.osu.edu/fin/journal/jofsites.htm.

- An academic society that could be of interest to you is IAFE (International Association of Financial Engineers, http://www.iafe.org). Another interesting academic society is the Bachelier Finance Society (http://www.bachelierfinance.com).

REFERENCES

1. P. Artzner, F. Delbaen, J.-M. Eber, and D. Heath. Coherent Measures of Risk. *Mathematical Finance*, 9:203–228, 1999.

2. E.J. Elton and M.J. Gruber. *Modern Portfolio Theory and Investment Analysis (5th ed.)*. Wiley, New York, 1995.

3. P. Embrechts, C. Kluppelberg, and T. Mikosch. *Modelling Extremal Events for Insurance and Finance*. Springer-Verlag, Berlin, 2000.

4. F.J. Fabozzi. *Bond Markets: Analysis and Strategies.* Prentice Hall, Upper Saddle River, NJ, 1996.

5. F.J. Fabozzi. *Fixed Income Mathematics: Analytical and Statistical Techniques (3rd ed.).* McGraw-Hill, New York, 1997.

6. F.J. Fabozzi and G. Fong. *Advanced Fixed Income Portfolio Management: The State of the Art.* McGraw-Hill, New York, 1994.

7. B.E. Gup and R. Brooks. *Interest Rate Risk Management: The Banker's Guide to Using Futures, Options, Swaps, and Other Derivative Instruments.* Irwin Professional Publishing, New York, 1993.

8. J.C. Hull. *Options, Futures, and Other Derivatives (4th ed.).* Prentice Hall, Upper Saddle River, NJ, 2000.

9. Jr. J.E. Ingersoll. *Theory of Financial Decision Making.* Rowman & Littlefield, Totowa, NJ, 1987.

10. P. Jorion. *Value at Risk: The New Benchmark for Controlling Derivatives Risk.* McGraw-Hill, New York, 1997.

11. J.G. Kallberg and W.T. Ziemba. Comparison of Alternative Utility Functions in Portfolio Selection Problems. *Management Science,* 29:1257–1276, 1983.

12. Y.K. Kwok. *Mathematical Models of Financial Derivatives.* Springer-Verlag, Berlin, 1998.

13. D.G. Luenberger. *Investment Science.* Oxford University Press, New York, 1998.

14. T. Mikosch. *Elementary Stochastic Calculus with Finance in View.* World Scientific Publishing, Singapore, 1998.

15. S. Neftci. *An Introduction to the Mathematics of Financial Derivatives (2nd ed.).* Academic Press, San Diego, CA, 2000.

16. J. Paroush and E.Z. Prisman. On the Relative Importance of Duration Constraints. *Management Science,* 43:198–205, 1997.

17. S.R. Pliska. *Introduction to Mathematical Finance: Discrete Time Models.* Blackwell Publishers, Malden, MA, 1997.

18. S.M. Sundaresan. *Fixed Income Markets and Their Derivatives.* South Western College Publishing, Cincinnati, OH, 1997.

19. N. Taleb. *Dynamic Hedging: Managing Vanilla and Exotic Options.* Wiley, New York, 1996.

20. S.R. Veale, editor. *Stocks, Bonds, Options, Futures: Investments and their Markets.* New York Institute of Finance / Prentice Hall, Paramus, NJ, 1987.

21. P. Wilmott. *Quantitative Finance (vols. I and II).* Wiley, Chichester, West Sussex, England, 2000.

## Part II

Numerical Methods

2

Basics of numerical analysis

The core of the MATLAB system implements a set of functions to cope with some classical numerical problems. Although there is no need for a deep knowledge of numerical analysis in order to use MATLAB, a grasp of the basics is useful in order to choose among competing methods and to understand what may go wrong with them. Indeed, numerical computation is affected by machine precision and error propagation, in ways that may result in quite unreasonable outcomes. The effect of finite precision arithmetic and the issue of numerical instability are outlined briefly in section 2.1. This material is essential, among other things, in understanding the pitfalls of pricing derivatives by solving PDEs.

Then we describe methods for solving systems of linear equations in section 2.2; MATLAB provides the user with both direct and iterative methods to this purpose, and it is important to understand the characteristics of the two classes of methods. Section 2.3 introduces the reader to the problems of approximating functions and interpolating data values. Solving nonlinear equations is the subject of section 2.4. Finally, numerical integration is dealt with in section 2.5.

There are (at least) two basic topics that have been omitted here: computing matrix eigenvalues and eigenvectors and solving ordinary differential equations. Both problems are solved by methods available in MATLAB, but since they will not be used in the rest of the book, we refer the reader to the references listed at the end of the chapter.

2.1   NATURE OF NUMERICAL COMPUTATION

Real analysis is based on real numbers. Unfortunately, dealing with real numbers on a computer is impossible. Each number is represented by a finite number of bits, taking the values 0 or 1. Hence, we have to settle for binary and finite precision arithmetic. The progress in computing hardware has improved the quality of the representation, since more bits may be used efficiently without resorting to low-level software tricks. Yet some representation error is unavoidable, and its effect may lead to unexpected results.

2.1.1   Working with a finite precision arithmetic

To get a feeling for the pitfalls of working with a finite arithmetic, it is best to consider some simple examples. Consider the following expression:

$$9 \cdot 8.1 + 8.1$$

Everyone would agree that this is just a complicated way to write $10 \times 8.1 = 81$. Let us try it on a computer, using MATLAB:

```
>> 9 * 8.1 + 8.1
ans =
   81.0000
```

Everything seems right. Now, there is a built-in function in MATLAB, `fix`, which can be used to round a number to the integer nearest to zero.[1] Let us try it on the expression above:

```
>> fix(9*8.1 + 8.1)
ans =
    80
```

Now something seems quite wrong. Actually, the point is that the first result is not what it looks like. This may be seen by changing the visualization format of numbers and trying again:

```
>> format long
>> 9 * 8.1 + 8.1
ans =
   80.99999999999999
```

The problem is that an innocent-looking number like 8.1 is not represented exactly on a computer. This is because a computer works with a finite binary representation, which can represent some numbers only approximately (more

---

[1] The reader is urged to explore the differences between `fix` and the similar functions `floor`, `ceil`, and `round`.

on this in subsection 2.1.2), even if their representation is finite in another system, like the decimal system we are used to.

We see that changing the base may complicate the matter. One would also expect some trouble with numbers, such as $1/3$, that have no finite representation in the decimal base. To see an example of this, refer back to example 1.8 on page 40, where we considered a naive capital budgeting model. We had to select a subset of available investment opportunities, e.g., projects, taking budget constraints into account, with the aim of maximizing profit. One limitation of that model is that uncertainty is not modeled; another issue is that in general there might be some interaction among different projects. For instance, it could be the case that a certain project, say project P_0, may be started only if projects $P_1, P_2, \ldots, P_N$ are started. This logical constraint is easily modeled using the binary decision variables, say x_i $(i = 0, 1, \ldots, N)$, set to 1 if project P_i is started and 0 otherwise. One possibility is to express the constraint in the following form:

$$x_0 \leq \frac{\sum_{i=1}^{N} x_i}{N}.$$

If we start all the N required activities, the right-hand side of this inequality is simply $N/N = 1$, so that we *may* start P_0, since the constraint boils down to $x_0 \leq 1$. If some required project is missing, the constraint amounts to something like $x_0 \leq \alpha < 1$, which together with the binary requirement $x_0 \in \{0, 1\}$, enforces $x_0 = 0$. In principle, the idea is fine; does it really work on a computer? Well, in many cases it does, but consider what happens with $N = 3$. Project P_0 will never be selected. In fact, in this case, you should read the constraints as

$$x_0 \leq \frac{1}{3}x_1 + \frac{1}{3}x_2 + \frac{1}{3}x_3,$$

but unfortunately, even if all the x_i are set to 1, due to the finite precision of the computer we have something like

$$x_0 \leq 0.3333333 + 0.3333333 + 0.3333333 = 0.9999999 < 1,$$

where the number of decimals depends on the numerical precision of the machine and on the software involved. Actually, sophisticated optimization software for integer programming does not incur this trouble, since some integrality tolerance is introduced, and 0.9999999 is considered just like 1. Similar considerations apply to any high-quality numerical software. However, if the problem is first written to a text file, which is then loaded by an optimization solver, it may be the case that the number of digits is too small.[2] So it is

---

[2]For instance, if you solve the model within a modeling system like AMPL, calling a solver like CPLEX, there is no trouble. But if you write an MPS file and load the file with CPLEX, the result will not be correct. MPS files are text files representing optimization models according to standard rules; they are read by many optimization software packages.

better to avoid the trouble in the first place, by rewriting the constraint like this:

$$Nx_0 \le \sum_{i=1}^{N} x_i,$$

or, even better, in the disaggregated form

$$x_0 \le x_i \qquad i = 1, \ldots, N.$$

Why this is the preferred form depends on how mixed-integer programming problems are solved by branch and bound methods, explained in section 3.5 (see also example 3.16).

Numerical errors may affect the precision in representing numbers, but this issue is not much trouble in itself; after all, a derivative price will not be quoted in millionths of a dollar. But how about the *propagation* of errors within a numerical algorithm? The effect may well be a negative price for an option, as we will see in chapter 8.

One typically troublesome situation is when you subtract two nearly equal numbers. To see why, consider the following example:

$$\begin{aligned}
x &= 0.3721478693 \\
y &= 0.3720230572 \\
x - y &= 0.0001248121.
\end{aligned}$$

If you represent the numbers by five significant digits only (rounding the last one), the actual result will be

$$\hat{x} - \hat{y} = 0.37215 - 0.37202 = 0.00013,$$

with a relative error of about 4% with respect to the correct result. In fact, it is good practice to avoid expressions like

$$\sqrt{x^2 + 1} - 1,$$

which could result in remarkable losses in significance for small values of x. In such cases, it is easy to rewrite the expression as follows:

$$\sqrt{x^2 + 1} - 1 = \left(\sqrt{x^2 + 1} - 1\right) \left(\frac{\sqrt{x^2 + 1} + 1}{\sqrt{x^2 + 1} + 1}\right) = \frac{x^2}{\sqrt{x^2 + 1} + 1}.$$

Here there is no subtraction involved, but In other cases, there is no easy way to avoid the difficulty. We may consider a classical example involving the solution of a system of linear equations.

Example 2.1 Let us consider a system of linear equations:

$$\mathbf{Hx} = \mathbf{b},$$

where **H** is a particular matrix known as the *Hilbert matrix*:

$$\mathbf{H} = \begin{bmatrix} 1 & \frac{1}{2} & \frac{1}{3} & \cdots & \frac{1}{n} \\ \frac{1}{2} & \frac{1}{3} & \frac{1}{4} & \cdots & \frac{1}{n+1} \\ \frac{1}{2} & \frac{1}{3} & \frac{1}{4} & \cdots & \frac{1}{n+1} \\ \frac{1}{3} & \frac{1}{4} & \frac{1}{5} & \cdots & \frac{1}{n+2} \\ \vdots & \vdots & \vdots & \ddots & \vdots \\ \frac{1}{n} & \frac{1}{n+1} & \frac{1}{n+2} & \cdots & \frac{1}{2n-1} \end{bmatrix}.$$

MATLAB provides you with a function, `hilb`, to build a Hilbert matrix. Let us try solving the system for $n = 20$; we cheat a little here, since we assume that the solution is known, and we build the term **b**; then we check if that solution is obtained by solving the system. Let the solution be

$$\mathbf{x} = [1\ 2\ 3\ \cdots\ n]^T.$$

Using MATLAB, we obtain something like[3]

```
>> H=hilb(20);
>> x=(1:20)';
>> b=H*x;
>> H\b
Warning: Matrix is close to singular or badly scaled.
         Results may be inaccurate. RCOND = 3.557198e-019.
ans =
    1.0000
    2.0006
    2.9771
    4.3845
    1.5693
   23.8973
  -49.5444
  113.7311
  -93.4233
   37.6248
    0.1412
  145.4572
 -210.2104
  202.1644
 -197.4117
  301.8510
```

[3]The actual result may depend on the MATLAB version you use.

```
-172.4143
  25.1207
  67.0509
   4.0339
```

We see that the result doesn't look quite as it should. This is the effect of propagation of numerical errors, giving rise to numerical instability. Indeed, the problem here is that the Hilbert matrix is close to singular; in fact, this is detected by MATLAB, which issues a warning message about the matrix condition number (see later). In section 2.2.2 we will see that some methods to solve systems of linear equations involve repeated subtractions, and this is the root of the difficulty.   ◻

2.1.2   Number representation, rounding, and truncation

The usual way we represent numbers is based on a decimal base. When writing 1492, we actually mean

$$1 \times 10^3 + 4 \times 10^3 + 9 \times 10^1 + 2 \times 10^0.$$

Similarly, when we have to represent the fractional part of a number, we use negative powers of the base 10:

$$0.42 \Rightarrow 4 \times 10^{-1} + 2 \times 10^{-2}.$$

Some numbers, such as $1/3 = 0.\overline{3}$, do not have a finite representation and should be thought as limits of an infinite series. However, on a computer we must use a binary base, since the hardware is based on a binary logic: for instance,

$$(21.5)_{10} \Rightarrow 2^4 + 2^2 + 2^0 + 2^{-1} = (10101.1)_2.$$

How can we convert numbers from a decimal to a binary base? Let us begin with an integer number N. It can be thought of as

$$N = (b_k \cdot 2^k) + (b_{k-1} \cdot 2^{k-1}) + \cdots + (b_1 \cdot 2^1) + (b_0 \cdot 2^0).$$

Dividing both sides by 2, we get

$$\frac{N}{2} = (b_k \cdot 2^{k-1}) + (b_{k-1} \cdot 2^{k-2}) + \cdots + (b_1 \cdot 2^0) + \frac{b_0}{2}.$$

Hence, the rightmost digit in the binary representation, b_0, is simply the remainder of the integer division of N by 2. We may think of N as

$$N = 2 \cdot Q + b_0,$$

where Q is the result of the integer division by 2. Repeating this step, we obtain all the digits of the binary representation. This suggests the algorithm

```
function b=DecToBinary(n)
n0 = n;
i=1;
while (n0 > 0)
    n1 = floor(n0/2);
    b(i) = n0 - n1*2;
    n0=n1;
    i = i+1;
end
b=fliplr(b);
```

Fig. 2.1 MATLAB code to obtain the binary representation of an integer number.

whose MATLAB code is illustrated in figure 2.1. The function DecToBinary takes an integer number n and returns a vector b containing the binary digits:[4]

```
>> DecToBinary(3)
ans =
     1     1
>> DecToBinary(8)
ans =
     1     0     0     0
>> DecToBinary(13)
ans =
     1     1     0     1
```

Similarly, the fractional part of a number is represented in a binary base as

$$R = \sum_{k=1}^{\infty} d_k 2^{-k}.$$

Some numbers, which can be represented finitely in a decimal base, cannot in a binary base: for instance,

$$7/10 = (0.7)_{10} = (0.1\overline{0110})_2.$$

Clearly, in such cases the infinite series is truncated, with a corresponding error. The binary representation of a fractional number R can be obtained by the following algorithm, which is similar to the previous one (int and frac denote the integer and the fractional part of a number, respectively):

---

[4]This is not the best implementation, as the output vector b is resized incrementally. We could compute the number of necessary bits and preallocate b.

1. Set $d_1 = \text{int}(2R)$ and $F_1 = \text{frac}(2R)$.

2. Recursively compute $d_k = \text{int}(2F_{k-1})$ and $F_1 = \text{frac}(2F_{k-1})$ for $k = 2, 3, \ldots$

Knowing how to change the base may seem useless, but we will see an application of these procedures in section 4.5, dealing with quasi-Monte Carlo simulation.

In practice, we have to represent both quite large and quite small numbers. Hence we resort to a floating-point representation like

$$x \approx \pm q \times 2^n,$$

where q is the mantissa and n is the exponent. The exact details of the representation depend on the standard chosen and the underlying hardware. In any case, since only a finite memory space is available to store the mantissa, we will have a *roundoff* error.

Rounding off is not the only source of error in numerical computation. Another one is truncation. This occurs, for instance, when we substitute a finite sum for an infinite sum. As an example, consider the following expression for the exponential function:

$$e^x = \sum_{k=0}^{\infty} \frac{x^k}{k!}.$$

When we truncate a sum like this, a truncation error occurs.

2.1.3   Error propagation and instability

Roundoff errors have been mitigated by the increase in the number of bits used to store numbers on modern computers. From a practical perspective, numbers are virtually represented exactly. Nevertheless, such errors may accumulate within the steps of an algorithm, possibly with disruptive effects, as we have seen in example 2.1. Hence, algorithms should be analyzed with respect to their numerical stability properties. We will see specific examples later in the context of finite difference methods for PDEs; here we just introduce some basic concepts within a general setting.

From an abstract point of view, a numerical problem may be considered as a mapping:

$$y = f(x),$$

which transforms the input data x into the output y. An algorithm is a computationally workable approach to computing that function; different algorithms may be used to solve the same numerical problem, possibly with different characteristics with respect to computational effort and stability properties. When using a computer, roundoff errors will be introduced in the representation of the input; we should check the effects on the output of a perturbation δx in the input data. Denoting the actual input by $\bar{x} = x + \delta x$, the output should be $f(\bar{x})$, whereas an algorithm will yield some answer, say y^*. An algorithm is stable if the relative error

$$\frac{\|f(\bar{x}) - y^*\|}{\|f(\bar{x})\|}$$

is of the same order of magnitude as the machine precision.

By comparing $f(\bar{x})$ with $f(x)$, we analyze a different issue, called the *conditioning* of the numerical problem. We should compare the error in the output with the error in the input; when the input error is small, the output error should be small, too. Ideally, it would be nice to have a bounding relationship like

$$\frac{\|f(x) - f(\bar{x})\|}{\|f(x)\|} \leq K \frac{\|x - \bar{x}\|}{\|x\|}, \tag{2.1}$$

where $\|\cdot\|$ is an appropriate norm.[5] The number K is called the *condition number* of the problem. Later, we investigate the condition number for the problem of solving a system of linear equations.

Note that the conditioning issue is linked to the numerical problem itself, not to the specific algorithm used to solve it. Putting the two concepts together, we will find a "good" answer to a specific problem when the problem is well conditioned and the algorithm is stable.

2.1.4 Order of convergence and computational complexity

Many numerical algorithms are based on an iterative scheme. Given an approximate solution $\mathbf{x}^{(k)}$, some transformation is applied to obtain an improved approximation $\mathbf{x}^{(k+1)}$. The minimal requirement of a good algorithm is that the sequence generated converges to the correct solution $\mathbf{x}^*$. Furthermore, one would hope that such convergence is reasonably fast. The speed of convergence may be quantified by a rate. We say that the sequence converges at a rate q if

$$\lim_{k \to \infty} \frac{\|\mathbf{x}^{(k+1)} - \mathbf{x}^*\|}{\|\mathbf{x}^{(k)} - \mathbf{x}^*\|^q} < \infty.$$

The larger the rate q, the better; quadratic convergence ($q = 2$) is preferred to linear convergence ($q = 1$). Sometimes, convergence depends on the initial estimate $\mathbf{x}^{(0)}$.

When we use an iterative algorithm, we may have no precise idea of the number of iterations we need to get a satisfactory solution. In other cases, some direct method will yield the answer. By *direct method* we mean a procedure which, after a known number of steps, gives the desired solution (if no difficulty due to instability arises). For direct methods, it may be possible to

---

[5]The reader should be familiar with the norm concept for vectors; anyway, it is recalled in section 2.2.1.

quantify the number of elementary operations (e.g., additions and multiplications) needed to get the answer; this measures the computational complexity of the algorithm. The amount of computation will be a function of the size of the problem. The number of operations may depend on implementation details, and the size of the problem may depend on the type of encoding used to represent the problem. In practice, it is not necessary to be overly precise in this measure. It is usually enough to have an idea of the rate of growth of the computational effort with respect to the increase in problem size. Furthermore, the computational complexity may depend on the specific problem instance at hand, where by *problem instance* we mean a specific problem with specific numerical data. Sometimes, it is possible to analyze the average complexity with respect to the universe of problem instances. Usually, it is easier to quantify the worst-case complexity.

Consider again the capital budgeting problem of example 1.8. Since there is a finite set of possible solutions, in principle one could find the optimal solution by enumerating all of them. However, since each project may or may not be financed, there may be up to 2^N solutions, where N is the number of competing projects and is the essential measure of the problem size. This number is actually only an upper bound on the number of solutions, since many will be infeasible with respect to the budget constraint. Yet we may say that the worst-case complexity of complete enumeration is on the order of 2^N [technically speaking, we say that the complexity is $O(2^N)$].[6] Clearly, an exponential growth like this is quite undesirable. Efficient algorithms are usually characterized by a polynomial growth of the computational effort; their complexity is something like $O(N^p)$ for some constant p.

2.2 SOLVING SYSTEMS OF LINEAR EQUATIONS

We have solved a small system of linear equations to cope with an elementary problem in example 1.6. The solution of systems of linear equations is an important problem per se; however, it is also instrumental for a variety of other problems. For instance, Newton's method for solving systems of nonlinear equations calls for the repeated solution of linear systems (see section 2.4.2); in chapter 5 we will also see how solving linear systems is needed in certain methods to cope with PDEs.

It is not our aim to dwell too deeply on this subject; we limit ourselves to the basic concepts needed to understand what MATLAB offers to solve linear equations. Example 2.1 shows that solving linear systems may be a difficult task in the case of an ill-conditioned matrix. A condition number can be associated with a numerical problem [see equation (2.1)] to measure

---

[6]A function $f(n)$ is $O(g(n))$ if $\lim_{n \to \infty} f(n)/g(n) < \infty$.

the potential difficulties due to numerical instability; for the case of linear equations, a condition number may be associated with a matrix.

Methods of solving linear equations can be broadly classified as direct or iterative. Direct methods have a clearly defined computational complexity, as they yield the result directly within a given number of steps; *iterative methods* build a sequence of solutions whose limit is (under some conditions) the desired solution. For iterative methods, the number of steps is not known a priori, as it depends on convergence speed. They are useful for some large systems characterized by sparse matrices (i.e., matrices with a small number of nonzero entries). Both classes are available in MATLAB, and there exist definite situations where application of one class is advantageous over application of the other.

2.2.1   Condition number for a matrix

To introduce the condition number for a matrix, we must first recall some definitions related to vector and matrix norms. The norm is a function mapping vectors $\mathbf{x} \in \mathbb{R}^n$ to real numbers $\|\mathbf{x}\|$ such that:

- $\|\mathbf{x}\| > 0$ for any $\mathbf{x} \neq \mathbf{0}$, and $\|\mathbf{x}\| = 0$ if and only if $\mathbf{x} = \mathbf{0}$.
- $\|c\mathbf{x}\| = |c| \, \|\mathbf{x}\|$ for any $c \in \mathbb{R}$.
- $\|\mathbf{x} + \mathbf{y}\| \leq \|\mathbf{x}\| + \|\mathbf{y}\|$ for any $\mathbf{x}, \mathbf{y} \in \mathbb{R}^n$.

These properties are the intuitive properties a measure of vector length should satisfy. The most natural way to define a vector length is through the Euclidean norm

$$\|\mathbf{x}\|_2 \equiv \sqrt{\sum_{i=1}^{n} |x_i|^2},$$

which extends the ordinary notion of distance between two points on the plane or in the space. However, there are different notions of vector length, which satisfy the conditions above for a vector norm. The most common ones are:

- $\|\mathbf{x}\|_\infty \equiv \max_{1 \leq i \leq n} |x_i|$.
- $\|\mathbf{x}\|_1 \equiv \sum_{i=1}^{n} |x_i|$.

Generally speaking, one may define a vector p-norm as

$$\|\mathbf{x}\|_p = \left(\sum_i |x_i|^p \right)^{1/p}.$$

A less familiar concept is the matrix norm, which is defined by requiring the same properties as above. In the case of square matrices, the norm function maps $\mathbb{R}^{n \times n}$ to $\mathbb{R}$. The required properties are:

- $\|\mathbf{A}\| > 0$ for any $\mathbf{A} \neq \mathbf{0}$, and $\|\mathbf{A}\| = 0$ if and only if $\mathbf{A} = \mathbf{0}$.

- $\|c\mathbf{A}\| = |c| \cdot \|\mathbf{A}\|$ for any $c \in \mathbb{R}$.

- $\|\mathbf{A} + \mathbf{B}\| \leq \|\mathbf{A}\| + \|\mathbf{B}\|$ for any $\mathbf{A}, \mathbf{B} \in \mathbb{R}^{n \times n}$.

Sometimes, the following additional condition is required:

$$\|\mathbf{AB}\| \leq \|\mathbf{A}\| \cdot \|\mathbf{B}\|.$$

Typical matrix norms are:

- $\|\mathbf{A}\|_\infty \equiv \max_{1 \leq i \leq n} \sum_{j=1}^{n} |a_{ij}|$.

- $\|\mathbf{A}\|_1 \equiv \max_{1 \leq j \leq n} \sum_{i=1}^{n} |a_{ij}|$.

- $\|\mathbf{A}\|_F \equiv \left(\sum_{i=1}^{n} \sum_{j=1}^{n} |a_{ij}|^2 \right)^{1/2}$, the Frobenius norm.

- $\|\mathbf{A}\|_2 \equiv \sqrt{\rho(\mathbf{A}^T \mathbf{A})}$, the *spectral norm*, where $\rho(\cdot)$ is the spectral radius of a matrix, i.e., $\rho(\mathbf{B}) \equiv \max\{|\lambda_k| : \lambda_k \text{ is an eigenvalue of } \mathbf{B}\}$.

Example 2.2 Vector and matrix norms are computed in MATLAB by the norm function.

```
>> v = [2 4 -1 3];
>> [norm(v,1) norm(v,2) norm(v,inf)]
ans =
   10.0000    5.4772    4.0000
```

The function takes two arguments: the vector and an optional parameter specifying the type of norm. The default value for the optional parameter is 2. A call like norm(v,p) corresponds to

sum(abs(v).^p)^(1/p).

The $\|\cdot\|_\infty$ norm is computed when the value of the optional parameter is inf. Similarly, a call like

```
>> A = [ 2 4 -1 ; 3 1 5 ; -2 3 -1];
>> [norm(A,inf) norm(A,1) norm(A,2) norm(A,'fro')]
ans =
    9.0000    8.0000    6.1615    8.3666
```

computes the four matrix norms we have defined, including the spectral and Frobenius norms. For the spectral norm, you may check the result by computing the square root of the eigenvalues of $\mathbf{A}^T \mathbf{A}$:

```
>> sqrt(eig(A' * A))
ans =
```

```
2.2117
5.2100
6.1615
```

and picking up the largest value.   ▯

When solving a system of linear equations, we deal with both matrices and vectors. Hence, it is natural to investigate the relationships between vector and matrix norms. We say that a matrix and a vector norms are compatible if

$$\|\mathbf{Ax}\| \leq \|\mathbf{A}\| \cdot \|\mathbf{x}\|$$

for any $\mathbf{A} \in \mathbb{R}^{n \times n}$ and $\mathbf{x} \in \mathbb{R}^n$. It can be shown that for any matrix norm that is compatible with a vector norm, we have

$$\rho(\mathbf{A}) \leq \|\mathbf{A}\|.$$

Given a vector norm, it is also possible to build a matrix norm as follows:

$$\|\mathbf{A}\| \equiv \sup_{\mathbf{x} \neq 0} \frac{\|\mathbf{Ax}\|}{\|\mathbf{x}\|} = \max_{\|\mathbf{x}\|=1} \|\mathbf{Ax}\|.$$

In this case we say that the matrix norm is *induced* by the vector norm. It is easy to see that in this case the two norms are compatible. For instance, it can be shown that the vector $\|\cdot\|_\infty$ norm induces the matrix $\|\cdot\|_\infty$ norm and that the Euclidean vector norm induces the spectral matrix norm (see, e.g., [11]).

Now we are ready to start analyzing the effect of numerical errors on the solution of a linear system. Consider the system

$$\mathbf{Ax} = \mathbf{b}$$

and suppose that we perturb $\mathbf{b}$ by adding a term $\delta\mathbf{b}$; such a perturbation may indeed occur due to rounding off. Then the solution will somehow be perturbed, too. We will have

$$\mathbf{A}(\mathbf{x} + \delta\mathbf{x}) = \mathbf{b} + \delta\mathbf{b},$$

which implies that

$$\mathbf{A}\delta\mathbf{x} = \delta\mathbf{b} \implies \delta\mathbf{x} = \mathbf{A}^{-1}\delta\mathbf{b}.$$

We would like to assess the error in the solution, $\delta\mathbf{x}$, as a function of the input error $\delta\mathbf{b}$. If we adopt compatible matrix and vector norms, we may write

$$\|\delta\mathbf{x}\| = \|\mathbf{A}^{-1}\delta\mathbf{b}\| \leq \|\mathbf{A}^{-1}\| \cdot \|\delta\mathbf{b}\|$$
$$\|\mathbf{b}\| = \|\mathbf{Ax}\| \leq \|\mathbf{A}\| \cdot \|\mathbf{x}\|.$$

Dividing term by term these two inequalities yields

$$\frac{\|\delta\mathbf{x}\|}{\|\mathbf{A}\|\,\|\mathbf{x}\|} \leq \|\mathbf{A}^{-1}\| \cdot \frac{\|\delta\mathbf{b}\|}{\|\mathbf{b}\|} \implies \frac{\|\delta\mathbf{x}\|}{\|\mathbf{x}\|} \leq \|\mathbf{A}\| \cdot \|\mathbf{A}^{-1}\| \cdot \frac{\|\delta\mathbf{b}\|}{\|\mathbf{b}\|},$$

which is analogous to (2.1). The condition number $K(\mathbf{A}) \equiv \|\mathbf{A}\| \cdot \|\mathbf{A}^{-1}\|$ gives an upper bound on the ratio of the relative error in the solution to the relative perturbation. Generally speaking, the higher the condition number, the more difficult it is to solve a linear system. Indeed, it can be shown that the condition number gives a measure of how close a matrix is to being singular.

Example 2.3 The cond function computes the condition number. An optional parameter may be provided to select a norm; the default value corresponds to the spectral norm.

```
>> cond(hilb(3))
ans =
  524.0568
>> cond(hilb(7))
ans =
  4.7537e+008
>> cond(hilb(10))
ans =
  1.6025e+013
```

Checking these numbers it is easy to see why solving a linear system involving the Hilbert matrix is a difficult task. □

2.2.2   Direct methods for solving systems of linear equations

Direct methods for solving linear equations are based on the idea of transforming the matrix into a suitable form. More precisely, the system is transformed into an equivalent one, i.e., a system with the same solution as the original one. The point is to obtain an equivalent form that is easy to solve. In fact, solving the system

$$\mathbf{Ax} = \mathbf{b}$$

is easy if $\mathbf{A}$ is an upper triangular matrix, i.e., if the system looks like the following:

$$
\begin{aligned}
a_{11}x_1 + a_{12}x_2 + \cdots + a_{1n}x_n &= b_1 \\
a_{22}x_2 + \cdots + a_{2n}x_n &= b_2 \\
\vdots &= \vdots \\
a_{nn}x_n &= b_n.
\end{aligned}
$$

In this case, it is easy to solve the system by back substitution, starting from the last variable x_n:

$$x_n = \frac{b_n}{a_{nn}}$$

$$x_k = \left(b_k - \sum_{j=k+1}^{n} a_{kj}x_j \right) \Big/ a_{kk} \qquad k = n-1, n-2, \ldots, 1.$$

A similar method could be used for lower triangular matrices. So we should come up with a systematic method to transform a linear system of equations into an equivalent triangular form. Gaussian elimination is such a procedure. In principle, the idea is rather simple; we must form linear combinations of equations in order to eliminate some coefficients from some equations. Since combining equations linearly does not change the solution, the resulting system is equivalent to the original one. Starting from the system in the form

$$
\begin{array}{ll}
(E_1) & a_{11}x_1 + a_{12}x_2 + \cdots + a_{1n}x_n = b_1 \\
(E_2) & a_{21}x_1 + a_{22}x_2 + \cdots + a_{2n}x_n = b_2 \\
& \vdots \qquad = \quad \vdots \\
(E_n) & a_{n1}x_1 + a_{n2}x_2 + \cdots + a_{nn}x_n = b_n
\end{array}
$$

we may try to obtain a column of zeros under the coefficient a_{11}. This is the first step in getting an equivalent triangular system. For each equation (E_k) $(k = 2, \ldots, n)$, we must apply the transformation

$$(E_k) \leftarrow (E_k) - \frac{a_{k1}}{a_{11}}(E_1),$$

which leads to the equivalent system:

$$
\begin{array}{ll}
a_{11}x_1 + a_{12}x_2 + \cdots + a_{1n}x_n = b_1 \\
a_{22}^{(1)}x_2 + \cdots + a_{2n}^{(1)}x_n = b_2^{(1)} \\
\vdots \qquad = \quad \vdots \\
a_{n2}^{(1)}x_2 + \cdots + a_{nn}^{(1)}x_n = b_n^{(1)}.
\end{array}
$$

Now we may repeat the procedure to obtain a column of zeros under the coefficient $a_{22}^{(1)}$, and so on, until the desired form is obtained, allowing for back substitution. We will not quantify the number of operations needed for the overall procedure, but it is evident that the algorithm has a quantifiable computational complexity. Actually, what we have explained is only the starting point of Gaussian elimination, as many things may go wrong with this naive procedure. A first point is that we must have $a_{11} \neq 0$ to carry out the first step of the Gaussian elimination; by the same token, we must have $a_{22}^{(1)} \neq 0$ and so on. Fortunately, if the original system is nonsingular, this may be accomplished by a suitable permutation of variables (columns) or equations (row). In practice, some further care is needed to minimize the effects of finite precision arithmetic. We have seen that subtraction is a potentially dangerous operation, because of the potential loss of significance. Suitable row and column permutations may help in keeping the trouble to a minimum; such operations are called *pivoting*. Scaling the size of the coefficients may be used, too. These points are well treated in any numerical analysis book, and the details are beyond the scope of this one.

It is also worth noting that Gaussian elimination may be seen as a way of factoring the matrix $\mathbf{A}$ into the product of a lower triangular matrix $\mathbf{L}$ and an upper triangular matrix $\mathbf{U}$. More precisely we have

$$\mathbf{PA} = \mathbf{LU},$$

where $\mathbf{P}$ is a permutation matrix, i.e., a matrix whose entries are either 0 or 1, such that only one element along each row and each column is nonzero. It may be seen that multiplying a generic matrix by such a matrix has the effect of exchanging rows. We may try to understand, at least intuitively, where the above factorization comes from. The permutation matrix $\mathbf{P}$ corresponds to the pivoting operations; if pivoting is not required for a matrix, then this matrix can be neglected. The upper triangular matrix $\mathbf{U}$ corresponds to the end result of Gaussian elimination we just described. The lower triangular matrix $\mathbf{L}$ corresponds to the transformations we must carry out to obtain the equivalent system in upper triangular form. These transformations are linear combinations of rows, which can be obtained by multiplying the original matrix by suitable elementary matrices; the matrix $\mathbf{L}$ is linked to the product of these elementary matrices. This factorization is called *LU-decomposition*.

Example 2.4 LU-decomposition is obtained in MATLAB by calling the `lu` function with a matrix argument.

```
>> A = [1 4 -2 ; -3 9 8; 5 1 -6];
>> [L,U,P] = lu(A)
L =
      1.0000        0        0
     -0.6000   1.0000        0
      0.2000   0.3958   1.0000
U =

      5.0000   1.0000  -6.0000
           0   9.6000   4.4000
           0        0  -2.5417
P =
        0      0      1
```

```
0    1    0
1    0    0
```

With such a factorization, solving a system like $\mathbf{Ax} = \mathbf{b}$ is equivalent to solving the two systems

$$\begin{aligned}\mathbf{Ly} &= \mathbf{Pb} \\ \mathbf{Ux} &= \mathbf{y}\end{aligned}$$

in cascade.

```
>> b = [1;2;3];
>> x = A\b
x =
    1.0820
    0.1967
    0.4344
>> x = U \ ( L \ (P*b))
x =
    1.0820
    0.1967
    0.4344
```

It is worth noting that an explicit LU-decomposition may be advantageous when it is necessary to solve a system repeatedly with different right-hand sides, as occurs in the solution of certain PDEs by finite difference methods.
                                                                              ▯

The LU-decomposition takes a special form when applied to a symmetric positive definite matrix; such matrices occur in many optimization problems, and a typical example is a covariance matrix. If $\mathbf{A}$ is a symmetric positive definite matrix, it can be shown that there exists a unique upper triangular matrix $\mathbf{L}$ such that $\mathbf{A} = \mathbf{L}^T\mathbf{L}$; this is called *Cholesky factorization*, as it decomposes $\mathbf{A}$ into the product of two factors.[7] The Cholesky factorization may be a suitable alternative to the usual Gaussian elimination for special matrices.

Example 2.5 The Cholesky factorization is computed in MATLAB by the chol function. For instance, let us define a matrix and check that it is positive definite:

```
>> A = [ 3 1 4 ; 1 5 3 ; 4 3 7 ]
```

---

[7]In many texts, a lower triangular matrix is considered, and the factorization is written as $\mathbf{A} = \mathbf{LL}^T$. It is easy to see that the two definitions are actually equivalent. We will stick to this one, since the MATLAB function chol returns an upper triangular matrix.

```
A =
     3     1     4
     1     5     3
     4     3     7
>> eig(A)
ans =
    0.3803
    3.5690
   11.0507
```

Given a known term **b**, we may factor **A** and solve the system.

```
>> b=(1:3)';
>> L=chol(A)
L =
    1.7321    0.5774    2.3094
         0    2.1602    0.7715
         0         0    1.0351
>> L \ (L' \ b)
ans =
   -1.0000
   -0.0000
    1.0000
```
                                                                    ⬜

In chapter 4 we will see that the Cholesky factorization is useful when we have to simulate random variables with a multivariate normal distribution.

2.2.3 Tridiagonal matrices

In certain applications, the matrix of a system of linear equations has a very specific form. One such case is the tridiagonal matrix, which may occur in the solution of option pricing problems by PDEs. A tridiagonal matrix has the following form:

$$
\mathbf{A} = \begin{bmatrix}
a_{11} & a_{12} & 0 & 0 & 0 & \cdots & 0 \\
a_{21} & a_{22} & a_{23} & 0 & 0 & \cdots & 0 \\
0 & a_{32} & a_{33} & a_{34} & 0 & \cdots & 0 \\
\vdots & \vdots & \vdots & \vdots & \vdots & \ddots & \vdots \\
0 & \cdots & \cdots & a_{n-2,n-3} & a_{n-2,n-2} & a_{n-2,n-1} & 0 \\
0 & \cdots & \cdots & 0 & a_{n-1,n-2} & a_{n-2,n-1} & a_{n-1,n} \\
0 & \cdots & \cdots & 0 & 0 & a_{n,n-1} & a_{nn}
\end{bmatrix}.
$$

This matrix has a banded form, and it is *sparse*; i.e., it has few nonzero entries. Without loss of generality, assume that $a_{i,j+1} \neq 0$. If $a_{j,j+1} = 0$, it is easy to see that the original system may be decomposed into two subsystems, since in such a case we have an upper block of lower triangular form. We may solve the system by a specially structured direct method. Consider the first equation:

$$a_{11}x_1 + a_{12}x_2 = b_1.$$

We may solve for x_2, in terms of x_1:

$$x_2 = c_2 + d_2x_1,$$

where $c_2 = b_1/a_{12}$ and $d_2 = -a_{11}/a_{12}$. By the same token, we may obtain an expression of x_3 in terms of x_1. In fact, given the second equation

$$a_{21}x_1 + a_{22}x_2 + a_{23}x_3 = b_2,$$

we may express x_3 as a function of x_1 and x_2. But since we know x_2 as a function of x_1, we may get an expression of the form

$$x_3 = c_3 + d_3x_1.$$

Going on the same way for all equations up to the $(n-1)$th one, we obtain expressions like $x_k = c_k + d_kx_1$, for all $k = 2, \ldots, n$. Finally, plugging the expressions for x_{n-1} and x_n into the last equation, we end up with

$$a_{n,n-1}x_{n-1} + a_{nn}x_n = a_{n,n-1}(c_{n-1} + d_{n-1}x_1) + a_{nn}(c_n + d_nx_1) = b_n,$$

which yields x_1, and, by substitution, all the other unknowns. The approach may be adapted in the case of similar banded matrices. It is also worth noting that memory savings may be obtained by storing only the nonzero matrix entries. In other cases, despite the sparseness of the matrix, it is difficult to adapt the direct substitution method. Then it might be advisable to avoid a process like Gaussian elimination, which would destroy the sparseness of the matrix, and adopt an iterative approach such as those described in the next section.

2.2.4   Iterative methods for solving systems of linear equations

In many situations we must solve a large system of linear equations, characterized by a sparse matrix. PDEs are a typical source of such systems. Storing a sparse matrix is a waste of memory, since many entries are zero; special techniques have been developed to avoid the problem. However, applying direct methods to a sparse matrix will destroy its characteristic. So we may try a different approach. One possibility is an iterative method, generating a sequence of vectors that converges to the solution desired. The process may be stopped when a reasonable accuracy has been achieved. Note that unlike direct methods, the number of steps required by an iterative algorithm is not known a priori, and its behavior should be characterized in terms of convergence speed, along the lines illustrated in section 2.1.4. The first issue to consider is how to characterize the conditions under which an iterative method converges; in fact, the method could simply blow up due to instability, giving rise to an unbounded sequence.

Here we illustrate the basic iterative approaches described in any numerical analysis text. It is worth emphasizing that MATLAB has efficient capabilities to represent sparse matrices and provides the user with a rich set of iterative methods, which are much more sophisticated than the ones we describe here. Nevertheless, we believe that the background behind relatively simple iterative methods will be a useful reading, for at least a couple of reasons. On the one hand, they have been proposed in the literature on financial engineering to solve PDEs (see, e.g., [20, pp. 895-901] for a comparison of LU-decomposition and successive overrelaxation in option pricing). Second, in chapter 5 we investigate the numerical stability of finite difference methods for solving PDEs, using the same concepts we use here to study the convergence of iterative methods.

Consider a generic operator $\mathbf{G}(\cdot)$ and assume that you want to find a fixed point of $\mathbf{G}$, i.e., a point satisfying the equation

$$\mathbf{x} = \mathbf{G}(\mathbf{x}).$$

A possible approach is to generate a sequence of approximations of the solution, according to the iteration scheme

$$\mathbf{x}^{(k+1)} = \mathbf{G}(\mathbf{x}^{(k)}), \tag{2.2}$$

starting from some initial approximation $\mathbf{x}^{(0)}$. This approach, called *fixed-point iteration*, may be used for both linear and nonlinear equations, and for many other problems as well. Now the question is if and when this scheme will converge to a fixed point of $\mathbf{G}$. The general answer lies in the *contraction mapping* concept, which is widely applied in many diverse settings. Let us investigate the idea in the case of the familiar system of linear equations $\mathbf{Ax} = \mathbf{b}$, which can be rewritten as

$$\mathbf{x} = (\mathbf{A} + \mathbf{I})\mathbf{x} - \mathbf{b} = \hat{\mathbf{A}}\mathbf{x} - \mathbf{b}.$$

We want to find a fixed point of the operator $\mathbf{G}(x) = \hat{\mathbf{A}}\mathbf{x} - \mathbf{b}$, and we could consider the iterative approach (2.2). Would such a scheme converge? To begin with, consider starting from a first guess $\mathbf{x}^{(0)}$, and trace the first iteration steps:

$$
\begin{aligned}
\mathbf{x}^{(1)} &= \hat{\mathbf{A}}\mathbf{x}^{(0)} - \mathbf{b} \\
\mathbf{x}^{(2)} &= \hat{\mathbf{A}}\mathbf{x}^{(1)} - \mathbf{b} = \hat{\mathbf{A}}^2\mathbf{x}^{(0)} - \hat{\mathbf{A}}\mathbf{b} - \mathbf{b} \\
\mathbf{x}^{(3)} &= \hat{\mathbf{A}}^3\mathbf{x}^{(0)} - \hat{\mathbf{A}}^2\mathbf{b} - \hat{\mathbf{A}}\mathbf{b} - \mathbf{b}
\end{aligned}
$$

$$\cdots$$

Intuition suggests that if the elements of the matrix $\hat{\mathbf{A}}^n$ grow without bound as $n \to \infty$, the iteration scheme will diverge. Indeed, it can be shown that convergence will occur only if all the eigenvalues of $\hat{\mathbf{A}}$ have an absolute value less than 1 (see below). Since this may well not be the case for an arbitrary system of equations, it is better to take a slightly different approach and split the matrix $\mathbf{A}$ as follows:

$$\mathbf{A} = \mathbf{D} + \mathbf{C},$$

which yields an equivalent system

$$\mathbf{D}\mathbf{x} = -\mathbf{C}\mathbf{x} + \mathbf{b}.$$

Then we may apply the iteration scheme

$$\begin{aligned}
\mathbf{d}^{(k)} &= -\mathbf{C}\mathbf{x}^{(k)} + \mathbf{b} \\
\mathbf{D}\mathbf{x}^{(k+1)} &= \mathbf{d}^{(k)}
\end{aligned} \tag{2.3}$$

in order to generate a sequence of approximations $\mathbf{x}^{(k)}$. In some sense, this is a generalization of the previous fixed-point approach, but the flexibility in choosing $\mathbf{D}$ may be exploited to improve convergence. To investigate the convergence issue further, we may write, as before,

$$\mathbf{x}^{(k+1)} = -\mathbf{D}^{-1}\mathbf{C}\mathbf{x}^{(k)} + \mathbf{D}^{-1}\mathbf{b}.$$

Letting $\mathbf{B} = -\mathbf{D}^{-1}\mathbf{C} = \mathbf{I} - \mathbf{D}^{-1}\mathbf{A}$, we may check how the absolute error $\mathbf{e}^{(k)} = \mathbf{x}^* - \mathbf{x}^{(k)}$ evolves, where $\mathbf{x}^*$ is the correct solution:

$$\mathbf{e}^{(k+1)} = \mathbf{x}^* - \mathbf{x}^{(k+1)} = (\mathbf{B}\mathbf{x}^* + \mathbf{D}^{-1}\mathbf{b}) - (\mathbf{B}\mathbf{x}^{(k)} + \mathbf{D}^{-1}\mathbf{b}) = \mathbf{B}(\mathbf{x}^* - \mathbf{x}^{(k)}) = \mathbf{B}\mathbf{e}^{(k)},$$

from which it is easy to see that

$$\lim_{k\to\infty} \mathbf{e}^{(k)} = \lim_{k\to\infty} \mathbf{B}^k \mathbf{e}^{(0)}.$$

It can be shown that

$$\lim_{k\to\infty} \mathbf{B}^k = \mathbf{0}$$

if and only if the spectral radius of $\mathbf{B}$ is strictly less than 1, i.e., if all of its eigenvalues have an absolute value less than 1. This implies that the approach will converge if and only if

$$\rho(\mathbf{I} - \mathbf{D}^{-1}\mathbf{A}) < 1.$$

To verify this condition, we should compute the eigenvalues of a possibly large matrix (actually, only the largest one in absolute value is needed). We may avoid this trouble by recalling that

$$\rho(\mathbf{B}) \le \|\mathbf{B}\|$$

for any matrix norm compatible with a vector norm. Hence, we may settle the convergence question, in the sense of characterizing sufficient but not necessary conditions for convergence, by considering easily computable matrix norms such as $\|\mathbf{B}\|_1$ or $\|\mathbf{B}\|_\infty$. From a practical point of view, the whole approach makes sense only if solving the linear equation (2.3) is easy. By a proper choice of $\mathbf{D}$, we obtain the methods described in the following.

Jacobi method   A particularly convenient choice for $\mathbf{D}$ is a diagonal matrix:

$$\mathbf{D} = \begin{pmatrix} a_{11} & 0 & 0 & \cdots & 0 \\ 0 & a_{22} & 0 & \cdots & 0 \\ 0 & 0 & a_{33} & \cdots & 0 \\ \vdots & \vdots & \vdots & \ddots & \vdots \\ 0 & 0 & 0 & \cdots & a_{nn} \end{pmatrix},$$

which is easily inverted provided that $a_{ii} \neq 0$; this condition may be obtained by proper row/column permutations if $\mathbf{A}$ is nonsingular. Choosing the $\|\cdot\|_\infty$ norm, we obtain a sufficient condition for convergence:

$$\|\mathbf{I} - \mathbf{D}^{-1}\mathbf{A}\|_\infty = \max_{1 \leq i \leq n} \sum_{\substack{j=1 \\ j \neq i}}^{n} \left| \frac{a_{ij}}{a_{ii}} \right| < 1,$$

which actually boils down to diagonal dominance, i.e.,

$$|a_{ii}| > \sum_{\substack{j=1 \\ j \neq i}}^{n} |a_{ij}| \qquad \forall i.$$

To implement the method, we must rewrite the initial equation as

$$x_i = \frac{b_i - \sum\limits_{\substack{j=1 \\ j \neq i}}^{n} a_{ij}x_j}{a_{ii}} \qquad i = 1, \ldots, n,$$

which leads immediately to the iteration scheme

$$x_i^{(k+1)} = \frac{b_i - \sum\limits_{\substack{j=1 \\ j \neq i}}^{n} a_{ij}x_j^{(k)}}{a_{ii}} \qquad i = 1, \ldots, n.$$

The iterations should be stopped when a satisfactory precision has been achieved. One possible condition to check is related to the absolute error. Having specified a tolerance parameter ϵ, we could stop the algorithm when

$$\|\mathbf{x}^{(k+1)} - \mathbf{x}^{(k)}\| < \epsilon.$$

Example 2.6 The Jacobi method is easily coded in MATLAB, as illustrated in figure 2.2; note that in MATLAB a vector implementation is preferred. Let us try it with the following data:

$$\mathbf{A} = \begin{bmatrix} 5 & -1 & 1 \\ 2 & 8 & -1 \\ -1 & 1 & 4 \end{bmatrix} \qquad \mathbf{b} = \begin{bmatrix} 10 \\ 11 \\ 3 \end{bmatrix}.$$

```
function z=Jacobi(A,b,x0,n)
D = diag(diag(A));
Dinv = diag(1./diag(A));
C = A - D;
B = - Dinv * C;
b1 = Dinv * b;
z = x0;
for i=1:n
   z = B * z + b1;
end
```

Fig. 2.2   Implementation of the Jacobi iterative method.

As the matrix **A** is diagonally dominant, convergence is expected. Since the matrix is small, we may also compute the spectral radius of **B**.

```
>> A = [5 -1 1; 2 8 -1; -1 1 4];
>> b = [10;11;3];
>> B = eye(3) - (diag(1./diag(A)) * A );
>> norm(B)
ans =
    0.4210
>> Jacobi(A,b,zeros(3,1),10)
ans =
    2.0000
    1.0000
    1.0000
```

We see that convergence is rather fast. Now, let us try perturbing the matrix a bit, increasing the value of the off-diagonal entries.

```
>> A = [5 -2 2; 2 8 -3; -2 1 4];
>> B = eye(3) - (diag(1./diag(A)) * A );
>> norm(B)
ans =
    0.6531
>> Jacobi(A,b,zeros(3,1),10)
ans =
    2.0029
    1.3935
    1.4064
>> Jacobi(A,b,zeros(3,1),20)
ans =
    2.0001
```

```
      1.4000
      1.3999
>> Jacobi(A,b,zeros(3,1),30)
ans =
      2.0000
      1.4000
      1.4000
```

We see that the convergence to the solution $(2, 1.4, 1.4)$ is slower. The problem is that the spectral radius is larger here.

```
>> A = [5 -2 4; 12 8 -3; -2 3 4];
>> B = eye(3) - (diag(1./diag(A)) * A );
>> norm(B)
ans =
      1.6448
>> Jacobi(A,b,zeros(3,1),10)
ans =
      3.1273
      7.3010
     -5.6352
>> Jacobi(A,b,zeros(3,1),30)
ans =
    418.4938
   -738.5725
     93.7298
>> Jacobi(A,b,zeros(3,1),100)
ans =
    1.0e+009 *
      0.1007
     -4.3863
      2.8764
```

The new matrix is not diagonally dominant; even more important, the spectral radius is larger than 1 and, as expected, the sequence is blowing up. □

Gauss-Seidel method The Gauss-Seidel method is a variant of the Jacobi method. The idea is to use the updated values of $x_i^{(k+1)}$ immediately, as soon as they are computed. The iteration scheme is therefore

$$x_i^{(k+1)} = \frac{b_i - \sum_{j=1}^{i-1} a_{ij} x_j^{(k+1)} - \sum_{j=i+1}^{n} a_{ij} x_j^{(k)}}{a_{ii}} \qquad i = 1, \ldots, n.$$

To analyze the convergence of this method, we may note that this corresponds to choosing as $\mathbf{D}$ the lower triangle of $\mathbf{A}$:

$$\mathbf{D} = \begin{pmatrix} a_{11} & 0 & 0 & \cdots & 0 \\ a_{21} & a_{22} & 0 & \cdots & 0 \\ a_{31} & a_{32} & a_{33} & \cdots & 0 \\ \vdots & \vdots & \vdots & \ddots & \vdots \\ a_{n1} & a_{n2} & a_{n3} & \cdots & a_{nn} \end{pmatrix}.$$

Then it can be shown that diagonal dominance is again a sufficient condition for convergence:

$$\|\mathbf{I} - \mathbf{D}^{-1}\mathbf{A}\|_\infty \leq \max_{1 \leq i \leq n} \sum_{\substack{j=1 \\ j \neq i}}^{n} \left| \frac{a_{ij}}{a_{ii}} \right| < 1.$$

Speeding up convergence: successive overrelaxation Consider the iteration scheme

$$\mathbf{x}^{(k+1)} = \mathbf{B}\mathbf{x}^{(k)} + \mathbf{d}.$$

Since we move from the current point $\mathbf{x}^{(k)}$ to the updated point $\mathbf{x}^{(k+1)}$, we may think of it as the addition of a displacement to the old approximation:

$$\mathbf{x}^{(k+1)} = \mathbf{x}^{(k)} + \mathbf{r}^{(k)}.$$

Even though this method will converge if $\rho(\mathbf{B}) < 1$, convergence will be slow if the spectral radius of $\mathbf{B}$ is close to 1 (see example 2.6). We could try to speed up convergence by modifying the iteration:

$$\mathbf{x}^{(k+1)} = \mathbf{x}^{(k)} + \omega\mathbf{r}^{(k)}.$$

Intuitively, if $\mathbf{r}^{(k)}$ is a good direction, we might think of accelerating the movement by setting $\omega > 1$. We must make sure that a poor choice of ω does not lead to instability. On the other hand, if the starting iteration is itself unstable, we might think that the difficulty stems from moving "too much" along the directions $\mathbf{r}^{(k)}$, which leads to oscillations and instability. In this case, we might think of dampening the oscillations with a suitable modification of the iteration scheme. To pursue this dampening, we may form a convex combination[8] of the new and the old point as follows:

$$\begin{aligned} \hat{\mathbf{x}}^{(k+1)} &= \omega\mathbf{x}^{(k+1)} + (1-\omega)\mathbf{x}^{(k)} \\ &= \omega(\mathbf{B}\mathbf{x}^{(k)} + \mathbf{d}) + (1-\omega)\mathbf{x}^{(k)} = \mathbf{B}_\omega\mathbf{x}^{(k)} + \omega\mathbf{d}. \end{aligned} \qquad (2.4)$$

---

[8] A convex combination of two points $\mathbf{x}_1$ and $\mathbf{x}_2$ is just a particular linear combination with nonnegative weights, such that their sum is 1: $\lambda\mathbf{x}_2 + (1-\lambda)\mathbf{x}_2$ for $\lambda \in [0,1]$.

This is actually a convex combination if $\omega \in (0,1)$. It is worth noting that it looks like well-known exponential smoothing methods for time series analysis, where the aim is just to dampen oscillations in the estimates. The iterative scheme is stable if $\rho(\mathbf{B}_\omega) < 1$. Moreover, by a suitable choice of ω, the spectral radius will be reduced, with a corresponding improvement in convergence speed.

The reasoning above suggests that we may try to pursue modifications of the iterative approaches we have just described. For instance, we may try the idea on the Gauss-Seidel scheme. In the basic method, each component of the displacement is

$$r_i^{(k)} = x_i^{(k+1)} - x_i^{(k)} = \frac{1}{a_{ii}}\left[b_i - \sum_{j=1}^{i-1} a_{ij} x_j^{(k+1)} - \sum_{j=i+1}^{n} a_{ij} x_j^{(k)} \right].$$

In order to analyze the effect of the correction $\mathbf{x}^{(k+1)} = \mathbf{x}^{(k)} + \omega \mathbf{r}^{(k)}$, let us rewrite the Gauss-Seidel scheme in a compact form, based on the following decomposition of $\mathbf{A}$:

$$\mathbf{A} = \mathbf{L} + \mathbf{D} + \mathbf{U},$$

where

$$\mathbf{L} = \begin{bmatrix} 0 & 0 & 0 & \cdots & 0 & 0 \\ a_{21} & 0 & 0 & \cdots & 0 & 0 \\ a_{31} & a_{32} & 0 & \cdots & 0 & 0 \\ \vdots & \vdots & \vdots & \ddots & \vdots & \vdots \\ a_{n-1,1} & a_{n-1,2} & a_{n-1,3} & \cdots & 0 & 0 \\ a_{n1} & a_{n2} & a_{n3} & \cdots & a_{n,n-1} & 0 \end{bmatrix}$$

$$\mathbf{D} = \begin{bmatrix} a_{11} & 0 & 0 & \cdots & 0 & 0 \\ 0 & a_{22} & 0 & \cdots & 0 & 0 \\ 0 & 0 & a_{33} & \cdots & 0 & 0 \\ \vdots & \vdots & \vdots & \ddots & \vdots & \vdots \\ 0 & 0 & 0 & \cdots & a_{n-1,n-1} & 0 \\ 0 & 0 & 0 & \cdots & 0 & a_{nn} \end{bmatrix}$$

$$\mathbf{U} = \begin{bmatrix} 0 & a_{12} & a_{13} & \cdots & a_{1,n-1} & a_{1n} \\ 0 & 0 & a_{23} & \cdots & a_{2,n-1} & a_{2n} \\ 0 & 0 & 0 & \cdots & a_{3,n-1} & a_{3n} \\ \vdots & \vdots & \vdots & \ddots & \vdots & \vdots \\ 0 & 0 & 0 & \cdots & 0 & a_{n-1,n} \\ 0 & 0 & 0 & \cdots & 0 & 0 \end{bmatrix}.$$

With this notation, the Gauss-Seidel scheme may be rewritten as

$$\mathbf{x}^{(k+1)} = \mathbf{D}^{-1}(\mathbf{b} - \mathbf{L}\mathbf{x}^{(k+1)} - \mathbf{U}\mathbf{x}^{(k)}).$$

```
function z=SpeedJacobi(A,b,x0,omega,n)
D = diag(diag(A));
Dinv = diag(1./diag(A));
C = A - D;
B = - Dinv * C;
b1 = Dinv * b;
z = x0;
for i=1:n
   z = omega*(B * z + b1) + (1-omega)*z;
end
```

Fig. 2.3 Implementation of a modified Jacobi iterative method.

By the same token

$$\mathbf{r}^{(k)} = \mathbf{D}^{-1}[\mathbf{b} - \mathbf{L}\mathbf{x}^{(k+1)} - (\mathbf{D} + \mathbf{U})\mathbf{x}^{(k)}].$$

Now the modified iteration is

$$(\mathbf{D} + \omega\mathbf{L})\mathbf{x}^{(k+1)} = [(1 - \omega)\mathbf{D} - \omega\mathbf{U}]\mathbf{x}^{(k)} + \omega\mathbf{b},$$

which will be stable if

$$\rho((\mathbf{D} + \omega\mathbf{L})^{-1}[(1 - \omega)\mathbf{D} - \omega\mathbf{U}]) < 1.$$

By proper selection of the parameter ω, we may reduce the spectral radius of the matrix, thus improving convergence. We speak of *overrelaxation* if $1 < \omega < 2$ and of *underrelaxation* if $0 < \omega < 1$. It can be shown that for other values of ω, the method cannot converge.

Example 2.7 Consider again the last matrix of example 2.6, for which Jacobi iteration blows up. We may try to stabilize the method by using the convex combination of equation (2.4). The corresponding code is illustrated in figure 2.3, and it is stable if

$$\rho[\omega\mathbf{B} + (1 - \omega)\mathbf{I}] < 1,$$

where $\mathbf{B} = \mathbf{I} - \mathbf{D}^{-1}\mathbf{A}$.

```
>> A = [5 -2 4; 12 8 -3; -2 3 4];
>> b = [10;11;3];
>> omega = 0.4;
>> B=eye(3) - diag(1./diag(A))*A;
>> norm(omega * B + (1-omega)*eye(3))
ans =
```

```
    0.9897
>> SpeedJacobi(A,b,zeros(3,1),omega,100)
ans =
    1.1107
    0.1549
    1.1891
>> A\b
ans =
    1.1107
    0.1549
    1.1891
```

We see that for $\omega = 0.4$ the method does converge to the solution, although the speed is not very satisfactory. Determining the parameter ω by trial and error and applying the idea to the Gauss-Seidel scheme may not be easy, but it does pay off in some cases.   ▯

2.3 FUNCTION APPROXIMATION AND INTERPOLATION

Function approximation and interpolation are two common tasks in numerical analysis. An example of function approximation occurs when you have a complicated function $f(\cdot)$, so complicated that you have no direct and easy way of computing its values $f(x)$. A typical example is the standard normal distribution function

$$N(x) = \frac{1}{\sqrt{2\pi}} \int_{-\infty}^{x} e^{-y^2/2} \, dy.$$

which occurs in the Black-Scholes pricing formula. Note that we would like an approximation valid over an extended range of values of the independent variable. In other cases, we are interested in the values in a neighborhood of some value x_0, in which case a Taylor expansion would suffice:

$$f(x) \approx f(x_0) + f'(x_0)(x - x_0) + \frac{1}{2}f''(x_0)(x - x_0)^2 + \cdots.$$

Indeed, this is behind the duration/convexity approximation used for bond portfolio immunization and the delta-gamma approximation used for VaR computation with derivatives (see examples 1.4 and 1.14 on pages 16 and 56, respectively).

In other situations, you have the values $f(x_i)$ of a function over a set of points x_i, $i = 1, \ldots, n$, and you would like a simple function which approximates $f(\cdot)$ in a well-defined sense. One way of setting up the problem is to define a suitable class of approximating functions and to look for the best solution in terms of a chosen metric. A possible choice for the class of approximating functions is represented by the class of polynomials of given degree

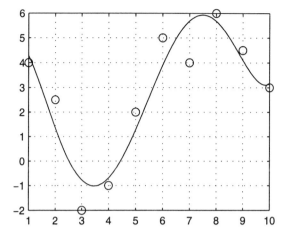

Fig. 2.4   Approximating a given data set by a fifth-degree polynomial.

m; let $P_m(x; \boldsymbol{\alpha})$ denote the polynomial with coefficients represented by the vector $\boldsymbol{\alpha}$. One reason behind this choice is that polynomials are continuous functions, as well as their derivatives, which lend themselves to easy differentiation and integration. One possible metric to select the best approximation is the least squares approximation, whereby you try to minimize the average square deviation of the approximating function from f on the selected points x_i. The approximation problem can be stated as

$$\min_{\boldsymbol{\alpha}} \quad \sum_{i=1}^{n} [f(x_i) - P_m(x_i; \boldsymbol{\alpha})]^2 \, .$$

Different objective functions could be used, basically corresponding to different ways of measuring the norm of the vector of the approximation errors. Another typical choice is the "min-max" metric, which is essentially the $\|\cdot\|_\infty$ norm:

$$\min_{\boldsymbol{\alpha}} \quad \max_{i=1,\dots,n} |f(x_i) - P_m(x_i; \boldsymbol{\alpha})| \, .$$

An issue we are neglecting here is the selection of the points x_i. This selection may have an influence on the quality of the approximation; a natural choice is to select equidistant points, but this need not be the best idea. The details are beyond the scope of the book, but it is worth noting that approximation theory is a subject area in itself, with strong links to both numerical analysis and optimization theory. We refer the reader to the references.

Example 2.8 Polynomial approximation in the least squares sense is easily accomplished in MATLAB by the `polyfit` function. Consider for example the following 10-point data set:

```
>> x=1:10;
>> y = [4 2.5 -2 -1 2 5 4 6 4.5 3];
```

Fitting a polynomial of degree 5 yields a vector of coefficients:

```
>> p=polyfit(x,y,5)
p =
    0.0060    -0.1600    1.4370    -4.7512    3.4640    4.3000
```

Using the function `polyval`, we may evaluate the approximated function outside the data set.

```
>> x2=1:0.05:10;
>> y2=polyval(p,x2);
>> plot(x,y,'o',x2,y2);
>> grid on
```

The result is shown in figure 2.4. Apparently, the fitting is not that good, and we could consider alternative approximating functions.   ▯

In other situations we require that the approximating functions actually pass through the data points. For instance, we might have a set of spot rates, derived from some bond market data, and we would like to obtain a whole set of spot rates, describing the whole term structure of interest rates. In such cases we speak of *interpolation*. A possible class of interpolating functions is again the set of polynomials of sufficient degree. Let us consider a set of support points (x_i, y_i), $i = 0, 1, \ldots, n$, where $y_i = f(x_i)$ and $x_i \neq x_j$ for $i \neq j$. It is easy to find a polynomial of degree (at most) n such that $P_n(x_i) = y_i$ for any i. We may rely on the *Lagrange polynomials* $L_i(x)$, defined as

$$L_i(x) = \prod_{\substack{j=0 \\ j \neq i}}^{n} \frac{x - x_j}{x_i - x_j}. \tag{2.5}$$

Note that these are polynomials of degree n and that

$$L_i(x_k) = \begin{cases} 1 & \text{if } i = k \\ 0 & \text{otherwise.} \end{cases}$$

Now an interpolating polynomial can be easily written as

$$P_n(x) = \sum_{i=0}^{n} y_i L_i(x).$$

In practice, no one should use this form for computational purposes, and some tricks are needed for the sake of computational efficiency. The next example shows the basic features of interpolating polynomials.

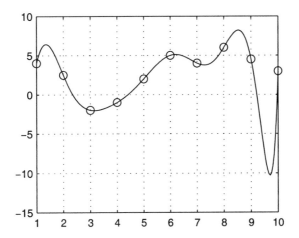

Fig. 2.5 Interpolating a given data set by a tenth-degree polynomial.

Example 2.9 We may try to improve the fitting of example 2.8 by increasing the degree of the polynomial. Since there are 10 data points, we may try interpolating with a polynomial of degree 10, passing through the support points.

```
>> p=polyfit(x,y,10);
>> y2=polyval(p,x2);
>> plot(x,y,'o',x2,y2);
>> grid on
```

The result is shown in figure 2.5. We may see that now the polynomial passes through the data set, which is also easily checked numerically:

```
>> polyval(p,x)
ans =
  Columns 1 through 7
    4.00    2.50   -2.00   -1.00    2.00    5.00    4.00
  Columns 8 through 10
    6.00    4.50    3.00
```

Unfortunately, we see that the interpolating polynomial has some undesirable oscillation behavior near the left end of the interval. This is not surprising: a polynomial of degree n may have up to $n - 1$ local minima and maxima and oscillations are to be expected. ▯

The oscillation of high-degree interpolating polynomials is a typical difficulty. One way to overcome the problem is to use more sophisticated functions, for both approximation and interpolation. Sometimes, rational functions fit nicely. We will not go in any detail, but we give a typical example of the use of rational functions in approximation.

```
function z = mynormcdf(x)
c = [ 0.31938153 , -0.356563782 , 1.781477937 , ...
        -1.821255978 , 1.330274429 ];
gamma = 0.2316419;
vx = abs(x);
k = 1./(1+gamma.*vx);
n = exp(-vx.^2./2)./sqrt(2*pi);
matk = ones(length(x),1) * k;
matexp = (ones(length(x),1)*(1:5))';
matv = matk.^matexp;
z = 1 - n.*(c*matv);
i = find(x < 0);
z(i) = 1-z(i);
```

Fig. 2.6 MATLAB code to approximate the cumulative normal distribution.

Example 2.10 There are various approximation formulas that can be used to evaluate the standard normal distribution function $N(x)$. One is the following:[9]

$$N(x) = \begin{cases} 1 - N'(x)(a_1k + a_2k^2 + a_3k^3 + a_4k^4 + a_5k^5) & \text{if } x \geq 0 \\ 1 - N(-x) & \text{if } x < 0, \end{cases}$$

where

$$N'(x) = \frac{1}{\sqrt{2\pi}}e^{-x^2/2} \qquad\qquad k = \frac{1}{1+\gamma x}$$

$$\gamma = 0.2316419 \qquad\qquad a_1 = 0.31938153$$

$$a_2 = -0.356563782 \qquad\qquad a_3 = 1.781477937$$

$$a_4 = -1.821255978 \qquad\qquad a_5 = 1.330274429.$$

The MATLAB code for this function is shown in figure 2.6; it is a little involved, as we have made sure it can operate on vector arguments (as should be the case with good MATLAB functions). This is not really the formula used in the MATLAB function normcdf, but we may compare the two approximations:

```
>> mynormcdf([-1 -0.5 0.5 1 1.5 ])
ans =
       0.1587      0.3085      0.6915      0.8413      0.9332
```

[9]This formula is proposed in [8, p. 252]. It is based on approximation 7.1.26 of the error function in [1], which in turn refers to [7]. If you have some archaeological instinct, you may go further.

```
>> normcdf([-1 -0.5 0.5 1 1.5 ])
ans =
    0.1587    0.3085    0.6915    0.8413    0.9332
```

Another trick to avoid oscillating polynomials is resorting to low-degree polynomials, interpolating the data points piecewise. The simplest idea is to use a piecewise linear interpolation. Given the $N + 1$ support points (or knots) (x_i, y_i), we may use N first-degree polynomials $S_i(x)$, each valid on the interval (x_i, x_{i+1}). An obvious requirement is that the resulting function is continuous, i.e., $S_i(x_{i+1}) = S_{i+1}(x_{i+1})$. Recalling the Lagrange polynomials defined in equation (2.5), we have

$$ S_i(x) = y_i \frac{x - x_{i+1}}{x_i - x_{i+1}} + y_{i+1} \frac{x - x_i}{x_{i+1} - x_i} \quad x \in [x_i, x_{i+1}]. $$

This type of interpolation is called *linear spline*. Whereas the interpolating function is continuous, its derivative is not. We may enforce the continuity of the derivatives of the spline by increasing the degree of the polynomials. The most common spline is obtained by "joining" N third-degree polynomials $S_i(x)$, with coefficients $s_{i0}, s_{i1}, s_{i2}, s_{i3}$, which must satisfy the following requirements:

$$
\begin{aligned}
S(x) &= S_i(x) = s_{i0} + s_{i1}(x - x_i) + s_{i2}(x - x_i)^2 + s_{i3}(x - x_i)^3 \\
&\qquad x \in [x_i, x_{i+1}] \qquad i = 0, 1, \ldots, N-1 \\
S(x_i) &= y_i \qquad i = 0, 1, \ldots, N \\
S_i(x_{i+1}) &= S_{i+1}(x_{i+1}) \qquad i = 0, 1, \ldots, N-2 \\
S_i'(x_{i+1}) &= S_{i+1}'(x_{i+1}) \qquad i = 0, 1, \ldots, N-2 \\
S_i''(x_{i+1}) &= S_{i+1}''(x_{i+1}) \qquad i = 0, 1, \ldots, N-2.
\end{aligned}
$$

The resulting spline $S(x)$ is called a *cubic spline*. We see that we require continuity for the spline itself and for its first and second derivatives. To specify a spline, we must give $4N$ coefficients. Passage through the support points gives $N + 1$ conditions; the continuity of the spline and the two derivatives enforces a total of $3(N - 1)$ conditions, yielding a total of $4N - 2$ conditions. Hence, we have two degrees of freedom which may be eliminated by enforcing further requirements. Usually, they involve some conditions at, or near, the endpoints x_0 and x_N. Among the most common conditions, we recall the following:

- $S''(x_0) = S''(x_N) = 0$, which leads to *natural splines*.

- $S'(x_0) = f'(x_0)$ and $S'(x_N) = f'(x_N)$, which may be used if we have a precise idea of the behavior of $f(x)$ near the endpoints.

- The *not-a-knot* condition, which is obtained by requiring that the third-order derivative $S'''(x)$ be continuous in x_1 and x_{N-1}.

Despite the appealing name, the first choice is usually avoided. Its importance stems from the following theorem, which we state without proof.[10]

THEOREM 2.1 *Let f'' be continuous in (a, b) and let $a = x_0 < x_1 < \cdots < x_N = b$. If S is the natural cubic spline interpolating f on the knots x_i, then*

$$\int_a^b [S''(x)]^2 \, dx \leq \int_a^b [f''(x)]^2 \, dx.$$

The importance of this theorem can be understood by recalling that the curvature of the curve described by the equation $y = f(x)$ is given by

$$|f''(x)| \cdot \{1 + f'(x)^2\}^{-3/2}.$$

Hence, if f' is sufficiently small, we see that $|f''(x)|$ approximates the curvature; hence, the natural spline is, in some sense, an approximation of minimal curvature over the interval (a, b). When nothing is known about the function, the not-a-knot condition is the recommendable choice. The name stems from the fact that in this case the first and last interior points, x_1 and x_{n-1}, do not behave as knots. In fact, if we require continuity of S''' in those points, we find that the first two and last two polynomials are the same.

To find the unknown coefficients, we have to set up a system of linear equations; the details are a bit tedious, and since they are implemented in a ready-to-use MATLAB function, they are omitted. Yet it is interesting to note that for most choices of free conditions, the resulting system has a tridiagonal form like that discussed in section 2.2.3; furthermore, it is symmetric and diagonally dominant, hence it is particularly easy to solve.

Example 2.11 Splines are so important that an entire toolbox is devoted to them. In the base MATLAB system, you have two functions that may be used for cubic spline interpolation. One is `interp1`, provided that you call it with the parameter `'spline'`; the other is `spline`. We may use a cubic spline for the same data points as in example 2.8 (see figure 2.7).

```
>> x=1:10;
>> y = [4 2.5 -2 -1 2 5 4 6 4.5 3];
>> x2=1:0.05:10;
>> y2=interp1(x,y,x2,'spline');
>> plot(x,y,'o',x2,y2);
>> grid on;
```

The same result is obtained by calling `spline`, which also returns a spline object; this object may be used for later evaluations by the function `ppval`.

---

[10]See, e.g., [11, pp. 380-381].

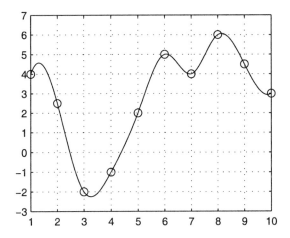

Fig. 2.7 Interpolating a given data set by a cubic spline.

```
>> pp=spline(x,y);
>> v=[1.4 2.8 3.5 4.6];
>> ppval(pp,v)
ans =
    4.5242   -1.4331    -2.0737      0.6520
>> interp1(x,y,v,'spline')
ans =
    4.5242   -1.4331    -2.0737      0.6520
```

The free condition is by default the not-a-knot condition defined above. If you provide the function **spline** with an y vector with two more components than the vector x, the first and last components are used to enforce a value for the spline slopes at the extreme points of the interval.   ⬜

Cubic splines are only the basic type of spline; many more have been proposed. A typical application in finance is in estimating term structures of interest rates given a limited set of market data related to bond prices (see, e.g., [3], [4], and the references therein).

2.4 SOLVING NONLINEAR EQUATIONS

Solving nonlinear equations is a common task in finance; the most elementary example is the computation of the internal rate of return (see example 1.1 on page 10), which calls for finding the roots of a polynomial. A polynomial equation is a particular case of general nonlinear equations. You might wish to find a solution of an equation in a single variable, such as

$$f(x) = 0$$

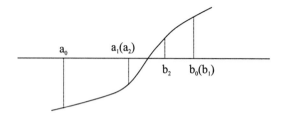

Fig. 2.8 Example of the bisection method.

or a system of equations in several variables, such as

$$\mathbf{F(x)} = \mathbf{0}.$$

MATLAB offers different functionalities to this purpose. We first outline the basic features of numerical methods for nonlinear equations, limiting the treatment to the bisection and Newton methods; then we illustrate the application of MATLAB functions to financial derivatives.

2.4.1 Bisection method

The bisection method is the simplest method for solving the scalar equation

$$f(x) = 0$$

without requiring anything more than the ability to evaluate, or estimate, the function f at a given point. This is an important feature, since in some cases we do not even have an analytical expression for the function f, and therefore we are not able to apply more sophisticated methods such as Newton's method, which calls for computation of the derivative of f. Suppose that we know two points a, b $(a < b)$ such that $f(a) < 0$ and $f(b) > 0$. Then, if the function is continuous, it is obvious that it must cross the zero axis somewhere in the interval $[a, b]$ (see figure 2.8). The same observation holds if the signs of the function in a and b are reversed. So $[a, b]$ is an interval encapsulating a root of the equation. Then we may try to reduce this interval by checking the sign of f in the midpoint of the interval, i.e.,

$$c = \frac{a + b}{2}.$$

If $f(c) = 0$, possibly within some prespecified tolerance, we are done. If $f(c) < 0$, we may conclude that a zero must be located somewhere in the interval $[c, b]$; otherwise, the interval to check is $[a, c]$. Going on this way, we build a sequence of smaller and smaller intervals bracketing the zero. When we reach a satisfactory approximation, we may stop the algorithm.

There are other methods, such as *regula falsi* and the secant method, which are described in the numerical analysis literature and that just like the bisection method, do not require the ability of computing the derivative of f.

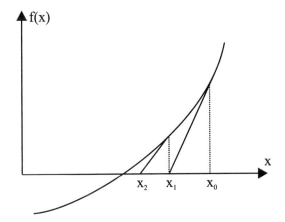

Fig. 2.9   Geometrical illustration of Newton's method.

2.4.2   Newton's method

Unlike bisection, Newton's method exploits more knowledge of the function f; in particular, it requires computing the first-order derivative of the function f. The method can be applied to solving a system of nonlinear equations, but let us first consider Newton's method for the scalar equation

$$f(x) = 0$$

and assume that $f \in C^2$. Consider a point $x^{(0)}$, which is not a solution of the equation since $f(x^{(0)}) \neq 0$. We would like to move by a step Δx, such that the new point $x = x^{(0)} + \Delta x$ solves the equation, i.e.,

$$f(x^{(0)} + \Delta x) = 0.$$

To obtain the displacement Δx, we may consider the Taylor expansion:

$$f(x^{(0)} + \Delta x) \approx f(x^{(0)}) + f'(x^{(0)}) \, \Delta x.$$

Solving this equation for Δx, we get

$$\Delta x = -\frac{f(x^{(0)})}{f'(x^{(0)})}.$$

Since the Taylor expansion is truncated, we will not find a root of the equation in one step, but we may use the idea to define a sequence of points:

$$x^{(k+1)} = x^{(k)} - \frac{f(x^{(k)})}{f'(x^{(k)})}.$$

Geometrically, the method uses the tangent of f in $x^{(k)}$ to improve the estimate of the solution, as shown in figure 2.9. Actually, there may be difficulties due to lack of convergence or division by zero; yet, using fixed-point arguments, it can be shown that the method converges under suitable hypotheses, and that the order of convergence is 2 (in the case of a single root).

Newton's method is immediately generalized to a vector equation such as

$$\mathbf{F}(\mathbf{x}) = \mathbf{0},$$

where $\mathbf{F} = [f_1 \ f_2 \ \cdots \ f_n]^T$. Given an approximation $\mathbf{x}^{(k)} = [x_1^{(k)} \ x_2^{(k)} \ \cdots \ x_n^{(k)}]^T$ of the root $\mathbf{x}^* = [x_1^* \ x_2^* \ \cdots \ x_n^*]^T$, we may write

$$f_1(\mathbf{x}^{(k)}) + (x_1^* - x_1^{(k)})\left(\frac{\partial f_1}{\partial x_1}\right)_{\mathbf{x}=\mathbf{x}^{(k)}} + \cdots + (x_n^* - x_n^{(k)})\left(\frac{\partial f_1}{\partial x_n}\right)_{\mathbf{x}=\mathbf{x}^{(k)}} \approx 0$$

$$f_2(\mathbf{x}^{(k)}) + (x_1^* - x_1^{(k)})\left(\frac{\partial f_2}{\partial x_1}\right)_{\mathbf{x}=\mathbf{x}^{(k)}} + \cdots + (x_n^* - x_n^{(k)})\left(\frac{\partial f_2}{\partial x_n}\right)_{\mathbf{x}=\mathbf{x}^{(k)}} \approx 0$$

$$\cdots$$

$$f_n(\mathbf{x}^{(k)}) + (x_1^* - x_1^{(k)})\left(\frac{\partial f_n}{\partial x_1}\right)_{\mathbf{x}=\mathbf{x}^{(k)}} + \cdots + (x_n^* - x_n^{(k)})\left(\frac{\partial f_n}{\partial x_n}\right)_{\mathbf{x}=\mathbf{x}^{(k)}} \approx 0,$$

which is simply a system of linear equations in which the matrix coefficients form the Jacobian matrix

$$\mathbf{J}^{(k)} = \mathbf{J}(\mathbf{x}^{(k)}) = \begin{bmatrix} \left(\frac{\partial f_1}{\partial x_1}\right)_{\mathbf{x}=\mathbf{x}^{(k)}} & \left(\frac{\partial f_1}{\partial x_2}\right)_{\mathbf{x}=\mathbf{x}^{(k)}} & \cdots & \left(\frac{\partial f_1}{\partial x_n}\right)_{\mathbf{x}=\mathbf{x}^{(k)}} \\ \left(\frac{\partial f_2}{\partial x_1}\right)_{\mathbf{x}=\mathbf{x}^{(k)}} & \left(\frac{\partial f_2}{\partial x_2}\right)_{\mathbf{x}=\mathbf{x}^{(k)}} & \cdots & \left(\frac{\partial f_2}{\partial x_n}\right)_{\mathbf{x}=\mathbf{x}^{(k)}} \\ \vdots & \vdots & \ddots & \vdots \\ \left(\frac{\partial f_n}{\partial x_1}\right)_{\mathbf{x}=\mathbf{x}^{(k)}} & \left(\frac{\partial f_n}{\partial x_2}\right)_{\mathbf{x}=\mathbf{x}^{(k)}} & \cdots & \left(\frac{\partial f_n}{\partial x_n}\right)_{\mathbf{x}=\mathbf{x}^{(k)}} \end{bmatrix}.$$

A sequence of solution estimates is built by solving the linear systems

$$\mathbf{J}^{(k)} \Delta\mathbf{x}^{(k)} = -\mathbf{F}(\mathbf{x}^{(k)})$$

and setting

$$\mathbf{x}^{(k+1)} = \mathbf{x}^{(k)} + \Delta\mathbf{x}^{(k)}.$$

2.4.3   Solving nonlinear equations in MATLAB

The simplest MATLAB function to solve a nonlinear equation is `fzero`, which is used to find the zero of a scalar function of a scalar argument. This function needs either an M-file or an inline expression, and possibly an estimate of the solution; if the equation has different roots and we give an initial estimate, `fzero` should return a root near the initial estimate. When the nonlinear equation is a polynomial equation, the `roots` function, which is able to return all the polynomial roots, including the complex ones, is preferred. Its use was illustrated in example 1.1 to compute the internal rate of return; similarly, it may be used in yield calculations for simple bonds. An example of solving a generic nonlinear equation occurs when the implied volatility of an option is needed.

Example 2.12 In section 1.4.2 we have seen how the Black-Scholes pricing formula may be used to price a vanilla European call or put given the value of some key parameters, including the volatility σ. In practice, the assumptions behind the Black-Scholes model are not met exactly in practice, and real prices will differ from the theoretical ones. Nevertheless, one might wonder for which value of the volatility σ the theoretical price would match the real one; in other words, we would like to apply the Black-Scholes model the other way around and estimate the *implied* volatility. This might be useful in order to estimate volatility as perceived by the market participants rather than using historical data; indeed, this approach has been advocated for VaR calculations.

This is easily accomplished in MATLAB. Consider a call option with strike price $54, expiring in five months, on a stock whose current price is $50, volatility 30%, when the risk-free interest rate is 7%. Its price is obtained as follows:

```
>> c=blsprice(50, 54, 0.07, 5/12, 0.3)
c =
    2.8466
```

Now let's go the other way around, and check which volatility would yield this price. We may define an inline function and find a zero using `fzero`:

```
>> f=inline('blsprice(50, 54, 0.07, 5/12, x) - 2.8466');
>> fzero(f,1)
Zero found in the interval: [0.094903, 1.64].
ans =
    0.3000
```

Alternatively, we could use an M-file to define the function.   ⌷

Since in the Black-Scholes formula we have the option price in analytical terms, one might wonder if it is better to use Newton's method rather than simpler methods such as bisection. This requires computing the derivative of the nonlinear function, but this effort could pay off in terms of efficiency. In fact, the Financial toolbox includes a function, `blsimpv`, which computes the implied volatility of a European call by Newton's method. Its performance may be compared with that of `fzero`.

```
>> tic, fzero(f,1), toc
Zero found in the interval: [0.094903, 1.64].
```

```
ans =
    0.3000
elapsed_time =
    1.2700
>> tic, blsimpv(50,54,0.07,5/12,2.8466), toc
ans =
    0.3000
elapsed_time =
    0.2200
```

We see that indeed Newton's method is more efficient. This does not imply that `fzero` must be disregarded, since the next example shows how it may be used even when the function whose root we are seeking is not known.

Example 2.13 We consider here a pay-later call option on a stock whose current price and volatility are $12 and 20%, respectively.[11] The risk-free rate is 10%; the strike price is $14; the expiration date is in 10 months. The feature of the pay-later option is that no premium is paid when the contract is entered. If the option is in the money at expiration, the option must be exercised and a premium is paid to the writer. Otherwise, the option expires worthless and no premium is due. How can we find the fair premium value?

The problem can be solved by setting up a binomial lattice giving the price of a call option when a premium P is subtracted from the payoff when the option is exercised (at expiration). This is accomplished by the MATLAB code shown in figure 2.10. Note that the function yields the price, given a premium P, using a time interval corresponding to one month. All we have to do is to find out for which value of P the initial price is zero.

```
>> fzero(inline('L11(X,12,14,0.1, 0.2, 1/12, 10)'),2)
Zero found in the interval: [1.9434, 2.0566].
ans =
    2.0432
```

We see how `fzero` could be used in all those cases in which an analytical pricing formula is not known. □

If you have to solve a vector nonlinear equation, you may use the `fsolve` function. As this function use a least squares approach, it is not included in the base MATLAB system but in the Optimization toolbox.

---

[11]This example is based on [12, chapter 13, exercise 11].

```
% exercise 11 chapter 13 from Luenberger, Investment Science
% use with fzero(inline('L11(X,12,14,0.1, 0.2, 1/12, 10)'),2)
function [ initprice , lattice ] = L11(premium,S0,X,r,sigma,deltaT,N)
u=exp(sigma * sqrt(deltaT));
d=1/u;
p=(exp(r*deltaT) - d)/(u-d);
lattice = zeros(N+1,N+1);
for j=0:N
   if (S0*(u^j)*(d^(N-j)) > X)
      lattice(N+1,j+1)=S0*(u^j)*(d^(N-j)) - X - premium;
   end
end
for i=N-1:-1:0
   for j=0:i
      lattice(i+1,j+1) = exp(r*deltaT) * ...
         (p * lattice(i+2,j+2) + (1-p) * lattice(i+2,j+1));
   end
end
initprice = lattice(1,1);
```

Fig. 2.10 MATLAB code to price a pay-later option by a binomial lattice.

2.5 NUMERICAL INTEGRATION

Consider the problem of approximating the value of a definite integral like

$$I[f] = \int_a^b f(x)\,dx$$

over a bounded interval $[a, b]$ for a function f of a single variable. Actually, the reader will find little direct use of such integrals in this book; nevertheless, when we must estimate an expected value of a function of a random variable, we are actually trying to estimate an integral. This happens when we price a derivative computing the expected value of the discounted payoff, with respect to a risk-neutral probability measure. Unfortunately, the integral we wish to estimate may be a multidimensional integral. The purpose of this section is just to present the basis of numerical integration and to motivate the development of alternative approaches.

Since the integration is a linear operator, it is natural to look for an approximation preserving this property. Using a finite number of values of f over a set of nodes x_j such that

$$a = x_0 < x_1 < \cdots < x_N = b,$$

we may define a quadrature formula such as

$$Q[f] = \sum_{j=0}^{n} w_j f(x_j).$$

A quadrature formula is characterized by the weights w_j and by the nodes x_j. In order to compute a suitable set of weights, we may consider the truncation error:

$$I[f] - Q[f].$$

A reasonable requirement is that the error should be zero for sufficiently simple functions such as polynomials. We may define a degree of precision of a certain quadrature formula as the maximum degree m such that the truncation error is zero for all the polynomials of degree m or less. In other words, if the original function is substituted by an interpolating polynomial, we should not commit any error in integrating the polynomial.

One way to derive quadrature formula is to consider equally spaced nodes:

$$x_j = a + jh \qquad j = 0, 1, 2, \ldots, N,$$

where $h = (b - a)/N$; also let $f_j = f(x_j)$. Given those $N + 1$ nodes, we may consider the interpolating polynomial $P_M(x)$ using Lagrange polynomials of degree M:

$$f(x) \approx P_M(x) = \sum_{j=0}^{N} f_j L_j(x).$$

Then we may compute the correct weights as follows:

$$\int_a^b f(x)\,dx \;\approx\; \int_a^b P_M(x)\,dx = \int_a^b \left[\sum_{j=0}^{N} f_j L_j(x)\right] dx$$

$$= \sum_{j=0}^{N} f_j \left[\int_a^b L_j(x)\,dx\right] = \sum_{j=0}^{N} w_j f_j.$$

Consider the case of two nodes only, $x_0 = a$ and $x_1 = b$. Here we are just interpolating f by a straight line:

$$P_1(x) = f_0 \frac{x - x_1}{x_0 - x_1} + f_1 \frac{x - x_0}{x_1 - x_0}.$$

A straightforward calculation yields

$$\int_{x_0}^{x_1} P_1(x)\,dx = h \frac{f_1 + f_0}{h}.$$

Actually, what we are saying is that we may approximate the area below the function using trapezoidal elements, as depicted in figure 2.11, and the formula

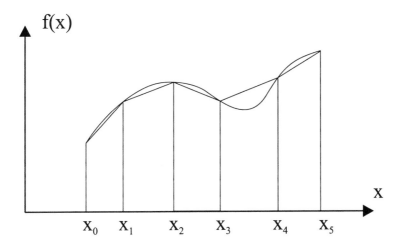

Fig. 2.11 Example of the trapezoidal quadrature formula.

above gives the area of one element. Applying the idea to more subintervals, we get the trapezoidal rule:

$$Q[f] = h \left[\frac{1}{2} f_0 + \sum_{j=1}^{N-1} f_j + \frac{1}{2} f_N \right].$$

A less straightforward calculation is needed for $M = 2$, and the result is Simpson's rule:

$$\int_{x_0}^{x_2} f(x)\, dx \approx \frac{h}{3}(f_0 + 4f_1 + f_2).$$

Going on this way, we obtain the entire set of Newton-Cotes quadrature formulas. As the reader may imagine, this is only a starting point. To begin with, nodes should be added dynamically until a prespecified accuracy is obtained; this leads to recursive quadrature formulas. Furthermore, the choice of the nodes may be improved by adapting it to the function characteristics; more nodes are needed where there is more variation, and less are needed where the function is more "constant"; this leads to adaptive quadrature formulas. All these improvements are included in MATLAB functions.

Example 2.14 Consider the integral

$$I = \int_0^{2\pi} e^{-x} \sin(10x)\, dx.$$

Integration by parts yields

$$I = -\frac{1}{101} e^{-x} \left[\sin(10x) + 10\cos(10x) \right] \Big|_0^{2\pi} \approx 0.0988.$$

Using the quad8 function, we get

```
>> f=inline('exp(-x).*sin(10*x)');
>> quad8(f,0,2*pi)
ans =
   0.0988
```
                                                                              ▯

We will not go into much detail about refined quadrature formulas. This is because, from a financial point of view, what we usually need is to compute high-dimensional integrals. Although a straightforward extension of quadrature formulas to more dimensions is possible, it would require a huge number of nodes, resulting in an excessive computational effort. In fact, MATLAB offers only a function, dblquad, to compute integrals in a two-dimensional region. This is why one usually resorts to alternative methods, such as Monte Carlo or quasi-Monte Carlo integration, which are discussed in chapter 4.

For further reading

In the literature

- The literature on numerical methods is quite extensive. One classical reference is [18]. Other references are [2], [11], and [17].

- An interesting book on numerical methods from an economist's point of view is [9].

- Splines are dealt with in depth in [5]. They are a widespread tool, both in engineering (e.g., in computer-aided design) and in economics. For a recent application in financial economics, see [10].

- A classical source for special function evaluation is [1].

- Approximation theory is the subject of [15] and [19].

- If you would like a "cookbook" collection of algorithms, [16] is a well-known reference providing many C-language codes implementing numerical methods (a Fortran version is available, too).

- Several numerical analysis books have been written based on MATLAB; see, e.g., [6] and [13].

- A few years ago I first learned about numerical methods on an old edition of [14]. This reference, being written in Italian, will not be useful to many readers; nevertheless, it has still been an influence in writing this chapter.

On the Web

- `http://www.netlib.org` is a web site offering many pointers to numerical analysis material.

- `http://www.mathworks.com/support/books` lists several MATLAB-based books, including basic numerical analysis texts.

REFERENCES

1. M. Abramowitz and I.A. Stegun, editors. *Handbook of Mathematical Functions.* Dover Publications, New York, 1972.

2. K.E. Atkinson. *An Introduction to Numerical Analysis (2nd ed.).* Wiley, Chichester, West Sussex, England, 1989.

3. L. Barzanti and C. Corradi. A Note on Interest Rate Term Structure Estimation Using Tension Splines. *Insurance Mathematics and Economics,* 22:139–143, 1998.

4. J.F. Carriere. Nonparametric Confidence Intervals of Instantaneous Forward Rates. *Insurance Mathematics and Economics,* 26:193–202, 2000.

5. C. de Boor. *A Practical Guide to Splines.* Springer-Verlag, New York, 1978.

6. L.V. Fausett. *Applied Numerical Analysis Using MATLAB.* Prentice Hall, Upper Saddle River, NJ, 1999.

7. C. Hastings. *Approximations for Digital Computers.* Princeton University Press, Princeton, NJ, 1955.

8. J.C. Hull. *Options, Futures, and Other Derivatives (4th ed.).* Prentice Hall, Upper Saddle River, NJ, 2000.

9. K.L. Judd. *Numerical Methods in Economics.* MIT Press, Cambridge, MA, 1998.

10. K.L. Judd, F. Kubler, and K. Schmedders. Computing Equilibria in Infinite-Horizon Finance Economies: The Case of One Asset. *Journal of Economic Dynamics and Control,* 24:1047–1078, 2000.

11. D. Kincaid and W. Cheney. *Numerical Analysis: Mathematics of Scientific Computing.* Brooks/Cole Publishing Company, Pacific Grove, CA, 1991.

12. D.G. Luenberger. *Investment Science.* Oxford University Press, New York, 1998.

13. J.H. Mathews and K.D. Fink. *Numerical Methods Using MATLAB (3rd ed.)*. Prentice Hall, Upper Saddle River, NJ, 1999.

14. G. Monegato. *Fondamenti di Calcolo Numerico*. CLUT, Torino, Italy, 1998.

15. M.J.D. Powell. *Approximation Theory and Methods*. Cambridge University Press, Cambridge, 1981.

16. W.H. Press, S.A. Teukolsky, W.T. Vetterling, and B.P. Flannery. *Numerical Recipes in C*. Cambridge University Press, Cambridge, 1993.

17. H.R. Schwarz. *Numerical Analysis: A Comprehensive Introduction*. Wiley, Chichester, West Sussex, England, 1989.

18. J. Stoer and R. Burlisch. *Introduction to Numerical Analysis*. Springer-Verlag, New York, 1980.

19. G.A. Watson. *Approximation Theory and Numerical Methods*. Wiley, Chichester, West Sussex, England, 1980.

20. P. Wilmott. *Quantitative Finance (vols. I and II)*. Wiley, Chichester, West Sussex, England, 2000.

3

Optimization methods

Covering in depth all optimization methods that could be useful in solving finance-related problems would require a few books (tough ones, by the way). The aim of this chapter is twofold but much less ambitious. We would like:

- On the one hand, to provide the reader with a minimal background required to grasp what MATLAB offers in the Optimization toolbox; in particular, one should know what she's doing when choosing one among the various methods that are available to cope with the same type of problem.

- On the other hand, we would like to present the basics of relatively advanced modeling and solving techniques, such as branch and bound methods for nonconvex optimization (global optimization and mixed-integer programming) and stochastic programming; although these methods are not available nor easily implemented in MATLAB, we feel that students should at least be exposed to the basics, since both mixed-integer and stochastic programming are likely to find increasing applications in portfolio management.

3.1 CLASSIFICATION OF OPTIMIZATION PROBLEMS

We have already met simple linear and quadratic programming models in sections 1.2 and 1.3. They are just two examples of a vast array of model classes which are dealt with by an equally vast array of methods. Hence, the starting point of this chapter should be a listing of the basic features by which an optimization model may be characterized.

3.1.1   Finite- vs. infinite-dimensional problems

In this chapter we are concerned with problems whose abstract form is

$$
\begin{aligned}
&\min && f(\mathbf{x}) \\
&\text{s.t.} && \mathbf{x} \in S \subseteq \mathbb{R}^n.
\end{aligned}
\tag{3.1}
$$

The objective function f is a scalar function quantifying the suitability of a solution $\mathbf{x}$, which is a vector of decision variables and must belong to a feasible set S, which is a subset of the set of vectors with n real components. Since the solution is expressed by a finite-dimensional vector, we speak of a *finite-dimensional problem*. There is no loss of generality in considering only minimization problems, since a maximization problem may be transformed into a minimization problem simply by changing the sign in the objective:

$$
\max f(x) \quad \Rightarrow \quad -\min[-f(x)].
$$

Indeed, all MATLAB functions in the Optimization toolbox assume a minimization problem. Solving an optimization problem like (3.1) means finding a point $\mathbf{x}^* \in S$ such that

$$
f(\mathbf{x}^*) \le f(\mathbf{x}) \qquad \forall \mathbf{x} \in S.
\tag{3.2}
$$

The point $\mathbf{x}^*$ is said to be a *global optimum* (the term *minimizer* is also used). Neither the existence nor the uniqueness of a global optimum should be taken for granted. To begin with, the problem may be unbounded, which is the case if there is a sequence of solutions $\mathbf{x}^{(k)} \in S$ such that

$$
\lim_{k \to \infty} f(\mathbf{x}^{(k)}) = -\infty.
$$

Furthermore, the problem may be infeasible, i.e., the feasible set S may be empty. Finally, the solution is not unique when condition (3.2) is satisfied by a set of alternative optima, which may be a discrete and finite set, or an infinite set. If the condition (3.2) holds only in a neighborhood of $\mathbf{x}^*$, we speak of a *local optimum*.

Example 3.1 A typical objective function that gives rise to local optima is a polynomial function; recall that the oscillatory behavior of high-order polynomials is the reason why they are not well suited to function interpolation (see example 2.9 on page 106). We may check this with a simple MATLAB snapshot. Consider a polynomial like

$$
f(x) = x^4 - 10.5x^3 + 39x^2 - 59.5x + 30
$$

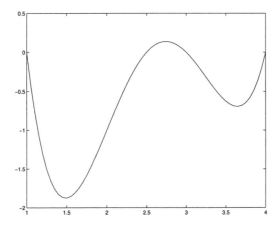

Fig. 3.1 Global and local optima for a polynomial.

and use MATLAB to plot it.

```
>> g = inline('polyval( [ 1 -10.5 39 -59.5 30], x)');
>> xvet=1:0.05:4;
>> plot(xvet,g(xvet))
```

The plot produced is illustrated in figure 3.1, from which it is clear that there are two local minimizers. One MATLAB function that will solve a minimization problem is **fminunc**; the "unc" stands for *unconstrained*, since we are not enforcing any requirement on the decision variable. This function requires an argument which is the initial point of the search process.

```
>> [x,fval] = fminunc(g, 0)
Warning: Gradient must be provided for trust-region method;
   using line-search method instead.
x =
    1.4878
fval =
   -1.8757
>> [x,fval] = fminunc(g, 5)
Warning: Gradient must be provided for trust-region method;
   using line-search method instead.
x =
    3.6437
fval =
   -0.6935
```

We see that depending on the starting point, we get the global or the local minimizer. The MATLAB output has been cut a little, but we see some

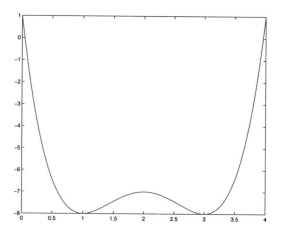

Fig. 3.2 Objective function with two global optima.

messages concerning trust region and line search; the meaning of these terms is illustrated in the following (this is all this chapter is about, after all). A different situation occurs in the following case:

```
>> f = inline('polyval( [ 1 -8 22 -24 1], x)');
>> xvet=0:0.05:4;
>> plot(xvet,f(xvet))
```

The plot is shown in figure 3.2. It may be seen that we have two alternative global minima.   ⬜

Example 3.2 It is easy to build problems which are, respectively:

1. Unbounded:

$$\begin{aligned}
\max \quad & x_1 + x_2 \\
\text{s.t.} \quad & x_1 + x_2 \geq 4 \\
& x_1, x_2 \geq 0.
\end{aligned}$$

2. Infeasible:

$$\begin{aligned}
\max \quad & x_1 + x_2 \\
\text{s.t.} \quad & x_1 + x_2 \geq 4 \\
& 0 \leq x_1, x_2 \leq 1.
\end{aligned}$$

3. Characterized by an infinite set of optima:

$$\begin{aligned}
\max \quad & x_1 + x_2 \\
\text{s.t.} \quad & x_1 + x_2 \leq 4 \\
& x_1, x_2 \geq 0.
\end{aligned}$$

The reader is urged to check this by drawing the feasible set and the level curves of the objective function.

Another important remark is that some problems may have no solution because they are posed the wrong way. Consider the innocent-looking example

$$\min \quad x$$
$$\text{s.t.} \quad x > 2.$$

This problem has no solution, as the feasible set is open, and the apparently obvious solution $x = 2$ is not feasible. In fact, there is not a minimum but only an infimum. This is why in any optimization software you only get constraints such as $\geq$ or $\leq$. $\qquad \qquad \qquad \qquad \qquad \qquad \qquad \qquad \qquad \square$

So far we have assumed that the feasible set is a subset of the space on n-dimensional vectors with real components. In *infinite-dimensional problems* the solution is represented by an infinite collection of decision variables. This is the case when the solution we are seeking is a function of time over a continuous interval. Consider, for instance, a continuous-time dynamic system represented by a vector differential equation:

$$\dot{\mathbf{x}}(t) = \mathbf{h}[\mathbf{x}(t), \mathbf{u}(t)],$$

where $\mathbf{x}$ is the vector of state variables and $\mathbf{u}$ is the vector of control inputs. An optimal control $\mathbf{u}(t)$, $t \in [0, T]$ for this system may be found by solving

$$\min \quad \int_0^T f[\mathbf{x}(t), \mathbf{u}(t)] \, dt + g[\mathbf{x}(T)]$$
$$\text{s.t.} \quad \dot{\mathbf{x}}(t) = \mathbf{h}[\mathbf{x}(t), \mathbf{u}(t)] \qquad \forall t \in [0, T]$$
$$\mathbf{x}(0) = \mathbf{x}_0$$
$$\mathbf{u}(t) \in \Omega \qquad \forall t \in [0, T],$$

where $[0, T]$ is the time horizon we are interested in, $\mathbf{x}_0$ is the (known) initial state of the system, and Ω is the set of admissible controls. The objective function includes both a *trajectory cost*, depending on both the states and the controls, and a *terminal cost*, depending on the terminal state $\mathbf{x}(T)$. It is also possible to specify some constraints on the terminal state.

There is a vast literature on optimal control models in finance. They are actually formulated within a stochastic setting (returns are random and modeled by stochastic differential equations as discussed in chapter 1) and solved by dynamic programming (see, e.g., [14]). Optimal control methods are an excellent tool to analyze relatively simple models and to derive valuable insights from a qualitative and theoretical point of view; however, it might be argued that, in general, complex and realistic problems are usually best formulated and solved as finite-dimensional models. This is an admittedly debatable point, as many would disagree, particularly when it comes to stochastic models for finance (see, e.g., [11] for an alternative view). Anyway, we do not deal with this class of models, essentially to keep the book to a reasonable size. It is worth noting that finite-dimensional models may be used to approximate infinite-dimensional problems by discretizing the continuous-time model. For instance, the infinite-dimensional problem above can be transformed into the finite-dimensional problem

$$\min \quad \sum_{k=1}^{K} f(\mathbf{x}_k, \mathbf{u}_k) + g(\mathbf{x}_K)$$

$$\text{s.t.} \quad \mathbf{x}_k = \mathbf{h}(\mathbf{x}_{k-1}, \mathbf{u}_k) \quad k = 1, \dots, K$$

$$\mathbf{u}_k \in \Omega \quad k = 1, \dots, K,$$

where the time horizon has been discretized in time intervals of width Δ and $\mathbf{x}_k = \mathbf{x}(k\Delta)$. Note that $\mathbf{x}_k$ is the state *at the end* of the kth period [i.e., the period between $(k-1)\Delta$ and $k\Delta$], whereas $\mathbf{u}_k$ is the control applied *during* the kth period.

3.1.2   Unconstrained vs. constrained problems

If $S \equiv \mathbb{R}^n$, we have an *unconstrained problem*; otherwise, we have a *constrained problem*. Needless to say, real-life problems are rarely unconstrained; yet methods for unconstrained optimization are the foundation for many constrained optimization methods. The set S is usually specified by enforcing the following types of constraints on the decision variables.

- Equality constraints:

$$h_i(\mathbf{x}) = 0 \quad i \in E,$$

  or in vector form:

$$\mathbf{h}(\mathbf{x}) = \mathbf{0}.$$

- Inequality constraints:

$$g_i(\mathbf{x}) \le 0 \quad i \in I.$$

  or in vector form:

$$\mathbf{g}(\mathbf{x}) \le \mathbf{0},$$

  having stipulated that a vector inequality is interpreted componentwise. The constraint $g_i(\mathbf{x}) \le 0$ is said to be active at the point $\hat{\mathbf{x}}$ if $g_i(\hat{\mathbf{x}}) = 0$, and inactive if $g_i(\hat{\mathbf{x}}) < 0$. A "greater than" constraint such as $g_k(\mathbf{x}) \ge 0$ can be rewritten immediately in the form $-g_k(\mathbf{x}) \le 0$. In MATLAB, inequality constraints are assumed in the "less than" form. Nonnegativity restrictions such as $x \ge 0$, also denoted by $x \in \mathbb{R}_+$, may be thought of as inequality constraints. However, simple bounding constraints of the form $l \le x \le u$ are usually dealt with in a special way by optimization algorithms; hence, inequality constraints and bounds are passed separately to optimization procedures.

3.1.3 Convex vs. nonconvex problems

Depending on the nature of the objective function f and of the feasible set S, problem (3.1) may or may not be easy. In particular, when there is only one local optimum which is also the global optimum, the problem should be expected to be relatively easy. The key concept here, and in most optimization theory as well, is convexity. Some background in convex analysis is given in supplement S3.1 at the end of the chapter.

Problem (3.1) is a *convex problem* if f is a convex function and S is a convex set. Problem (3.1) is a *concave problem* if f is a concave function and S is a convex set. Assuming that the optimization problem has a finite solution, the following properties can be shown.

PROPERTY 3.1 *In a convex problem a local optimum is also a global optimum.*

PROPERTY 3.2 *In a concave problem the global optimum lies on the boundary of the feasible region S.*

To get a feeling for the second property, the reader is urged to solve the following problem graphically:

$$\min \quad -(x-2)^2 + 3$$
$$\text{s.t.} \quad 1 \le x \le 4.$$

Ideally, we would like to come up with a set of necessary and sufficient conditions for global optimality. Regrettably, what we have, in general, are just either sufficient or necessary conditions for local or global optimality. However, when the problem is unconstrained and the function is convex, it is easy to find a convenient characterization of a global minimizer.

THEOREM 3.3 *If the function f is convex and differentiable on $\mathbb{R}^n$, the point $\mathbf{x}^*$ is a global minimizer of f if and only if it satisfies the stationarity condition:*

$$\nabla f(\mathbf{x}^*) = \mathbf{0}.$$

Proof. If f is convex and differentiable, then we have

$$f(\mathbf{x}) \ge f(\mathbf{x}_0) + \nabla f^T(\mathbf{x}_0)(\mathbf{x} - \mathbf{x}_0) \qquad \forall \mathbf{x}, \mathbf{x}_0.$$

But if $\mathbf{x}^*$ is a stationarity point,

$$f(\mathbf{x}) \ge f(\mathbf{x}^*) + \nabla f^T(\mathbf{x}^*)(\mathbf{x} - \mathbf{x}^*) = f(\mathbf{x}^*) + \mathbf{0}^T(\mathbf{x} - \mathbf{x}^*) = f(\mathbf{x}^*) \qquad \forall \mathbf{x},$$

which simply says that $\mathbf{x}^*$ is a global optimum. ☐

The stationarity condition is a *first-order condition*; for generic functions, *second-order conditions* involving the Hessian matrix are required to guarantee that a stationary point is actually a (local) minimizer. The stationarity condition is easily extended to the case of a convex nondifferentiable function.

THEOREM 3.4 *If the function f is convex on $\mathbb{R}^n$, the point $\mathbf{x}^*$ is a global minimizer of f if and only if the subdifferential of f at $\mathbf{x}^*$ includes the zero vector:*

$$\mathbf{0} \in \partial f(\mathbf{x}^*).$$

Proof. As discussed in supplement S3.1, a convex function f is subdifferentiable at any point;[1] that is, at any point $\mathbf{x}_0$ there is a set of subgradients, which is called the *subdifferential*. A subgradient at $\mathbf{x}_0$ is a vector $\boldsymbol{\gamma}$ such that

$$f(\mathbf{x}) \geq f(\mathbf{x}_0) + \boldsymbol{\gamma}^T(\mathbf{x} - \mathbf{x}_0) \qquad \forall \mathbf{x}.$$

It is easy to see that if $\mathbf{0}$ belongs to the subdifferential at $\mathbf{x}^*$, we have $f(\mathbf{x}) \geq f(\mathbf{x}^*)$ for any $\mathbf{x}$. It is worth noting that this theorem is a generalization of the previous one, as if the function is differentiable in $\mathbf{x}^*$, the subdifferential includes only the gradient, and this condition boils down to stationarity. □

It should be noted that a set $S = \{\mathbf{x} \in \mathbb{R}^n \mid g_i(\mathbf{x}) \leq 0,\ i \in I\}$ is convex if the functions g_i are convex. To see this for a single function $g(\mathbf{x})$, assume that $\mathbf{x}_1, \mathbf{x}_2 \in S$. Convexity of g implies that

$$g[\lambda \mathbf{x}_1 + (1 - \lambda)\mathbf{x}_2] \leq \lambda g(\mathbf{x}_1) + (1 - \lambda)g(\mathbf{x}_2) \leq 0 \qquad \forall \lambda \in [0, 1].$$

Since the intersection of convex sets is a convex set, the result is valid for an arbitrary number of convex functions. The equality-constrained case is more critical. Since an inequality constraint $h_i(\mathbf{x}) = 0$ can be thought of as two inequalities,

$$h_i(\mathbf{x}) \leq 0 \qquad -h_i(\mathbf{x}) \leq 0,$$

we see that it will describe a convex set only if the function h_i is both convex and concave. This will be the case only if h_i is affine, i.e., it is of the form

$$\mathbf{a}_i^T \mathbf{x} = b_i.$$

3.1.4 Linear vs. nonlinear problems

A finite-dimensional problem is called a *linear programming* (LP) *problem* when both the constraints and the objective are expressed by affine functions. The general form of a linear programming problem is

$$\min \quad \sum_{j=1}^{n} c_j x_j$$

$$\text{s.t.} \quad \sum_{j=1}^{n} a_{ij} x_j = b_i \qquad \forall i \in E$$

---

[1]Strictly speaking, this is true only for the *interior* of the domain over which the function is convex.

$$\sum_{j=1}^{n} d_{ij} x_j \leq e_i \qquad \forall i \in I,$$

which can be written in matrix form as

$$
\begin{aligned}
\min \quad & \mathbf{c}^T \mathbf{x} \\
\text{s.t.} \quad & \mathbf{A}\mathbf{x} = \mathbf{b} \\
& \mathbf{D}\mathbf{x} \leq \mathbf{e}.
\end{aligned}
$$

Linear programming problems have two important features; they are both convex and concave problems. Thus, a local optimum is also a global one, and it lies on the boundary of the feasible solution; actually, it turns out that the feasible set is a polyhedron and that there is an optimal solution which corresponds to one of its vertices.

Example 3.3 Here is an example of an LP problem:

$$
\begin{aligned}
\min \quad & 2x_1 + 3x_2 + 3x_3 \\
\text{s.t.} \quad & x_1 + 2x_2 = 3 \\
& x_1 + x_3 \geq 3 \\
& x_1, x_2, x_3 \geq 0.
\end{aligned}
$$
☐

If either condition is not met, i.e., if the objective function or a constraint is expressed by a nonlinear function, we have a nonlinear programming problem.

Example 3.4 The following are examples of nonlinear programming problems:

$$
\begin{aligned}
\min \quad & 2x_1 + 3x_2 + 3x_3 \\
\text{s.t.} \quad & x_1 + x_2^2 = 3 \\
& x_1 + x_3 \geq 3 \\
& x_1, x_2, x_3 \geq 0.
\end{aligned}
$$

$$
\begin{aligned}
\min \quad & 2x_1 + 3x_2 x_3 \\
\text{s.t.} \quad & x_1 + 2x_2 = 3 \\
& x_1 x_3 \geq 3 \\
& x_1, x_2, x_3 \geq 0.
\end{aligned}
$$

$$
\begin{aligned}
\min \quad & 2x_1^2 + 3x_2^2 + 3x_1 x_3 \\
\text{s.t.} \quad & x_1 + 2x_2 = 3 \\
& x_1 + x_3 \geq 3 \\
& x_1, x_2, x_3 \geq 0.
\end{aligned}
$$

The last problem is characterized by a quadratic objective function and by linear constraints. This kind of problem is called a *quadratic programming problem*. Quadratic programming problems are the simplest nonlinear programming problems, provided that the objective function is convex. If the quadratic part of the objective is related to a covariance matrix, as happens for mean-variance efficient portfolios, the objective function is convex, as the covariance matrix is positive semidefinite (see theorem 3.12 in supplement S3.1.1).   □

3.1.5   Continuous vs. discrete problems

Linear and quadratic programming problems are rather easy to solve, as they are convex problems. In some decision problems, it is necessary to enforce integrality constraints on some decision variables:

$$\mathbf{x} \in \mathbb{Z}_{+}^{n},$$

where $\mathbb{Z}_+ = \{0, 1, 2, \ldots\}$ is the set of nonnegative integers (models involving negative integer variables are quite rare). If the integrality constraint applies to all of the decision variables, we have a *pure integer program*; otherwise, we have a *mixed-integer program*. Such a restriction makes the problem much harder, mainly because a discrete feasible region is not convex. While nonlinear integer programming techniques are known, robust commercial tools are available only for mixed-integer linear programs.

Quite often, an integrality restriction has the form $x \in \{0, 1\}$, which is used when we have to model all-or-nothing decisions. One such case is the knapsack problem we met in example 1.8. It is useful to point out a few situations that require the introduction of binary decision variables.

Logical constraints   Consider a set of N activities, perhaps investment opportunities. Starting an activity or not is modeled by a corresponding binary decision variable x_i, $i = 1, \ldots, N$. You might wish to enforce some logical constraints involving subsets of activities. Here are a few examples:

- Exactly one activity within a subset S must start (exclusive "or"):

$$\sum_{j \in S} x_j = 1.$$

- At least one activity within a subset S must start (inclusive "or"):

$$\sum_{j \in S} x_j \geq 1.$$

- At most one activity within a subset S may start:

$$\sum_{j \in S} x_j \leq 1.$$

- If activity j is started, then activity k must start, too:

$$x_j \leq x_k.$$

All the constraints above may be generalized to more complex situations, which are relevant, for instance, if you want to enforce qualitative constraints on a portfolio of investments.

Fixed-charge problem and semicontinuous decision variables   We obtain LP models when we assume, among other things, that the costs of carrying out a set of activities depend linearly on the activity levels. In some cases, the cost structure is more complex; the *fixed-charge problem* is one such case. We are given a set of activities, indexed by $i = 1, \ldots, N$. The level of activity i is measured by a nonnegative continuous variable x_i; the activity levels are subject to a set of constraints, formally expressed as $\mathbf{x} \in S$. Each activity has a cost proportional to the level x_i and a fixed cost f_i which is paid whenever $x_i > 0$. The fixed cost does not depend on the activity level. It is interesting to note that the cost function is in this case discontinuous at the origin, but a simple modeling trick allows us to build a mixed-integer model.

Assume that we know an upper bound M_i on the level of activity i, and introduce a set of binary variables y_i such that

$$y_i = \begin{cases} 1 & \text{if } x_i > 0 \\ 0 & \text{otherwise.} \end{cases}$$

We can build the following model:

$$\begin{aligned} \min \quad & \sum_{i=1}^{N} (c_i x_i + f_i y_i) \\ \text{s.t.} \quad & x_i \leq M_i y_i \qquad \forall i \\ & \mathbf{x} \in S \\ & y_i \in \{0, 1\} \qquad \forall i. \end{aligned} \qquad (3.3)$$

Equation (3.3) is a common way to model fixed-charge costs. If $y_i = 0$, necessarily $x_i = 0$; if $y_i = 1$, then we obtain $x_i \leq M_i$, which is a nonbinding constraint if M_i is large enough.

Another common requirement on the level of an activity is that if it is undertaken, its level should be in the interval $[m_i, M_i]$. This means that the feasible set is nonconvex:

$$x_i \in \{0\} \cup [m_i, M_i].$$

Using the same trick as above, we may just write

$$x_i \geq m_i y_i \qquad x_i \leq M_i y_i.$$

These constraints define a *semicontinuous decision variable*. Semicontinuous variables may be used when the amount of an asset in a portfolio must be above a minimum threshold if the asset is included in the portfolio.

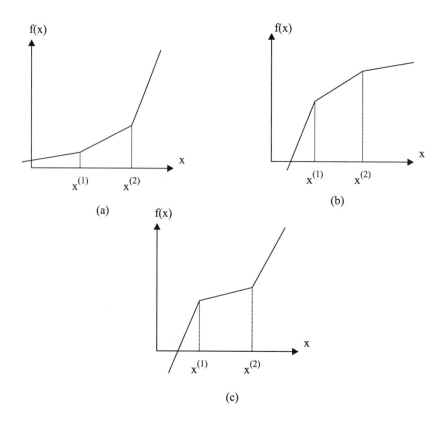

Fig. 3.3 Piecewise linear functions: (a) convex, (b) concave, (c) neither convex nor concave.

Piecewise linear functions   Sometimes we have to model a nonlinear dependency between two variables; to name one case, transaction costs may depend in a nonobvious way on the trading volume. Although it is possible to adopt nonlinear programming methods to cope with this case, it may be advisable to avoid the issue by approximating the nonlinear function by a piecewise linear function; in other words, we may try a linear interpolation (see section 2.3). Piecewise linear functions may arise quite naturally in applications. A few examples are shown in figure 3.3, where the points $x^{(i)}$ are the breakpoints separating the linearity intervals. There are different reasons for doing so. If the nonlinear function occurs in an equality constraints, the problem is nonconvex; the practical implication is that a nonlinear optimizer may get stuck in a local minimum. The same happens if the objective function is nonlinear and nonconvex. Here we show that these cases may be transformed into mixed-integer programming problems which are nonconvex but can be solved by branch and bound methods yielding a global optimum. Furthermore, it may be the case that the model involves integer decision variables, in which

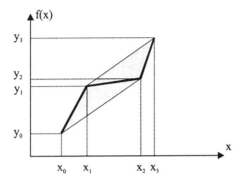

Fig. 3.4   Modeling a piecewise linear function.

case it may be preferable to keep the model linear, as nonlinear mixed-integer programming problems may be overly difficult to solve.

Consider a function like

$$f(x) = \begin{cases} c_1 x & 0 \le x \le x^{(1)} \\ c_2(x - x^{(1)}) + c_1 x^{(1)} & x^{(1)} \le x \le x^{(2)} \\ c_3(x - x^{(2)}) + c_1 x^{(1)} + c_2(x^{(2)} - x^{(1)}) & x^{(2)} \le x \le x^{(3)}. \end{cases}$$

If $c_1 < c_2 < c_3$ (increasing marginal costs), then $f(x)$ is convex (figure 3.3a); if $c_1 > c_2 > c_3$ (decreasing marginal costs), the function is concave (figure 3.3b); for arbitrary slopes c_i the function is neither convex nor concave (figure 3.3c).

The convex case is easy and it can be coped with by continuous LP models. The function $f(x)$ can be converted to a linear form by introducing three auxiliary variables y_1, y_2, y_3 and substituting:

$$x = y_1 + y_2 + y_3$$
$$0 \le y_1 \le x^{(1)}$$
$$0 \le y_2 \le (x^{(2)} - x^{(1)})$$
$$0 \le y_3 \le (x^{(3)} - x^{(2)}).$$

Then we can express

$$f(x) = c_1 y_1 + c_2 y_2 + c_3 y_3,$$

since $c_1 < c_2$, y_2 is positive in the optimal solution only if y_1 is set to its upper bound. Similarly, y_3 is activated only if both y_1 and y_2 are saturated to their upper bounds. If the function is not convex, this is not guaranteed, and we must come up with a modeling trick based on binary decision variables.

To get a clue on how a general piecewise linear function may be modeled, assume that the function is described by the knots (x_i, y_i), $y_i = f(x_i)$, $i = 0, 1, 2, 3$, as in figure 3.4. Any point on the line from (x_i, y_i) to (x_{i+1}, y_{i+1})

can be expressed as a convex combination:

$$
\begin{aligned}
x &= \lambda x_i + (1 - \lambda)x_{i+1} \\
y &= \lambda y_i + (1 - \lambda)y_{i+1},
\end{aligned}
$$

where $0 \le \lambda \le 1$. Now what about forming a convex combination of the four knots?

$$
x = \sum_{i=0}^{3} \lambda_i x_i
$$

$$
y = \sum_{i=0}^{3} \lambda_i y_i
$$

$$
\sum_{i=0}^{3} \lambda_i = 1 \qquad \lambda_i \ge 0.
$$

This is not really what we want, since this is the convex hull of the four knots (the shaded area in figure 3.4; see supplement S3.1). However, we are close; we have just to allow only pairs of adjacent coefficients λ_i to be greater than zero. This is accomplished by introducing a binary decision variable s_i, $i = 1, 2, 3$, for each line segment $(i - 1, i)$:

$$
x = \sum_{i=0}^{3} \lambda_i x_i
$$

$$
y = \sum_{i=0}^{3} \lambda_i y_i
$$

$$
0 \le \lambda_0 \le s_1
$$

$$
0 \le \lambda_1 \le s_1 + s_2
$$

$$
0 \le \lambda_2 \le s_2 + s_3
$$

$$
0 \le \lambda_3 \le s_3
$$

$$
\sum_{i=1}^{3} s_i = 1 \qquad s_i \in \{0, 1\}.
$$

In practice, optimization software packages provide the user with an easier but equivalent way to express piecewise linear functions.

3.1.6 Deterministic vs. stochastic problems

All the model classes we have considered so far assume on the one hand that there is no uncertainty in the data, and on the other that a sensible analytical model can be built. In some cases, building an analytical model is out of the question, because of both the randomness and the complexity involved. As an example, consider a set of rules for portfolio rebalancing; say these rules depend on a set of parameters and that you would like to find the optimal set of parameters. It may be the case that a thorough testing of the rules may be carried out only by running a set of simulated experiments. This means that a simulator acts as a black box mapping a vector of decision variables $\mathbf{x}$ into an estimate of an objective function $f(\mathbf{x}) = \mathrm{E}[U(\mathbf{x})]$, possibly related to an expected utility. In this case, you have to integrate stochastic simulation and optimization methods, as described in section 4.6.

Even if an analytical model can be built but the data are actually random variables, using a deterministic model is rather questionable. As a specific example, consider a stochastic LP model:

$$
\begin{aligned}
\text{``min''} \quad & \mathbf{c}(\omega)^T \mathbf{x} \\
\text{s.t.} \quad & \mathbf{A}(\omega)\mathbf{x} \geq \mathbf{b}(\omega) \\
& \mathbf{x} \geq \mathbf{0}.
\end{aligned}
\tag{3.4}
$$

Here the data $\mathbf{c}(\omega)$, $\mathbf{A}(\omega)$, and $\mathbf{b}(\omega)$ depend on random events ω. The "min" notation is used to point out that this problem actually does not make sense, since minimizing a random variable has no meaning. We could define a sensible objective function by taking an expected value:

$$
\mathrm{E}[\mathbf{c}(\omega)^T \mathbf{x}] = \mathrm{E}[\mathbf{c}(\omega)]^T \mathbf{x}.
$$

By the same token, we could consider the idea of solving a deterministic problem based on the expected values of all the uncertain data. Unfortunately, this approach may not be robust, as the solution we devise may turn out to be of questionable quality or even infeasible. Let us address the feasibility issue first.

Finding a solution $\mathbf{x}$ such that the constraints (3.4) are always satisfied may be impossible, or it could lead to a poor solution. A possible approach is to relax the constraints a bit and to accept the fact that in some cases the constraints could not be met; we might just ask that this undesirable event is unlikely. This leads to *chance-constrained models* such as

$$
\begin{aligned}
\min \quad & \mathbf{c}^T \mathbf{x} \\
\text{s.t.} \quad & \mathbf{A}\mathbf{x} \geq \mathbf{b} \\
& P\{\mathbf{G}(\omega)\mathbf{x} \geq \mathbf{h}(\omega)\} \geq \alpha \\
& \mathbf{x} \geq \mathbf{0},
\end{aligned}
$$

where we have separated the deterministic constraints from those involving uncertainty. Such models trade off the cost of the solution with its reliability, or robustness. Another fundamental point, which is not addressed by a chance-constrained model, is that when uncertainty is involved, the decision process is often dynamic. This means that we take a set of decisions here-and-now, based on limited information, but then we may adjust the decisions when the uncertainty is revealed. Of course, adjusting the decisions will imply some additional costs, and we would like to take good decisions minimizing the immediate costs as well as the expected value of the adjustment costs we will pay in the future. This idea leads to stochastic programming models with recourse. As an example, we may consider a two-stage stochastic linear programming model, which is usually stated as follows. The first-stage problem, involving the decisions $\mathbf{x}$ that we must take here and now, is

$$
\begin{aligned}
\min \quad & \mathbf{c}^T\mathbf{x} + \mathrm{E}[h(\mathbf{x}, \omega)] \\
\text{s.t.} \quad & \mathbf{A}\mathbf{x} = \mathbf{b} \\
& \mathbf{x} \geq \mathbf{0}.
\end{aligned}
$$

The first-stage problem involves a set of deterministic constraints and the expected cost of adjusting the solution at the second stage. The second-stage problem, involving the adjustments, or recourse variables $\mathbf{y}$, defines the function $h(\mathbf{x}, \omega)$:

$$
\begin{aligned}
h(\mathbf{x}, \omega) \equiv \min \quad & \mathbf{q}(\omega)^T \mathbf{y} \\
\text{s.t.} \quad & \mathbf{W}(\omega)\mathbf{y} = \mathbf{r}(\omega) - \mathbf{T}(\omega)\mathbf{x} \\
& \mathbf{y} \geq \mathbf{0}.
\end{aligned}
$$

There are a few things to point out in the second-stage problem.

- We have written the problem in its most general form, allowing randomness in all the parameters, but this need not be the case. For instance, if the recourse matrix $\mathbf{W}$ is deterministic, we have a *fixed recourse problem*. Some algorithms may only be applied if the recourse is fixed and if the recourse cost vector $\mathbf{q}$ is deterministic as well; other solution algorithms have no such limitations.

- The overall problem can be thought of as a nonlinear programming problem involving a recourse function $H(\mathbf{x}) \equiv \mathrm{E}[h(\mathbf{x}, \omega)]$. Such a function may seem intractable, as it involves the multidimensional integration of a function implicitly defined through an optimization problem. However, it may be shown that in the relevant cases the recourse function is convex. So, even if we do not know how to express $H(\mathbf{x})$ in a simple analytical form, we may still be able both to evaluate (or estimate) its value and to find a subgradient at a given point $\mathbf{x}$. On the contrary, chance-constrained problems are not convex problems in general.

- Depending on its structure, the second-stage problem may have a feasible solution for any first-stage vector $\mathbf{x}$ and for any random event ω, or not. In the second case, the second-stage problem implicitly defines some further constraints on $\mathbf{x}$.

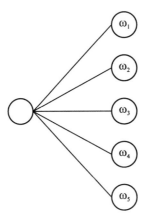

Fig. 3.5   Scenario tree for a two-stage stochastic optimization problem.

- The approach may be easily generalized to multiple stages. We will see how in chapter 6.

An important modeling decision is if we should treat the uncertainty as expressed by a continuous or discrete probability distribution. Although there are approximation algorithms able to cope with a continuous distribution, it's no surprise that a common practice is to approximate the original data through a suitable discrete distribution. In some cases this is all we can do, since the available information is just a set of plausible future scenarios, handed by an expert of the problem domain. Each possible realization of the uncertain data is called a *scenario*. It is most useful to depict the decision problem as illustrated in figure 3.5. The root node of the tree represents the present state of the world, from which different future states branch, corresponding to possible realizations of the uncertain data. We have to take first-stage decisions here and now, i.e., in the root of the tree; then, when the uncertainty is revealed, we will have the chance to take second-stage decisions to adapt to the circumstances; each possible contingency is represented by a leaf node in the tree. The overall problem entails taking a good first-stage decision, which should be robust, in that it should leave room for not too costly adaptations at the second-stage. Assume that we have a set of scenarios, indexed by $s \in S$, each with associated probability p_s. Then the two-stage stochastic LP problem boils down to a large-scale LP problem:

$$
\begin{aligned}
\min \quad & \mathbf{c}^T \mathbf{x} + \sum_{s \in S} p_s \mathbf{q}_s^T \mathbf{y}_s \\
\text{s.t.} \quad & \mathbf{A}\mathbf{x} = \mathbf{b} \\
& \mathbf{T}_s \mathbf{x} + \mathbf{W}_s \mathbf{y}_s = \mathbf{r}_s \qquad \forall s \in S \\
& \mathbf{x}, \mathbf{y}_s \geq \mathbf{0}.
\end{aligned}
$$

This problem could be, in principle, tackled simply by standard LP techniques; however, its size and its peculiar structure suggest the adoption of more specific approaches, one of which is described in section 3.7. Now a natural question is: Since solving a stochastic LP looks like a nontrivial task, why bother? Shouldn't we simply take the expected values of the data and solve a much simpler deterministic problem? Indeed, in some cases, solving a stochastic LP is a wasted effort. To characterize the cases in which the added effort is worthwhile, we may consider the VSS (*value of the stochastic solution*) concept.

Let us define the individual scenario problem

$$\text{min} \quad z(\mathbf{x}, \omega) = \mathbf{c}^T \mathbf{x} + \text{min}\{\mathbf{q}_\omega \mathbf{y} \mid \mathbf{W}_\omega = \mathbf{r}_\omega - \mathbf{T}_\omega \mathbf{x}, \mathbf{y} \geq \mathbf{0}\}$$
$$\text{s.t.} \quad \mathbf{A}\mathbf{x} = \mathbf{b}$$
$$\mathbf{x} \geq \mathbf{0}.$$

Note that the scenario problem assumes the knowledge of the future event ω. The recourse problem we have just considered amounts to solving

$$\text{RP} = \min_\mathbf{x} \text{E}_\omega[z(\mathbf{x}, \omega)].$$

Solving a deterministic problem, based on the expected values $\bar{\omega} = \text{E}[\omega]$ of the data, corresponds to the expected value problem:

$$\text{EV} = \min_\mathbf{x} z(\mathbf{x}, \bar{\omega}),$$

which yields a solution $\bar{\mathbf{x}}(\bar{\omega})$. However, this solution should be checked in the real context; this means that we should evaluate the expected cost of using the EV solution, which calls for some adjustments anyway:

$$\text{EEV} = \text{E}_\omega[z(\bar{\mathbf{x}}(\bar{\omega}), \omega)].$$

The VSS is defined as[2]
$$\text{VSS} = \text{EEV} - \text{RP}.$$

It can be shown that VSS ≥ 0. A large VSS value suggests that solving the stochastic problem is well worth the effort; a small value suggests the opportunity to take the much simpler deterministic approach. As expected, it turns out that finance is a typical field in which the stochastic character of the problem cannot be neglected. Furthermore, by a proper choice of the recourse function, different risk attitudes of the decision makers may be represented. We pursue this task in chapter 6, where we deal with some portfolio optimization models.

---

[2]A related but different concept is the expected value of perfect information (EVPI); see, e.g., [4, chapter 4].

3.2   NUMERICAL METHODS FOR UNCONSTRAINED OPTIMIZATION

In principle, an unconstrained problem $\min_{\mathbf{x} \in \mathbb{R}^n} f(\mathbf{x})$ may be solved by look-ing for a stationary point. Some care is needed for the nonconvex case, since second-order information should be checked; furthermore, what we get in general is a local optimizer; indeed, almost all the nonlinear programming libraries commercially available are aimed at local optimization. The station-arity condition yields a set of nonlinear equations which could be solved to spot candidate optima; in fact, there are a few links between unconstrained optimization and the numerical solution of nonlinear equations.

In optimization, one avoids direct solution of the nonlinear equations. The computational approaches are generally based on the generation of a sequence of points $\mathbf{x}^{(k)}$, converging to a local optimum $\mathbf{x}^*$. In order to drive the search process in the right direction, one should find, for each point $\mathbf{x}^{(k)}$ in the sequence, a descent direction, i.e., a vector $\mathbf{s}^{(k)} \in \mathbb{R}^n$ such that

$$f(\mathbf{x}^{(k)} + \alpha \mathbf{s}^{(k)}) < f(\mathbf{x}^{(k)})$$

for some $\alpha > 0$. If we consider the function $h(\alpha) = f(\mathbf{x} + \alpha \mathbf{s})$, a descent direction is characterized by

$$\left. \frac{dh}{d\alpha} \right|_{\alpha=0} = [\nabla f(\mathbf{x})]^T \mathbf{s} < 0.$$

It may be convenient to consider true direction vectors, i.e., unit norm vectors such that $\| \mathbf{s} \| = 1$. A general iteration scheme is, after initialization with a starting guess $\mathbf{x}^{(0)}$:

1. Find a descent direction $\mathbf{s}^{(k)}$.

2. Find a step length $\alpha^{(k)} \in \mathbb{R}_+$.

3. Update $\mathbf{x}^{(k+1)} = \mathbf{x}^{(k)} + \alpha^{(k)} \mathbf{s}^{(k)}$.

The scheme is iterated until some convergence criterion is met. There are a variety of choices, which lead to different algorithms, some of which are outlined briefly in the following. It should be noted that this approach can be extended to deal with constrained optimization problems. An easy case is when we have to solve

$$\min_{\mathbf{x} \in \mathbb{R}_+^n} f(\mathbf{x}).$$

Here it is sufficient to slightly modify the updating rule as follows:

$$\mathbf{x}^{(k+1)} = \max \left\{ 0, \mathbf{x}^{(k)} + \alpha^{(k)} \mathbf{s}^{(k)} \right\},$$

which should be interpreted componentwise; if some component becomes neg-ative, set it at zero. This operation essentially amounts to *projecting* $\mathbf{x}^{(k+1)}$ onto the feasible set $\mathbb{R}_+^n$ (projection can be exploited for more general feasible sets, with computational difficulties depending on their nature).

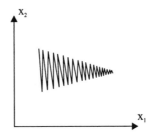

Fig. 3.6   Zigzagging in the steepest descent procedure.

3.2.1   Steepest descent method

One seemingly obvious choice for the descent direction is

$$\mathbf{s}^{(k)} = -\frac{\nabla f^{(k)}}{\|\nabla f^{(k)}\|},$$

which yields the steepest descent or gradient method. The step length α may be chosen by solving the one-dimensional problem

$$\min_{\alpha \geq 0} h(\alpha) = f(\mathbf{x}^{(k)} + \alpha \mathbf{s}^{(k)}).$$

This one-dimensional problem is easier than the original problem, as it is a scalar optimization problem. It can be solved by a variety of *line search methods*. One possibility, which works for convex functions, is using a quadratic fit. Assume that we have three points $0 \leq \alpha_1 < \alpha_2 < \alpha_3$, such that

$$h(\alpha_1) > h(\alpha_2) \qquad h(\alpha_2) < h(\alpha_3).$$

An initial set of points is easily found. Now we may fit a quadratic curve passing through the three points; minimization of the quadratic curve is easily accomplished, under convexity assumption, by setting its derivative to zero. This yields another point, α^*. Assume that $\alpha^* > \alpha_2$. If $h(\alpha^*) \geq h(\alpha_2)$, we proceed with the new set of points $(\alpha_1, \alpha_2, \alpha^*)$; otherwise, we proceed with $(\alpha_2, \alpha^*, \alpha_3)$. Actually, there is a rich set of line search methods, involving, e.g., cubic interpolation and other tricks of the trade; some may be selected by setting MATLAB option parameters.

Despite its apparent appeal, the steepest descent method may suffer from poor convergence near the minimizer. In some cases, pathological behavior called "zigzagging" is observed. The zigzagging phenomenon is illustrated in figure 3.6.[3]

---

[3]To really see zigzagging, the reader is urged to try the Optimization toolbox demos. Just type demo, which opens a window in which you should select the Optimization toolbox. Then try the "minimization of the banana function" demo. You may also try the command line demo.

3.2.2 The subgradient method

It is obvious that the gradient method cannot be applied to a nondifferentiable function. In supplement S3.1.1 we note that the subgradient is a generalization of the gradient concept to the case of nonsmooth functions. Hence, assuming we can compute a subgradient $\gamma^{(k)}$ for a convex function f at any point $\mathbf{x}^{(k)}$, we may wonder if a scheme like

$$\mathbf{x}^{(k+1)} = \mathbf{x}^{(k)} - \alpha^{(k)} \gamma^{(k)}$$

could work. The answer is not easy, since there is no guarantee that by changing the sign of the subgradient we find a descent direction. However, if some condition is enforced on the step lengths $\alpha^{(k)}$, it can be shown that the subgradient method converges to the optimal solution. An intuitive justification runs as follows.

Consider a point $\mathbf{x}_0$ and let γ_0 be a subgradient of f at $\mathbf{x}_0$. Then, by definition of a subgradient:

$$f(\mathbf{x}) \geq f(\mathbf{x}_0) + \gamma_0^T (\mathbf{x} - \mathbf{x}_0) \qquad \forall \mathbf{x}.$$

By applying this inequality to the optimal solution $\mathbf{x}^*$ and rearranging, we obtain

$$-\gamma_0^T (\mathbf{x}^* - \mathbf{x}_0) \geq f(\mathbf{x}_0) - f(\mathbf{x}^*) \geq 0.$$

Note that the vector $\mathbf{x}^* - \mathbf{x}_0$ is the direction along which we should move to reach the optimal solution from $\mathbf{x}_0$. The inequality above shows that this vector forms an angle less than 90 degrees with $-\gamma_0$. Hence, the subgradient, changed in sign, need not be a descent direction, but at least it points to the "right" half-space, where the optimal solution lies.

3.2.3 Newton and the trust region methods

The convergence problems in the gradient method are essentially due to the fact that the gradient method uses a *first-order* local approximation of f ignoring curvature information. The situation could be improved by using a second-order approximation:

$$f(\mathbf{x} + \delta) \approx f(\mathbf{x}) + [\nabla f(\mathbf{x})]^T \delta + \frac{1}{2} \delta^T \mathbf{H}(x) \delta,$$

where $\mathbf{H}$ is the Hessian matrix. If $\mathbf{H}$ is positive definite, the function is locally strictly convex and we may find a minimizer for the quadratic approximation simply by solving the linear system

$$\mathbf{H}(\mathbf{x})\delta = -\nabla f(\mathbf{x}).$$

This method has better convergence properties as well as higher computational costs. However, we are in trouble if the Hessian is not positive definite.

Another approach is to restrict the step α taken along the direction given by the gradient. The rationale is that the first-order approximation is valid only in a neighborhood of the current iterate $\mathbf{x}^{(k)}$. To find the displacement $\boldsymbol{\delta}$, we could consider the restricted minimization subproblem:

$$\min_{\boldsymbol{\delta}} \quad f(\mathbf{x}^{(k)}) + [\nabla f(\mathbf{x}^{(k)})]^T \boldsymbol{\delta}$$

$$\text{s.t.} \quad \|\boldsymbol{\delta}\| \leq h^{(k)}.$$

Exploiting this idea leads to *trust region methods*, which are actually used in MATLAB for large-scale problems. One way to limit the trust region is to compare the predicted improvement in the objective function (according to the approximating function) with the actual improvement. A large difference implies that the approximation is not reliable and that the step size should be reduced.

3.2.4 No-derivatives algorithms: quasi-Newton method and simplex search

One problem with Newton's method is that the Hessian matrix is required. Since providing the software with this information requires a good deal of work, alternative approaches have been developed in order to approximate this matrix based on function evaluations only. This leads to quasi-Newton methods. The same observation applies to providing the gradient of the objective function. One idea is to approximate the gradient by finite differences like

$$\left. \frac{\partial f(\mathbf{x})}{\partial x_i} \right|_{\mathbf{x}=\hat{\mathbf{x}}} \approx \frac{f(\hat{\mathbf{x}} + h_i \mathbf{1}_i) - f(\hat{\mathbf{x}})}{h_i}$$

or

$$\left. \frac{\partial f(\mathbf{x})}{\partial x_i} \right|_{\mathbf{x}=\hat{\mathbf{x}}} \approx \frac{f(\hat{\mathbf{x}} + h_i \mathbf{1}_i) - f(\hat{\mathbf{x}} - h_i \mathbf{1}_i)}{2h_i},$$

where $\mathbf{1}_i$ is the ith unit vector. This leads to no-derivatives methods.

In some circumstances, you would not be able to compute the gradient anyway; one case is when the objective function is not known, but it is implicitly computed by a simulation model; another is when there are discontinuities in the objective function. In such cases, it is useful to adopt methods that rely strictly on function evaluations. One such approach is the simplex search method developed by Nelder and Mead.[4] The rationale behind the method is illustrated in figure 3.7 for a minimization problem in $\mathbb{R}^2$. A simplex in $\mathbb{R}^n$ is simply the convex hull of a set of $n + 1$ affinely independent points

---

[4]This method should not be confused with the celebrated simplex method for linear programming.

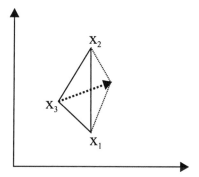

Fig. 3.7 Reflection of the worst value point in the Nelder-Mead simplex search procedure.

$x_1, \ldots, x_{n+1}$.[5] In two dimensions, a simplex is simply a triangle, whereas in three dimensions it is a tetrahedron. The simplex search method works by building and transforming a set of $n + 1$ points rather than generating a sequence of single points; the point with the worst value of the objective is spotted and replaced by another point. For instance, consider the three vertices of the triangle in figure 3.7 and assume that $f(x_3)$ is the worst objective value; then it seems reasonable to move away from x_3 simply by reflecting it through the center of the face formed by the other points. This is easily accomplished algebraically. Assume that x_{n+1} is the worst point; then we compute the centroid of the best n points as

$$ c = \frac{1}{n} \sum_{i=1}^{n} x_i, $$

and we try a new point of the form

$$ x_r = c + \alpha(c - x_{n+1}). $$

The reflection coefficient $\alpha > 0$ is adjusted depending on the circumstances. If x_r turns out to be even worse than x_{n+1}, we may argue that the step was too long, and the simplex should be contracted. If x_r turns out to be the new best point, we have found a good direction and the simplex should be expanded. Different tricks have been devised in order to improve the convergence of the method.

---

[5]Affine independence here means that the vectors $(x_2 - x_1), \ldots, (x_{n+1} - x_1)$ are linearly independent. For $n = 2$ this means that the three points do not lie on the same line. For $n = 3$ this means that the four points do not lie on the same plane.

3.2.5   Unconstrained optimization in MATLAB

Consider the unconstrained optimization problem

$$\min f(x_1, x_2) = (x_1 - 2)^4 + (x_1 - 2x_2)^2.$$

Clearly, $f(x_1, x_2) \geq 0$ and $f(2, 1) = 0$; hence $(2, 1)$ is a globally optimal solution. The gradient of f is given by

$$\nabla f(x_1, x_2) = \left[\begin{array}{c} 4(x_1 - 2)^3 + 2(x_1 - 2x_2) \\ -4(x_1 - 2x_2) \end{array} \right].$$

It is easy to see that $\nabla f(2, 1) = \mathbf{0}$.

In the Optimization toolbox we have two functions that can be used for unconstrained optimization:

- fminsearch, which implements a variant of the simplex search method.

- fminunc, which actually implements a variety of methods, which are selected according to a set of options controlled by the user.

Both functions require an M-file or an inline function to evaluate the objective and an initial estimate of the solution. An optional parameter may be used to set the desired options through the optimset function.

Let us first try the simplex search procedure, giving an initial estimate $\mathbf{x}_0 = \mathbf{0}$:

```
>> f = inline('(x(1) - 2)^4 + (x(1) - 2 * x(2))^2');
>> x=fminsearch(f,[0 0])
```

```
Optimization terminated successfully:
 the current x satisfies the termination criteria using OPTIONS.TolX
 of 1.000000e-004 and F(X) satisfies the convergence criteria
 using OPTIONS.TolFun of 1.000000e-004
```

```
x =
     2.0000    1.0000
>> f(x)
ans =
  2.9563e-017
```

Now we may try fminunc:

```
>> x=fminunc(f,[0 0])
Warning: Gradient must be provided for trust-region method;
   using line-search method instead.
```

```
Optimization terminated successfully:
 Current search direction is a descent direction, and magnitude of
 directional derivative in search direction less than 2*options.TolFun
```

```
x =
    1.9744    0.9868
>> f(x)
ans =
  9.9734e-007
```

The result is not exact really. The point is that the function is rather "flat" around the minimizer; in fact the objective function is close to zero in the solution reported by MATLAB. We could change the tolerance parameters in order to improve the solution, but this could make no sense in practice. We may also note that MATLAB complains about the lack of gradient information, so that it cannot apply a trust region method. This is not much trouble, as the gradient may be estimated numerically. However, we could ask MATLAB not to use the default "large-scale" algorithm, which is a trust region method, but a "medium-scale" algorithm.

```
>> options=optimset('largescale', 'off');
>> x=fminunc(f,[0 0], options)

Optimization terminated successfully:
 Current search direction is a descent direction, and magnitude of
 directional derivative in search direction less than 2*options.TolFun

x =
    1.9744    0.9868
```

Alternatively, we may provide a function to compute the gradient and tell MATLAB to use it within a large-scale algorithm, possibly with a stricter tolerance:

```
>> gradf = inline('[4*(x(1)-2)^3+2*(x(1)-2*x(2)) , -4*(x(1)-2*x(2))]');
>> options=optimset('gradobj','on', 'largescale','on', 'tolfun',1e-13);
>> x=fminunc(f, gradf, [0 0], options)
Optimization terminated successfully:
 Relative function value changing by less than OPTIONS.TolFun

x =
    1.9947    0.9973
```

Computing a gradient analytically is clearly an error-prone activity. To help with this task, it is possible to ask MATLAB to compare the gradient we provide with a numerical estimate. All we have to do is to reset the options and to set the **derivativecheck** option on. Here we may try this functionality, providing MATLAB with an incorrect expression for the gradient implemented in the inline function **gradf1**:

```
>> options = optimset;
```

```
>> options=optimset('gradobj', 'on', 'largescale', 'off', ...
   'derivativecheck', 'on');
>> gradf1 = inline('[6*(x(1)-2)^3+2*(x(1)-2*x(2)) , -4*(x(1)-2*x(2))]');
>> x=fminunc(f, gradf1, [0 0], options)
Maximum discrepancy between derivatives  = 16
Warning: Derivatives do not match within tolerance
Derivative from finite difference calculation:

finite_diff_deriv =
  -32.0000
        0
User-supplied derivative, [6*(x(1)-2)^3+2*(x(1)-2*x(2)),-4*(x(1)-2*x(2))]
analytic_deriv =
   -48
     0
Difference:
ans =
  -16.0000
        0
Strike any key to continue or Ctrl-C to abort
```

Indeed, we see that a warning is issued by the system, spotting probable trouble with our analytical gradient.

3.3 METHODS FOR CONSTRAINED OPTIMIZATION

Consider a general constrained optimization problem, such as

$$\begin{aligned}
\min \quad & f(\mathbf{x}) \\
\text{s.t.} \quad & h_i(\mathbf{x}) = 0 \qquad i \in E \\
& g_i(\mathbf{x}) \le 0 \qquad i \in I.
\end{aligned}$$

In this section we assume that all the involved functions have suitable differentiability properties. For a constrained problem, stationarity is not a necessary condition anymore, since the optimal solution may be a nonstationary point on the boundary of the feasible set (this means that there are descent directions, but they all lead outside the feasible region). One possible approach to cope with this difficulty is trying to transform the problem in such a way that stationarity condition may be applied again; this leads to the penalty function approach (section 3.3.1). Another idea is to develop optimality conditions which include some form of stationarity, plus some additional requirements; this leads to the Kuhn-Tucker conditions (section 3.3.2). Kuhn-Tucker conditions generalize the Lagrange multiplier method for equality-constrained problems, and they are linked to a body of optimization theory called duality theory (section 3.3.3), which leads both to theoretical insights and to practical algorithms. Another important observation is that a constrained problem is relatively easy when all the involved function are affine; indeed, linear programming is a very well developed branch of optimization theory (section 3.4). So it may be interesting to develop algorithms which somehow transform a nonlinear problem to a linear problem. This may be accomplished easily if the constraints are linear and the objective function is convex; Kelley's cutting planes algorithm (section 3.3.4) is based on this idea, and it is the conceptual basis of some methods for stochastic problems. In general, it is reasonable to assume that a linearly constrained problem has some specific features that may be exploited in a computational algorithm. The active set method (section 3.3.5) is one such strategy; it also worth noting that in the earlier versions of the Optimization toolbox the active set method was the basis of the functions for both linear and quadratic programming.

Due to its introductory nature, this book has been written sacrificing the mathematical rigor. This is particularly true for this part of the book, as optimization theory is a tough subject in which simplistic approaches may lead to disasters. Hence, the serious reader is urged not to take what we illustrate in the following as a foolproof set of recipes; it is a good starting point, but the references at the end of the chapter should be consulted for a more thorough treatment.

3.3.1 Penalty function approach

Penalty functions are based on the idea of relaxing constraints through the addition of a suitable term to the objective function. Consider a problem with equality constraints:

$$\begin{aligned} \min \quad & f(\mathbf{x}) \\ \text{s.t.} \quad & h_i(\mathbf{x}) = 0 \qquad i \in E. \end{aligned}$$

It is possible to approximate this problem by an unconstrained one such as

$$\min \Phi(\mathbf{x}, \sigma) = f(\mathbf{x}) + \sigma \sum_{i \in E} h_i^2(\mathbf{x}).$$

This function penalizes both positive and negative values of h_i. If σ is large enough, the optimization algorithm will, in some sense, first drive the solution toward the feasible region by minimizing the penalty term; then it will try to minimize the objective f. Actually, convergence difficulties will arise if we try solving the unconstrained problem with a large value of the penalty coefficient σ. So it is advisable to solve a sequence of unconstrained problems using the optimal solution of each subproblem as the initial solution of the next one:

1. Choose a sequence $\{\sigma^{(k)}\} \to \infty$.

2. Find the minimizer $\mathbf{x}^*(\sigma^{(k)})$ of $\Phi(\mathbf{x}, \sigma)$.

3. Stop if $h_i(\mathbf{x}^*)$ is sufficiently small for all i.

The case of inequality constraints

$$\begin{array}{ll} \min & f(\mathbf{x}) \\ \text{s.t.} & g_i(\mathbf{x}) \leq 0 \qquad i \in I. \end{array}$$

can be tackled by a similar approach. In this case, however, we must only penalize positive values of the constraint functions g_i. Using the notation $[y]^+ = \max\{y, 0\}$, we may use a penalty function like

$$f(\mathbf{x}) + \sigma \sum_{i \in I} \left[g_i^+(\mathbf{x}) \right]^2$$

or

$$f(\mathbf{x}) + \sigma \sum_{i \in I} g_i^+(\mathbf{x})$$

for increasing values of σ. The first penalty function is differentiable, whereas the second one is not, as you may see in figure 3.8a; however, the second function may be advantageous from the numerical point of view, as there is no need to use too large values of the penalty coefficients. Indeed, one of the driving forces between the development of nonsmooth optimization algorithms was the use of exact penalty functions.

In both cases, we are actually using an *exterior penalty function*. The name stems from the fact that the feasible set is approached from outside for increasing values of σ, as illustrated in figure 3.8a. If the optimal solution is on the boundary of the feasible set (which is usually the case, since some inequality constraints are active), a feasible solution is obtained only in the limit. In some cases, this is quite natural, as the constraints may be soft or "elastic" and express some desirable feature rather than a hard requirement. In other cases, we would like to be able to stop the algorithm whenever we want and still come up with a strictly feasible solution. To overcome this problem, an *interior penalty approach* can be pursued, by introducing a suitable *barrier function*. One example is

$$B(\mathbf{x}) = -\sum_{i \in I} \frac{1}{g_i(\mathbf{x})}.$$

The barrier function goes to infinity when $\mathbf{x}$ tends to the boundary of the feasible region from inside. Then an unconstrained problem,

$$\min f(\mathbf{x}) + \sigma B(\mathbf{x}),$$

is solved for decreasing values of σ, until the term $\sigma B(\mathbf{x})$ is small enough. As shown in figure 3.8b, in this case we approach the optimal solution on the boundary staying within the feasible region; this may be an advantage, provided that we have a way to start the iterations with a feasible point.

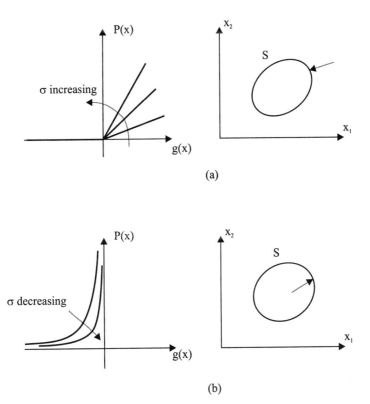

Fig. 3.8 Exterior (a) and interior (b) penalties.

From figure 3.8 it should also be clear that both exterior and interior penalty functions are numerically feasible ways of approximating the ideal penalty:

$$P_i(x) = \begin{cases} 0 & g_i(\mathbf{x}) \leq 0 \\ +\infty & g_i(\mathbf{x}) > 0. \end{cases}$$

Example 3.5 Consider the problem

$$\begin{aligned} \min \quad & (x - 1.5)^2 + (y - 0.5)^2 \\ \text{s.t.} \quad & x, y \leq 1, \end{aligned}$$

whose optimal solution is clearly $x^* = 1$, $y^* = 0.5$. An interior penalty function could be

$$(x - 1.5)^2 + (y - 0.5)^2 + \frac{\sigma}{1 - x} + \frac{\sigma}{1 - y}.$$

Using MATLAB graphics, we may easily plot the level curves of the penalty function for different values of the parameter σ. We need to define an inline function and to use the functions `meshgrid`, to define the grid of points on which we want to evaluate the function, and `contour`, to plot a set of level curves.

```
>> f = inline('(x-1.5).^2+(y-0.5).^2+sigma./(1-x)+sigma./(1-y)')
f =
     Inline function:
     f(sigma,x,y) = (x-1.5).^2+(y-0.5).^2+sigma./(1-x)+sigma./(1-y)
>> [x y] = meshgrid(0.01 : 0.01 : 0.99);
>> subplot(2,2,1)
>> contour(f(0.1,x,y),30)
>> subplot(2,2,2)
>> contour(f(0.01,x,y),30)
>> subplot(2,2,3)
>> contour(f(0.001,x,y),30)
>> subplot(2,2,4)
>> contour(f(0.0001,x,y),30)
```

The three plots are shown in figure 3.9. We see that the optimal solution of the unconstrained problem tends to the optimal solution of the original one from the inside.   ⬚

The penalty function approach is conceptually very simple, and some convergence properties can be proved. However, severe numerical difficulties may arise, for instance, when σ gets very large in the case of an exterior penalty. Nevertheless, penalty functions are most useful in providing a starting point for other, more sophisticated methods. They may be integrated with the Lagrangian methods described below, giving rise to the *augmented Lagrangian methods*, and they are one of the ingredients of the increasingly popular interior point methods for linear programming.

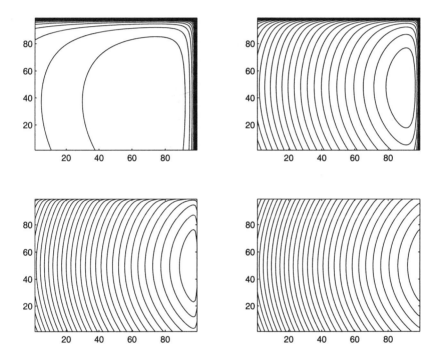

Fig. 3.9 Plots of the level curves for the interior penalty function of example 3.5 for $\sigma = 0.1$, $\sigma = 0.01$, $\sigma = 0.001$, $\sigma = 0.0001$.

3.3.2   Kuhn-Tucker conditions

Consider a general constrained problem (P_{EI}):

$$
\begin{aligned}
\min \quad & f(\mathbf{x}) \\
\text{s.t.} \quad & h_i(\mathbf{x}) = 0 \qquad i \in E \\
& g_i(\mathbf{x}) \le 0 \qquad i \in I.
\end{aligned}
$$

The stationarity of f plays no role in proving optimality here, but the stationarity of a related function does. Consider the Lagrangian function

$$
\mathcal{L}(\mathbf{x}, \boldsymbol{\lambda}, \boldsymbol{\mu}) = f(\mathbf{x}) + \sum_{i \in E} \lambda_i h_i(\mathbf{x}) + \sum_{i \in I} \mu_i g_i(\mathbf{x}). \tag{3.5}
$$

The stationarity of $\mathcal{L}$ does play a role in the following conditions.

THEOREM 3.5 (Kuhn-Tucker conditions) *Assume that the functions* f, h_i, g_i *in* (P_{EI}) *are continuously differentiable, and that* $\mathbf{x}^*$ *is feasible and satisfies a constraint qualification condition. Then a necessary condition for the local optimality of* $\mathbf{x}^*$ *is that there exist numbers* λ_i^* *$(i \in E)$ and* $\mu_i^* \ge 0$ *$(i \in I)$ such that*

$$
\begin{aligned}
& \nabla f(\mathbf{x}^*) + \sum_{i \in E} \lambda_i^* \nabla h_i(\mathbf{x}^*) + \sum_{i \in I} \mu_i^* \nabla g_i(\mathbf{x}^*) = \mathbf{0} \\
& \mu_i^* g_i(\mathbf{x}^*) = 0 \qquad \forall j \in I.
\end{aligned}
$$

The first condition is the stationarity of the Lagrangian function; if the set of inequality constraints is empty, these conditions boil down to the older Lagrange method to deal with equality-constrained problems. The numbers λ_i and μ_i are called *Lagrangian multipliers*; note that the multipliers for inequality constraints are restricted in sign. For reasons that will be clear in the next section, the multipliers are also called *dual variables* (as opposed to the primal variables $\mathbf{x}$). The Kuhn-Tucker conditions are, in a sense, rather weak, as they are only necessary conditions for local optimality, and they further require differentiability properties and some additional qualification condition on the constraints (to be clarified in example 3.6). They are, however, necessary and sufficient for global optimality in the convex case.

We will not prove these conditions, as a rigorous proof is beyond the scope of the book; informally, they can be derived by characterizing a local optimum as a point such that an improvement in the objective function can only be obtained by going outside the feasible region. It is worth noting that the stationarity condition says that the gradient of the objective function can be expressed as a linear combination of the gradients of the objectives; this clarifies a little what we mean by *constraint qualification*; if the gradients of the constraints are not linearly independent at $\mathbf{x}^*$, it might be the case that we cannot use them as a basis to express ∇f. So it may happen that the

Kuhn-Tucker conditions are not satisfied by a point that is actually a local minimizer.

Example 3.6 To understand the point behind the constraint qualification condition, consider the problem:

$$\begin{aligned}
\min \quad & x_1 + x_2 \\
\text{s.t.} \quad & h_1(\mathbf{x}) = x_2 - x_1^3 = 0 \\
& h_2(\mathbf{x}) = x_2 = 0.
\end{aligned}$$

It is easy to see that the feasible set is the single point $(0,0)$, which is the (trivial) optimal solution. If we try applying the Kuhn-Tucker conditions, we first build the Lagrangian function

$$\mathcal{L}(x_1, x_2, \lambda_1, \lambda_2) = x_1 + x_2 + \lambda_1(x_2 - x_1^3) + \lambda_2 x_2.$$

Writing the stationarity yields the system

$$\begin{aligned}
\frac{\partial \mathcal{L}}{\partial x_1} &= 1 - 3\lambda_1 x_1^2 = 0 \\
\frac{\partial \mathcal{L}}{\partial x_2} &= 1 + \lambda_1 + \lambda_2 = 0 \\
\frac{\partial \mathcal{L}}{\partial \lambda_1} &= x_2 - x_1^3 = 0 \\
\frac{\partial \mathcal{L}}{\partial \lambda_2} &= x_2 = 0,
\end{aligned}$$

which has no solution (the first equation requires that $x_1 \neq 0$, which is not compatible with the last two equations). This is due to the fact that the gradients of the two constraints are parallel at the origin:

$$\nabla h_1(0,0) = \left[\begin{array}{c} -3x_1^2 \\ 1 \end{array} \right]_{\mathbf{x}=0} = \left[\begin{array}{c} 0 \\ 1 \end{array} \right]$$

$$\nabla h_2(0,0) = \left[\begin{array}{c} 0 \\ 1 \end{array} \right]_{\mathbf{x}=0} = \left[\begin{array}{c} 0 \\ 1 \end{array} \right]$$

and are not a basis able to express the gradient of f:

$$\nabla f(0,0) = \left[\begin{array}{c} 1 \\ 1 \end{array} \right]_{\mathbf{x}=0} = \left[\begin{array}{c} 1 \\ 1 \end{array} \right].$$

Different constraint qualification conditions have been proposed in the literature. Sufficient conditions to avoid troubles are that the gradients of the active constraints are linearly independent or that the constraints are all linear. We will not pursue this issue any further, but we recommend a book like [19] as a warning against easy cookbook recipes in optimization.

The second set of conditions is known as *complementary slackness conditions*. They may be interpreted by noting that if a constraint is inactive at $\mathbf{x}^*$, i.e., if $g_i(\mathbf{x}^*) < 0$, the corresponding multiplier must be zero; by the same token, if the multiplier μ_i^* is strictly positive, the corresponding constraints must be active (which roughly means that it could be substituted by an equality constraint without changing the optimal solution). The complementary slackness conditions could be used, in principle, to find a feasible point and a set of multipliers satisfying the Kuhn-Tucker conditions.

Example 3.7 Consider the convex problem

$$\begin{aligned}
\min \quad & x_1^2 + x_2^2 \\
\text{s.t.} \quad & x_1 \geq 0 \\
& x_2 \geq 3 \\
& x_1 + x_2 = 4.
\end{aligned}$$

First write the Lagrangian function:

$$\mathcal{L}(\mathbf{x}, \boldsymbol{\mu}, \lambda) = x_1^2 + x_2^2 - \mu_1 x_1 - \mu_2(x_2 - 3) + \lambda(x_1 + x_2 - 4).$$

A set of numbers satisfying the Kuhn-Tucker conditions can be found by solving the following system:

$$\begin{aligned}
& 2x_1 - \mu_1 + \lambda = 0 \\
& 2x_2 - \mu_2 + \lambda = 0 \\
& x_1 \geq 0, \qquad x_2 \geq 3 \\
& x_1 + x_2 = 4 \\
& \mu_1 x_1 = 0, \qquad \mu_1 \geq 0 \\
& \mu_2(x_2 - 3) = 0, \qquad \mu_2 \geq 0.
\end{aligned}$$

We may proceed with a case-by-case analysis exploiting the complementary slackness conditions. If a multiplier is strictly positive, the corresponding inequality is active, which helps us in finding the value of a decision variable.

Case 1 ($\mu_1 = \mu_2 = 0$). In this case, the inequality constraints are dropped from the Lagrangian function. From the stationarity conditions we obtain the system

$$\begin{aligned}
& 2x_1 + \lambda = 0 \\
& 2x_1 + \lambda = 0 \\
& x_1 + x_2 - 4 = 0.
\end{aligned}$$

This yields a solution $x_1 = x_2 = 2$, which violates the second inequality constraint.

Case 2 ($\mu_1, \mu_2 \neq 0$). The complementary slackness conditions immediately yield $x_1 = 0, x_2 = 3$, violating the equality constraint.

Case 3 $(\mu_1 \neq 0, \mu_2 = 0)$**.** We obtain

$$x_1 = 0$$
$$x_2 = 4$$
$$\lambda = -2x_2 = -8$$
$$\mu_1 = \lambda = -8.$$

The Kuhn-Tucker conditions are not satisfied since the value of μ_1 is negative.

Case 4 $(\mu_1 = 0, \mu_2 \neq 0)$**.** We obtain

$$x_2 = 3$$
$$x_1 = 1$$
$$\lambda = -2$$
$$\mu_2 = 4,$$

which satisfy all the necessary conditions.

Since this is a convex problem, we have obtained the global optimum. Note how nonzero multipliers correspond to the active constraints, whereas the inactive constraint $x_1 \geq 0$ is associated to a multiplier $\mu_1 = 0$. The same result can easily be obtained through MATLAB. The quadprog function deals with quadratic programming problems such as

$$\min \quad \frac{1}{2}\mathbf{x}^T \mathbf{H} \mathbf{x} + \mathbf{f}^T \mathbf{x}$$
$$\text{s.t.} \quad \mathbf{A}\mathbf{x} \leq \mathbf{b}$$
$$\mathbf{A}_{eq}\mathbf{x} = \mathbf{b}_{eq}$$
$$\mathbf{l} \leq \mathbf{x} \leq \mathbf{u}.$$

For our example, some entries of the problem are empty. Note also that simple bounds are treated apart in practice and that the quadratic term in the objective function must be written in a specific way, as it involves a $1/2$ factor and it assumes a symmetric Hessian matrix $\mathbf{H}$.

```
>> H = 2*eye(2);
>> f = [0 0];
>> Aeq = [1 1];
>> beq = 4;
>> lb = [0 3];
>> [x,f,exitflag,output,lambda] = quadprog(H,f,[],[],Aeq,beq,lb);
Optimization terminated successfully.
>> x
x =
     1
     3
```

```
>> lambda.eqlin
ans =
   -2.0000
>> lambda.lower
ans =
     0
     4
```

The output arguments include the optimal decision variables, the optimal value of the objective function, an exit flag containing information about the termination of the algorithm, additional output information, and the multipliers included in the structure lambda. The multipliers in our case are associated to the linear equalities and to the lower bounds on the decision variables.   ⬚

Clearly, the approach we have taken in the example is not practical. Some alternative way must be found to spot the optimal multipliers. This leads to duality theory, which is the topic of next section. Before proceeding, it is also useful to get an intuitive grasp of the meaning of the Lagrangian multipliers.

Example 3.8 Consider the parameterized problem

$$\min \quad x_1^2 + x_2^2$$
$$\text{s.t.} \quad x_1 + x_2 = b.$$

The stationarity conditions on the Lagrangian function,

$$x_1^2 + x_2^2 + \lambda(x_1 + x_2 - b),$$

immediately yield $x_1^* = x_2^* = b/2$ and $\lambda^* = -b$. Now, ask how slight changes in the parameter b will affect the optimal value $f^* = b^2/2$:

$$\frac{df^*}{db} = b = -\lambda^*.$$

This suggests that, neglecting the sign, the dual variables are linked to the sensitivity of the optimal value with respect to perturbations in the right hand side of the constraints.   ⬚

The intuition suggested by the example is correct, provided we assume that the derivative makes sense. Consider an equality-constrained problem and apply a small perturbation to the constraints

$$h_i(\mathbf{x}) = \epsilon_i \qquad i \in E.$$

Applying the Lagrangian approach to the perturbed problem we get a new solution $\mathbf{x}^*(\boldsymbol{\epsilon})$ and a new multiplier vector $\boldsymbol{\lambda}^*(\boldsymbol{\epsilon})$, both depending on $\boldsymbol{\epsilon}$. The Lagrangian function for the perturbed problem is

$$\mathcal{L}(\mathbf{x}, \boldsymbol{\lambda}, \boldsymbol{\epsilon}) = f(\mathbf{x}) + \sum_{i \in E} \lambda_i (h_i(\mathbf{x}) - \epsilon_i). \tag{3.6}$$

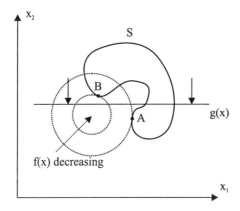

Fig. 3.10 Counterexample showing that a constraint may be necessary even if it has a null multiplier.

Equality constraints must be satisfied by the optimal solution of the perturbed problem. Hence:

$$f^* = f(\mathbf{x}^*(\boldsymbol{\epsilon})) = \mathcal{L}(\mathbf{x}^*(\boldsymbol{\epsilon}), \boldsymbol{\lambda}^*(\boldsymbol{\epsilon}), \boldsymbol{\epsilon}). \tag{3.7}$$

We can evaluate the derivative of the optimal value with respect to each component of $\boldsymbol{\epsilon}$,

$$\frac{df^*}{d\epsilon_i} = \frac{d\mathcal{L}}{d\epsilon_i} = \underbrace{\nabla_{\mathbf{x}}\mathcal{L}^T \frac{\partial \mathbf{x}}{\partial \epsilon_i} + \nabla_{\boldsymbol{\lambda}}\mathcal{L}^T \frac{\partial \boldsymbol{\lambda}}{\partial \epsilon_i}}_{=0} + \frac{\partial \mathcal{L}}{\partial \epsilon_i} = \frac{\partial \mathcal{L}}{\partial \epsilon_i} = -\lambda_i, \tag{3.8}$$

where we have used the stationarity condition of $\mathcal{L}$. As to inequality constraints, they are either inactive or active in $\mathbf{x}^*$: in the first case, they play no role for small enough perturbations; in the second one, they essentially act as equality constraints. It may be tempting to conclude that if a constraint is associated to a null multiplier, then it can be dropped without changing the optimal solution. The counterexample shown in figure 3.10 shows that this is not the case. Here we have a convex quadratic objective, to which the two concentric level curves are associated; the feasible region is the portion of the "bean" S below the constraint $g(\mathbf{x}) \leq 0$, which is actually an upper bound on x_2. The optimal solution is the point A, and the constraint $g(\mathbf{x}) \leq 0$ is inactive at that point; however, if we eliminate the constraint, the optimal solution is B (it remains true that A is a locally optimal solution). The issue here is that the overall problem is not convex.

3.3.3 Duality theory

In preceding sections we have shown that the stationarity of the Lagrangian function plays a crucial role in constrained optimization. Stationarity is linked to an optimality condition for either minimization or maximization. It is rather intuitive that we should minimize the Lagrangian function with respect to the primal variables, but what about the dual variables? This is an important point if we want to devise a numerical way to find optimal values for both the primal and dual variables. In this section we show that interesting results are obtained by *maximizing* a dual function with respect to the dual variables, leading to duality theory.

Consider the inequality-constrained problem

$$(P) \qquad \min \quad f(\mathbf{x})$$
$$\text{s.t.} \quad g_i(\mathbf{x}) \leq 0 \qquad i \in I \tag{3.9}$$
$$\mathbf{x} \in S \subseteq \mathbb{R}^n.$$

This problem is called the *primal problem*. Note that the set S is any subset of $\mathbb{R}^n$, possibly a *discrete* one; furthermore, in this section we do not assume the differentiability nor the convexity of the objective function. The results obtained are therefore extremely general.

Consider the Lagrangian function obtained by *dualizing* constraints (3.9):

$$\mathcal{L}(\mathbf{x}, \boldsymbol{\mu}) = f(\mathbf{x}) + \sum_{i \in I} \mu_i g_i(\mathbf{x}).$$

For a given multiplier vector $\boldsymbol{\mu}$, the minimization of the Lagrangian function with respect to $\mathbf{x} \in S$ is called the *relaxed problem*; the solution of the relaxed problem defines a function $w(\boldsymbol{\mu})$, called the *dual function*:

$$w(\boldsymbol{\mu}) = \min_{\mathbf{x} \in S} \mathcal{L}(\mathbf{x}, \boldsymbol{\mu}).$$

Consider the *dual problem*:

$$(D) \qquad \max_{\boldsymbol{\mu} \geq 0} w(\boldsymbol{\mu}) = \max_{\boldsymbol{\mu} \geq 0} \left\{ \min_{\mathbf{x} \in S} \mathcal{L}(\mathbf{x}, \boldsymbol{\mu}) \right\}. \tag{3.10}$$

The following theorem holds.

THEOREM 3.6 (Weak duality theorem) *For any $\boldsymbol{\mu} \geq 0$ the dual function is a lower bound of the optimum $f(\mathbf{x}^*)$ of the primal problem (P), i.e.,*

$$w(\boldsymbol{\mu}) \leq f(\mathbf{x}^*) \qquad \forall \boldsymbol{\mu} \geq 0.$$

Proof. Let us adopt the notation $\nu(P)$ to denote the optimal value of the objective function for an optimization problem P. Under the hypothesis $\boldsymbol{\mu} \geq 0$, it is easy to see that

$$\nu(P) \geq \nu \begin{pmatrix} \min & f(\mathbf{x}) \\ \text{s.t.} & \mathbf{x} \in S \\ & \boldsymbol{\mu}^T \mathbf{g}(\mathbf{x}) \leq 0 \end{pmatrix} \tag{3.11}$$

$$\geq \quad \nu \left(\begin{array}{ll} \min & f(\mathbf{x}) + \boldsymbol{\mu}^T \mathbf{g}(\mathbf{x}) \\ \text{s.t.} & \mathbf{x} \in S \\ & \boldsymbol{\mu}^T \mathbf{g}(\mathbf{x}) \leq 0 \end{array} \right) \qquad (3.12)$$

$$\geq \quad \nu \left(\begin{array}{ll} \min & f(\mathbf{x}) + \boldsymbol{\mu}^T \mathbf{g}(\mathbf{x}) \\ \text{s.t.} & \mathbf{x} \in S \end{array} \right). \qquad (3.13)$$

Inequality (3.11) is justified by the fact that the points satisfying the set of constraints $g_i(\mathbf{x}) \leq 0$ for all i also satisfy the aggregate constraint $\boldsymbol{\mu}^T \mathbf{g}(\mathbf{x}) \leq 0$ if $\boldsymbol{\mu} \geq \mathbf{0}$, but not vice versa. In other words, the feasible set of the first problem is a subset of the feasible set of the second one. Clearly, when we relax the feasible set, the optimal value cannot increase. Inequality (3.12) holds since the third problem involves the same feasible set as the second problem, but we have added a negative term to the objective function. Finally, inequality (3.13) holds since the fourth problem is a relaxation of the third one (we delete a constraint).

◻

We obtain a very general but weak relationship. Under suitable conditions (essentially convexity), a stronger property holds, known as *strong duality*:

$$\nu(D) = w(\boldsymbol{\mu}^*) = f(\mathbf{x}^*) = \nu(P).$$

The convexity assumption does not hold, in particular, for the case of a discrete set; therefore, in general, duality yields only a lower bound for discrete optimization problems. The following theorem is useful in establishing when the dual problem yields an optimal solution of the primal problem.

THEOREM 3.7 *If there is a pair* $(\mathbf{x}^*, \boldsymbol{\mu}^*)$, *where* $\mathbf{x}^* \in S$ *and* $\boldsymbol{\mu}^* \geq \mathbf{0}$, *satisfying the following conditions:*

1. $f(\mathbf{x}^*) + \boldsymbol{\mu}^{*T} \mathbf{g}(\mathbf{x}^*) = \min_{\mathbf{x} \in S} \{ f(\mathbf{x}) + \boldsymbol{\mu}^{*T} \mathbf{g}(\mathbf{x}) \}$;

2. $\boldsymbol{\mu}^{*T} \mathbf{g}(\mathbf{x}^*) = \mathbf{0}$;

3. $\mathbf{g}(\mathbf{x}^*) \leq \mathbf{0}$;

then $\mathbf{x}^*$ *is a global optimum for the primal problem* (P).

In other words, the optimal solution $\mathbf{x}^*$ of the relaxed problem for a multiplier vector $\boldsymbol{\mu}^*$ is a global optimum for the primal problem if the pair $(\mathbf{x}^*, \boldsymbol{\mu}^*)$ is primal feasible, dual feasible, and it satisfies the complementary slackness conditions. Note that these are *sufficient* conditions for global optimality.

Weak duality also holds in the equality-constrained case. Consider the optimal solution $\mathbf{x}^*$ of the primal problem:

$$\begin{array}{ll} \min & f(\mathbf{x}) \\ \text{s.t.} & h_i(\mathbf{x}) = 0 \qquad i \in E \\ & \mathbf{x} \in S, \end{array}$$

and the optimal solution $\bar{\mathbf{x}}$ of the relaxed problem:

$$\min_{\mathbf{x}\in S}\left\{f(\mathbf{x})+\boldsymbol{\lambda}^T\mathbf{h}(\mathbf{x})\right\}.$$

For any multiplier vector $\boldsymbol{\lambda}$ (not restricted in sign), it is easy to see that

$$f(\bar{\mathbf{x}})+\boldsymbol{\lambda}^T\mathbf{h}(\bar{\mathbf{x}})\le f(\mathbf{x}^*)+\boldsymbol{\lambda}^T\mathbf{h}(\mathbf{x}^*)=f(\mathbf{x}^*).$$

Unfortunately, convexity does not hold easily for equality constraints. In fact, it holds only for linear equality constraints such as $\mathbf{a}_i^T\mathbf{x}=b_i$. Hence, strong duality with equality constraints holds only in the linear programming case (see section 3.4.3).

Example 3.9 Consider the problem

$$\begin{array}{ll} \min & x_1^2+x_2^2 \\ \text{s.t.} & x_1+x_2\ge 4 \\ & x_1,x_2\ge 0. \end{array}$$

The optimal value is 8, corresponding to the optimal solution $(2,2)$. Since this is a convex problem, we can apply strong duality. The dual function is

$$\begin{aligned} w(\mu) &= \min_{x_1,x_2}\{x_1^2+x_2^2+\mu(-x_1-x_2+4);\ \text{s.t.}\ x_1,x_2\ge 0\} \\ &= \min_{x_1}\{x_1^2-\mu x_1;\ \text{s.t.}\ x_1\ge 0\}+\min_{x_2}\{x_2^2-\mu x_2;\ \text{s.t.}\ x_2\ge 0\}+4\mu. \end{aligned}$$

Since $\mu\ge 0$, the optima with respect to x_1,x_2 are obtained for

$$x_1^*=x_2^*=\frac{\mu}{2}.$$

Hence,

$$w(\mu)=-\frac{1}{2}\mu^2+4\mu.$$

The maximum of the dual function is reached for $\mu^*=4$, and we have $w(4)=f^*=8$. □

In example 3.9 we have found an explicit representation of the dual function. In general, the maximization of the dual function must be tackled by a numerical method; in practice, an iterative procedure is adopted (assuming the inequality-constrained case):

1. Assign an initial value $\boldsymbol{\mu}^{(0)}\ge\mathbf{0}$; set $k\leftarrow 0$.

2. Solve the relaxed problem with multipliers $\boldsymbol{\mu}^{(k)}$.

3. Given the solution $\hat{\mathbf{x}}^{(k)}$ of the relaxed problem, compute a search direction $\mathbf{s}^{(k)}$ and a step length $\alpha^{(k)}$, and update the multipliers (making sure they stay nonnegative):

$$\boldsymbol{\mu}^{(k+1)}=\max\left\{\mathbf{0},\boldsymbol{\mu}^{(k)}+\alpha^{(k)}\mathbf{s}^{(k)}\right\}.$$

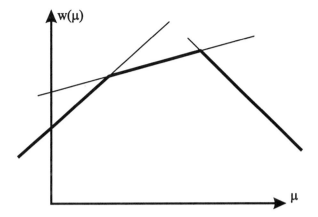

Fig. 3.11   Nondifferentiable dual function.

Then set $k \leftarrow k + 1$, and go to step 2.

In order to find a search direction, one would be tempted to compute a gradient of the dual function. Unfortunately, the dual function need not be everywhere differentiable, as we can see from the following example.

Example 3.10 Consider the discrete optimization problem

$$\begin{aligned} \min \quad & \mathbf{c}^T\mathbf{x} \\ \text{s.t.} \quad & \mathbf{a}^T\mathbf{x} \geq b \\ & \mathbf{x} \in S = \left\{\mathbf{x}^1, \mathbf{x}^2, \ldots, \mathbf{x}^m\right\}, \end{aligned}$$
(3.14)
(3.15)

where $\mathbf{c}, \mathbf{a}, \mathbf{x} \in \mathbb{R}^n$, $b \in \mathbb{R}$ and S is a *discrete* set. Dualizing constraint (3.14) with a multiplier $\mu \geq 0$, we obtain the dual function:

$$w(\mu) = \min_{j=1,\ldots,m} \left\{(b - \mathbf{a}^T\mathbf{x}^j)\mu + \mathbf{c}^T\mathbf{x}^j\right\}.$$

It is easy to see that the dual function is the lower envelope of a family of affine functions, as shown in figure 3.11. We have a nondifferentiability point when the relaxed problem has multiple optimal solutions. □

From example 3.10 we may conclude that there is no differentiability guarantee for the dual function; however, the dual function for this case is concave. In fact, we may easily prove that the dual function is always concave.

THEOREM 3.8 *The dual function $w(\boldsymbol{\mu})$ is a concave function.*

Proof. We must show that for any multiplier vectors $\boldsymbol{\mu}_1$ and $\boldsymbol{\mu}_2$,

$$w[\lambda\boldsymbol{\mu}_1 + (1 - \lambda)\boldsymbol{\mu}_2] \geq \lambda w(\boldsymbol{\mu}_1) + (1 - \lambda)w(\boldsymbol{\mu}_2) \qquad \lambda \in [0, 1].$$

Let us denote by $\hat{\mathbf{x}}_1$ and $\hat{\mathbf{x}}_2$ the optimal solutions of the relaxed subproblems with multipliers $\boldsymbol{\mu}_1$ and $\boldsymbol{\mu}_2$, respectively. We have

$$
\begin{aligned}
w(\boldsymbol{\mu}_1) &= f(\hat{\mathbf{x}}_1) + \boldsymbol{\mu}_1^T \mathbf{g}(\hat{\mathbf{x}}_1) \le f(\mathbf{x}_\lambda) + \boldsymbol{\mu}_1^T \mathbf{g}(\mathbf{x}_\lambda) \\
w(\boldsymbol{\mu}_2) &= f(\hat{\mathbf{x}}_2) + \boldsymbol{\mu}_2^T \mathbf{g}(\hat{\mathbf{x}}_2) \le f(\mathbf{x}_\lambda) + \boldsymbol{\mu}_2^T \mathbf{g}(\mathbf{x}_\lambda),
\end{aligned}
$$

where $\mathbf{x}_\lambda$ is the optimal solution corresponding to the multiplier vector $\lambda\boldsymbol{\mu}_1 + (1-\lambda)\boldsymbol{\mu}_2$. The result is obtained by multiplying the first inequality by λ, the second one by $1 - \lambda$, and summing.   ☐

Since maximizing a concave function is equivalent to minimizing a convex function, this is a reassuring result. In fact, we may apply a subgradient algorithm (see section 3.2.2) provided that we are able to find a subgradient of the dual function for any value of the multipliers.

THEOREM 3.9 *Let $\hat{\mathbf{x}}$ be an optimal solution of the relaxed problem for a multiplier vector $\hat{\boldsymbol{\mu}}$. Then $\mathbf{g}(\hat{\mathbf{x}})$ is a subgradient of the dual function at $\hat{\boldsymbol{\mu}}$.*

Proof. To show that $\mathbf{g}(\hat{\mathbf{x}}) \in \partial w(\hat{\boldsymbol{\mu}})$ we must show that for any $\boldsymbol{\mu}$, we have

$$
w(\boldsymbol{\mu}) \le w(\hat{\boldsymbol{\mu}}) + \mathbf{g}(\hat{\mathbf{x}})^T(\boldsymbol{\mu} - \hat{\boldsymbol{\mu}}).
$$

Here the inequality is reversed with respect to the definition of a subgradient for a convex function, since w is concave. We know that $\hat{\mathbf{x}}$ is the optimal solution of the relaxed problem for $\hat{\boldsymbol{\mu}}$:

$$
w(\hat{\boldsymbol{\mu}}) = f(\hat{\mathbf{x}}) + \hat{\boldsymbol{\mu}}^T \mathbf{g}(\hat{\mathbf{x}}) \tag{3.16}
$$

but not for a generic $\boldsymbol{\mu}$:

$$
w(\boldsymbol{\mu}) = \min_{\mathbf{x} \in S} \left\{ f(\mathbf{x}) + \boldsymbol{\mu}^T \mathbf{g}(\mathbf{x}) \right\} \le f(\hat{\mathbf{x}}) + \boldsymbol{\mu}^T \mathbf{g}(\hat{\mathbf{x}}). \tag{3.17}
$$

Subtracting equation (3.16) from inequality (3.17), we get

$$
w(\boldsymbol{\mu}) - w(\hat{\boldsymbol{\mu}}) \le \mathbf{g}^T(\hat{\mathbf{x}})(\boldsymbol{\mu} - \hat{\boldsymbol{\mu}}),
$$

and the result follows.   ☐

Theorem 3.9 allows us to solve the dual problem (3.10) by a subgradient algorithm. A remarkable point is that we are able to optimize a function, even if it is not known in explicit form; the same consideration applies to stochastic programming with recourse. A sequence of relaxed problems is solved, updating the dual variables as follows:

$$
\boldsymbol{\mu}^{(k+1)} = \max\left\{ \mathbf{0}, \boldsymbol{\mu}^{(k)} + \alpha^{(k)} \mathbf{g}(\hat{\mathbf{x}}^{(k)}) \right\}
$$

where $\hat{\mathbf{x}}^{(k)}$ is the solution of the kth relaxed problem. Note that this solution need not be feasible for the original (primal) problem. Provided that strong duality holds, the method converges to the optimal solution of the original problem. When only weak duality applies, we obtain a lower bound on the optimal value of the primal problem (which may be valuable in itself), and probably a near-feasible solution, from which a feasible near-optimal solution may be obtained with some problem-dependent procedure. It should be noted that duality theory in itself does not generally yield numerically efficient algorithms directly. Nevertheless, it may be fruitfully exploited for specially structured problems; in fact, we have seen in example 3.9 that dualizing certain constraints may decompose an optimization problem into independent subproblems; certain model formulations lend themselves to a decomposition by dualization of the interaction constraints. Furthermore, duality theory is a fundamental theoretical tool paving the way for important algorithmic developments.

3.3.4 Kelley's cutting plane algorithm

Assume that we have to solve a convex problem $\min_{x \in S} f(\mathbf{x})$, where the objective function f is actually not known in analytical form. This may seem absurd, but just think of a stochastic programming problem with recourse; we may be able to evaluate or estimate the value of the recourse function at a given point, but we have no explicit form for that function; the same consideration applies to the dual function. Suppose that for a given point $\mathbf{x}^k$ we are not only able to compute the function value $f(\mathbf{x}^k) = \alpha_k$, but also a subgradient γ_k, which does exist if the function is convex on the set S. In other words, we are able to find an affine function such that

$$f(\mathbf{x}^k) = \alpha_k + \gamma_k^T \mathbf{x}^k \tag{3.18}$$
$$f(\mathbf{x}) \geq \alpha_k + \gamma_k^T \mathbf{x} \qquad \forall \mathbf{x} \in S. \tag{3.19}$$

The availability of such a support hyperplane suggests the possibility of approximating f from below, through the upper envelope of support hyperplanes as illustrated in figure 3.12. The Kelley's cutting plane algorithm exploits this idea by building and improving a lower bounding function until some convergence criterion is met.

Step 0. Let $\mathbf{x}^1 \in S$ be an initial feasible solution; initialize the iteration counter $k \leftarrow 0$, the upper bound $u_0 = f(\mathbf{x}^1)$, the lower bound $l_0 = -\infty$, and the lower bounding function $\beta_0(\mathbf{x}) = -\infty$.

Step 1. Increment the iteration counter $k \leftarrow k + 1$. Find a subgradient of f at $\mathbf{x}^k$, such that equation (3.18) and condition (3.19) hold.

Step 2. Update the upper bound

$$u_k = \min\{u_{k-1}, f(\mathbf{x}^k)\}$$

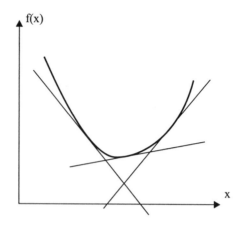

Fig. 3.12   Example of Kelley's cutting plane algorithm.

and the lower bounding function

$$\beta_k(\mathbf{x}) = \max\{l_{k-1}(\mathbf{x}), \alpha_k + \boldsymbol{\gamma}_k^T \mathbf{x}\}.$$

Step 3. Solve the problem

$$l_k = \min_{\mathbf{x} \in S} \beta_k(\mathbf{x}),$$

and let $\mathbf{x}^{k+1}$ be the optimal solution.

Step 4. If $u_k - l_k < \epsilon$, stop: $\mathbf{x}^{k+1}$ is a satisfactory approximation of the optimal solution; otherwise, go to step 1.

It is worth noting that if the feasible set S is polyhedral, then all the subproblems we solve are LP problems. The Kelley's cutting plane algorithm is the conceptual basis of some algorithms for stochastic programming, such as the L-shaped decomposition described in section 3.7.

3.3.5 Active set method

Although duality theory is a powerful tool, both in theory and practice, dual algorithms have the general drawback that a feasible solution is generally obtained only in the limit. A natural aim of many constrained optimization algorithms is to stay within the feasible region. This is particularly easy to accomplish if the problem is linearly constrained. Consider the problem

$$\begin{array}{ll} \min & f(\mathbf{x}) \\ \text{s.t.} & \mathbf{A}\mathbf{x} = \mathbf{b}, \end{array}$$

where the matrix $\mathbf{A} \in \mathbb{R}^{m,n}$, $m < n$, is assumed of full row rank for simplicity. Given a feasible solution $\hat{\mathbf{x}}$, how can we characterize descent directions $\boldsymbol{\delta}$ such that the new solution $\hat{\mathbf{x}} + \alpha\boldsymbol{\delta}$ remains feasible for some $\alpha > 0$? Since both solutions must be feasible,

$$\mathbf{A}(\hat{\mathbf{x}} + \alpha\boldsymbol{\delta}) = \mathbf{b} + \alpha\mathbf{A}\boldsymbol{\delta} = \mathbf{b} \quad \Rightarrow \quad \mathbf{A}\boldsymbol{\delta} = \mathbf{0}.$$

Technically speaking, the vector $\boldsymbol{\delta}$ must lie in the null space of the matrix $\mathbf{A}$; since this is a linear space, there must be a basis for it. Let $\mathbf{Z} \in \mathbb{R}^{n,(n-m)}$ be a matrix whose columns are a basis for this space; then we have

$$\mathbf{AZ} = \mathbf{0},$$

and the direction $\boldsymbol{\delta}$ is a linear combination of the columns of $\mathbf{Z}$:

$$\boldsymbol{\delta} = \mathbf{Zd}$$

The basis consists of $(n-m)$ vectors. To see why, consider that the m equality constraints eliminate m degrees of freedom for the n decision variables. Then we may move in some space with $(n-m)$ degrees of freedom. The first-order Taylor expansion for a perturbed point along the feasible direction is

$$f(\hat{\mathbf{x}} + \epsilon\mathbf{Zd}) \approx f(\hat{\mathbf{x}}) + \epsilon\mathbf{d}^T\mathbf{Z}^T\nabla f(\hat{\mathbf{x}}).$$

A descent direction is obtained when $\mathbf{d}^T\mathbf{Z}^T\nabla f(\hat{\mathbf{x}}) < 0$; furthermore, the first-order necessary optimality condition is

$$\mathbf{Z}^T\nabla f(\mathbf{x}^*) = \mathbf{0}. \tag{3.20}$$

The vector $\mathbf{Z}^T\nabla f$ is called the *reduced gradient*, and we see that a stationarity condition must be required for this reduced gradient. By the way, the condition (3.20) implies that the gradient ∇f is a linear combination of the rows of $\mathbf{A}$. This means that

$$\nabla f(\mathbf{x}^*) = \mathbf{A}^T\boldsymbol{\lambda},$$

which could also be obtained by the Lagrangian multipliers approach.

Provided that we are able to spot a suitable matrix $\mathbf{Z}$, an algorithm is readily devised, as we must simply spot descent directions and select the step length α in order to reduce the objective function while keeping the iterates feasible. One possible choice of $\mathbf{Z}$ is obtained by exploiting QR *factorization*. This factorization, which is implemented by the qr function in MATLAB, allows us to write

$$\mathbf{A}^T = \mathbf{Q}\begin{bmatrix} \mathbf{R} \\ \mathbf{0} \end{bmatrix} = \begin{bmatrix} \mathbf{Q}_1 & \mathbf{Q}_2 \end{bmatrix}\begin{bmatrix} \mathbf{R} \\ \mathbf{0} \end{bmatrix} = \mathbf{Q}_1\mathbf{R},$$

where $\mathbf{Q} \in \mathbb{R}^{n,n}$ is an orthogonal matrix (i.e., its columns are orthogonal vectors), and $\mathbf{R} \in \mathbb{R}^{n,n-m}$ is upper triangular. The choice $\mathbf{Z} = \mathbf{Q}_2$ satisfies our requirements, since the orthogonality of $\mathbf{Q}$ implies that

$$\mathbf{A} = \mathbf{R}^T\mathbf{Q}_1^T \quad \Rightarrow \quad \mathbf{AZ} = \mathbf{R}^T\mathbf{Q}_1^T\mathbf{Q}_2 = \mathbf{0}.$$

Different choices of $\mathbf{Z}$ and different approaches in selecting the descent direction and the step length result in a variety of methods which are described in the literature. It should also be mentioned that second-order conditions should be checked if f is not convex.

The approach may be extended to linear inequalities. To cope with a problem like

$$\begin{aligned} \min \quad & f(\mathbf{x}) \\ \text{s.t.} \quad & \mathbf{Ax} \leq \mathbf{b}, \end{aligned}$$

a possible idea is to restrict the attention to the active constraints, i.e., the constraints which are satisfied at equality. In principle, if we knew which constraints are active in the optimal solution, we could treat the problem like an equality-constrained problem. The active set strategy works on a pool of active constraints, trying to identify which constraints must be brought in and out of the active set. Roughly speaking, if we see that a relaxed constraint would get violated by a move along the feasible direction, it should be added to the set. Similarly, an inactive constraints can be dropped. The details of the method are not so easy, but it is enough to know the qualitative aspects of its working and that it is actually implemented and used in MATLAB functions for both quadratic and linear programming (see section 3.4.5 to appreciate this point).

3.4 LINEAR PROGRAMMING

A general LP problem can be expressed as

$$\begin{aligned} \min \quad & \mathbf{c}^T \mathbf{x} \\ \text{s.t.} \quad & \mathbf{a}_i^T \mathbf{x} = b_i \quad i \in E \\ & \mathbf{a}_i^T \mathbf{x} \geq b_i \quad i \in I, \end{aligned}$$

where $\mathbf{c}, \mathbf{a}_i, \mathbf{x} \in \mathbb{R}^n$, $b_i \in \mathbb{R}$. When dealing with solution algorithms for LP problems, it is convenient to assume that the problem has a specific form. An LP problem is said to be in *canonical form* if it involves only inequality constraints, and all the decision variables are restricted in sign. A canonical form for a maximization problem is

$$\begin{aligned} \max \quad & \mathbf{c}^T \mathbf{x} \\ \text{s.t.} \quad & \mathbf{Ax} \leq \mathbf{b} \\ & \mathbf{x} \geq \mathbf{0}, \end{aligned}$$

where $\mathbf{c}, \mathbf{x} \in \mathbb{R}^n$, $\mathbf{b} \in \mathbb{R}^m$, $\mathbf{A} \in \mathbb{R}^{m,n}$. We denote the ith row (corresponding to the ith constraint) of $\mathbf{A}$ by $\mathbf{a}_i^T$ and the jth column (corresponding to the jth variable) by $\mathbf{A}^j$. An LP problem is said to be in *standard form* if it involves only equality constraints:

$$
\begin{aligned}
\min \quad & \mathbf{c}^T \mathbf{x} \\
\text{s.t.} \quad & \mathbf{A} \mathbf{x} = \mathbf{b} \\
& \mathbf{x} \geq \mathbf{0},
\end{aligned}
$$

with the same notation as in the case of the canonical form. Clearly, we must have $m < n$, so that the system of linear equations is underdetermined and there are multiple solutions.

The reader might think that the canonical and standard forms are somewhat restrictive; in fact, this is not true, since a generic LP problem can be reduced to either form using the following transformations:

- If a variable x_j is not restricted in sign, it can be rewritten as $x_j = x_j^+ - x_j^-$, where $x_j^+, x_j^- \geq 0$.

- An inequality constraint
$$
\mathbf{a}_i^T \mathbf{x} \geq b_i
$$
can be transformed into an equality constraint by introducing a slack variable $s_i \geq 0$:
$$
\mathbf{a}_i^T \mathbf{x} - s_i = b_i.
$$

- An equality constraint
$$
\mathbf{a}_i^T \mathbf{x} = b_i
$$
can be transformed into two inequality constraints:
$$
\mathbf{a}_i^T \mathbf{x} \geq b_i \qquad -\mathbf{a}_i^T \mathbf{x} \geq -b_i.
$$

We know from supplement S3.1 that the feasible set of a LP problem is convex and polyhedral. Furthermore, the problem is both convex and concave. This implies that an optimal solution (if any exists) may be found on the boundary of the feasible set; more specifically, it will be a vertex of the feasible set. This is easy to see by expressing the feasible region S as the convex hull of its extreme points $\mathbf{X}^k$, $k = 1, \ldots, I$. Strictly speaking, if S is unbounded, we should also consider its extreme rays; however, if we assume that the optimal value is finite, there is no loss of generality by discarding the possibility of going to infinity along a ray. Denoting by C^k the cost of the extreme point $\mathbf{X}^k$, we may transform the LP problem

$$
\begin{aligned}
\min \quad & \mathbf{c}^T \mathbf{x} \\
\text{s.t.} \quad & \mathbf{x} \in S
\end{aligned}
$$

into the equivalent problem

$$
\min \quad \sum_{k=1}^{I} \lambda_k C^k \mathbf{X}^k
$$

$$\text{s.t.} \quad \sum_{i=1}^{I} \lambda_k = 1$$

$$\lambda_k \geq 0.$$

This problem has just one constraint, but a possibly huge number of variables; nevertheless, it is easy to see that an optimal solution can be found as the least cost extreme point.

If the problem is cast in standard form, the extreme points correspond to special solutions of the system of linear equations $\mathbf{Ax} = \mathbf{b}$; this is explained briefly in section 3.4.1 and is the basis of the simplex algorithm to which section 3.4.2 is devoted. Applying the duality principles to LP problems produces an interesting theory, outlined in section 3.4.3. The simplex algorithm is certainly the best known method for LP problems, but it is not what you get in MATLAB. The Optimization toolbox provides the user with two options: for medium-scale problems, a version of the active set method is implemented; for large-scale problems, an interior point method is available. Some ideas behind interior point methods are described in section 3.4.4. It is interesting to note that the simplex algorithm is not, in the worst case, a polynomial complexity algorithm, whereas polynomial complexity may be proved for interior point methods. In fact, interior point methods are faster on many problem instances, but not always.

3.4.1   Geometric and algebraic features of linear programming

Given an LP problem, one of the three following cases occurs:

1. The feasible set is empty, and the problem has no solution.

2. The optimal solution is unbounded; this case may occur only if the feasible set is an unbounded polyhedron, and we may keep improving the objective value by going to infinity along an extreme ray.

3. The problem has a finite optimal solution, corresponding to an extreme point of the feasible set; note that we have an infinite set of optimal solutions if the level curves of the objective function are parallel to a face of the polyhedron (see example 3.2).

Since there is a finite number of extreme points in a polyhedron, one way to solve an LP problem is to explore the set of extreme points of the feasible set without considering the interior points. Furthermore, a local minimizer will also be a global one; hence, if we find an extreme point such that no adjacent extreme point improves the objective function, then we have found the optimal solution.

To implement this idea, the geometrical intuition must be translated into algebraic terms. To this end, it is convenient to work on the standard LP form. For convenience, let us assume that the matrix $\mathbf{A} \in \mathbb{R}^{m,n}$, $m < n$, has full row rank. This assumption is not necessary in practice, as redundant equations are easily spot and eliminated. It is useful to consider a solution of the system $\mathbf{Ax} = \mathbf{b}$ as a way to express the vector $\mathbf{b}$ as a linear combination of the columns of $\mathbf{A}$:

$$\mathbf{Ax} = \begin{bmatrix} \mathbf{A}^1 \mathbf{A}^2 \cdots \mathbf{A}^n \end{bmatrix} \begin{bmatrix} x_1 \\ x_2 \\ \vdots \\ x_n \end{bmatrix} = \sum_{j=1}^{n} x_j \mathbf{A}^j = \mathbf{b}.$$

This system has infinite solutions, but not all of them are feasible with respect to the requirement $\mathbf{x} \geq \mathbf{0}$. Furthermore, we would like to work on feasible solutions which are extreme points of the feasible set. This is easily accomplished by considering only solutions in which at most m components x_j are strictly positive and the remaining $n - m$ variables are zero. Such solutions are called *basic solutions*; the name derives from the fact that the m column vectors associated with the m possibly nonnull variables are sufficient to express the m-dimensional vector $\mathbf{b}$. Any basic solution is associated with a basis of $\mathbb{R}^m$ consisting of m columns of $\mathbf{A}$. The m variables corresponding to the columns selected are called *basic variables*; the others are called *nonbasic variables*. A basic solution with nonnegative components is called a *basic feasible solution*.

Example 3.11 Consider the following system of linear equations:

$$\begin{bmatrix} -1 & 1 & 1 & -1 & 0 \\ 0 & 1 & 0 & 4 & 0 \\ 0 & 0 & 2 & 2 & 1 \end{bmatrix} \begin{bmatrix} x_1 \\ x_2 \\ x_3 \\ x_4 \\ x_5 \end{bmatrix} = \begin{bmatrix} 1 \\ 3 \\ 1 \end{bmatrix}$$

A basic solution is

$$x_1 = 2, \; x_2 = 3, \; x_3 = x_4 = 0, \; x_5 = 1,$$

which corresponds to the basis formed by the columns $\mathbf{A}^1, \mathbf{A}^2, \mathbf{A}^5$. This solution is also feasible. If we take the basis formed by $\mathbf{A}^2, \mathbf{A}^3, \mathbf{A}^5$, we obtain the basic solution

$$x_1 = 0, \; x_2 = 3, \; x_3 = -2, \; x_4 = 0, \; x_5 = 5,$$

which is not feasible since $x_3 < 0$.   ☐

Basic feasible solutions are fundamental because it can be shown that they actually correspond to the extreme points of the feasible set. Furthermore, given a current extreme point, the adjacent extreme point may be obtained by exchanging one basic variable with a nonbasic one; this means that we may move from a vertex to another one by driving one basic variable out of the basis and driving one nonbasic variable into the basis.

3.4.2   Simplex method

The simplex method is an iterative algorithm; given a current extreme point (or basic feasible solution, or basis), it looks for an adjacent extreme point such that the objective function is improved, and it stops when no improving adjacent extreme point is found.

Assume that we have a basic feasible solution $\mathbf{x}$; we will consider later how to obtain an initial basic feasible solution. We can partition the vector $\mathbf{x}$ into two subvectors: the subvector $\mathbf{x}_B \in \mathbb{R}^m$ of the basic variables and the subvector $\mathbf{x}_N \in \mathbb{R}^{n-m}$ of the nonbasic variables. Using a suitable permutation of the variable indexes, we may rewrite the system of linear equations

$$\mathbf{A}\mathbf{x} = \mathbf{b}$$

as

$$[\mathbf{A}_B \mathbf{A}_N] \begin{bmatrix} \mathbf{x}_B \\ \mathbf{x}_N \end{bmatrix} = \mathbf{A}_B \mathbf{x}_B + \mathbf{A}_N \mathbf{x}_N = \mathbf{b}, \tag{3.21}$$

where $\mathbf{A}_B \in \mathbb{R}^{m,m}$ is nonsingular and $\mathbf{A}_N \in \mathbb{R}^{m,n-m}$. If $\mathbf{x}$ is basic feasible, it may be written as

$$\mathbf{x} = \begin{bmatrix} \mathbf{x}_B \\ \mathbf{x}_N \end{bmatrix} = \begin{bmatrix} \hat{\mathbf{b}} \\ \mathbf{0} \end{bmatrix},$$

where

$$\hat{\mathbf{b}} = \mathbf{A}_B^{-1} \mathbf{b} \geq \mathbf{0}.$$

The objective function value corresponding to $\mathbf{x}$ is

$$\hat{f} = [\mathbf{c}_B^T \ \mathbf{c}_N^T] \begin{bmatrix} \hat{\mathbf{b}} \\ \mathbf{0} \end{bmatrix} = \mathbf{c}_B^T \hat{\mathbf{b}}. \tag{3.22}$$

Now we must find out if it is possible to improve the current solution by slightly changing the basis, i.e., by replacing one basic variable with a nonbasic one. To assess the potential benefit of introducing a nonbasic variable into the basis, we may eliminate the basic variables in equation (3.22). Using equation (3.21), we may express the basic variables as

$$\mathbf{x}_B = \mathbf{A}_B^{-1}(\mathbf{b} - \mathbf{A}_N \mathbf{x}_N) = \hat{\mathbf{b}} - \mathbf{A}_B^{-1} \mathbf{A}_N \mathbf{x}_N; \tag{3.23}$$

then we rewrite the objective function value

$$\mathbf{c}^T \mathbf{x} = \mathbf{c}_B^T \mathbf{x}_B + \mathbf{c}_N^T \mathbf{x}_N = \mathbf{c}_B^T (\hat{\mathbf{b}} - \mathbf{A}_B^{-1} \mathbf{A}_N \mathbf{x}_N) + \mathbf{c}_N^T = \hat{f} + \hat{\mathbf{c}}_N^T \mathbf{x}_N,$$

where

$$\hat{\mathbf{c}}_N^T = \mathbf{c}_N^T - \mathbf{c}_B^T \mathbf{A}_B^{-1} \mathbf{A}_N. \tag{3.24}$$

The quantities $\hat{\mathbf{c}}_N$ are called *reduced costs*, as they measure the marginal variation of the objective function with respect to the nonbasic variables. If $\hat{\mathbf{c}}_N \geq \mathbf{0}$, it is not possible to improve the objective function; in this case, bringing a nonbasic variable into the basis at some positive value can only increase the cost. Therefore, the current basis is optimal if $\hat{c}_N \geq 0$. If, on the contrary, there exists a $q \in N$ such that $\hat{c}_q < 0$, it is possible to improve the objective function by bringing x_q into the basis. A simple strategy is to choose q such that

$$\hat{c}_q = \min_{j \in N} \hat{c}_j.$$

This selection does not necessarily result in the best performance of the algorithm; we should consider not only the rate of change in the objective function, but also the value attained by the new basic variable. Furthermore, it may happen that the entering variable is stuck to zero and does not change the value of the objective. In such a case, there is danger of cycling on a set of bases; ways to overcome this difficulty are well explained in the literature.

When x_q is brought into the basis, a basic variable must "leave" in order to maintain $\mathbf{A}x = \mathbf{b}$. To spot the leaving variable we can reason as follows. Given the current basis, we can use it to express both $\mathbf{b}$ and the column $\mathbf{A}^q$ corresponding to the entering variable:

$$\mathbf{b} = \sum_{i=1}^{m} x_{B(i)} \mathbf{A}^{B(i)} \tag{3.25}$$

$$\mathbf{A}^q = \sum_{i=1}^{m} d_i \mathbf{A}^{B(i)}, \tag{3.26}$$

where $B(i)$ is the index of the ith basic variable $(i = 1, \ldots, m)$ and

$$\mathbf{d} = \mathbf{A}_B^{-1} \mathbf{A}^q.$$

If we multiply equation (3.26) by a number θ and subtract it from equation (3.25), we obtain

$$\mathbf{b} = \sum_{i=1}^{m} \left(x_{B(i)} - \theta d_i \right) \mathbf{A}^{B(i)} + \theta \mathbf{A}^q. \tag{3.27}$$

From equation (3.27) we see that θ is the value of the entering variable in the new solution, and that the value of the current basic is affected in a way depending on the sign of d_i. If $d_i \leq 0$, $x_{B(i)}$ remains nonnegative when x_q increases. But if there is an index i such that $d_i > 0$, then we cannot increase x_q at will, since there is a limit value for which a currently basic variable becomes zero. This limit value is attained by the entering variable x_q, and the first current basic variable which gets zero leaves the basis

$$x_q = \min_{\substack{i=1,\ldots,m \\ d_i > 0}} \frac{\hat{b}_i}{d_i}.$$

If $\mathbf{d} \leq \mathbf{0}$, there is no limit on the increase of x_q, and the optimal solution is unbounded.

In order to start the iterations, a starting basis is needed. One possibility is to introduce a set of auxiliary artificial variables $\mathbf{z}$ in the constraints:

$$\mathbf{Ax} + \mathbf{z} = \mathbf{b}$$
$$\mathbf{x}, \mathbf{z} \geq \mathbf{0}.$$

Assume also that the equations have been rearranged in such a way that $\mathbf{b} \geq \mathbf{0}$; then a basic feasible solution is trivially $\mathbf{z} = \mathbf{b}$. By minimizing the inadmissibility form

$$\phi = \min \sum_{i=1}^{m} z_i,$$

we may find a basic feasible solution if $\phi = 0$; otherwise, the original problem is not feasible.

At this point, one should wonder what is the connection between the simplex method for LP problems and the simplex search we have hinted at in section 3.2.4. Actually, the simplex method works on a simplex in the reduced space of the nonbasic variables. In this space, the origin corresponds to the current basic solution, as the nonbasic variables are zero; the remaining extreme points of the simplex correspond to the adjacent bases. The simplex method checks, in the reduced space, if any of these extreme points improves the objective function.

3.4.3   Duality in linear programming

We dealt with duality in nonlinear programming in section 3.3.3. Duality in LP can be developed without considering the more general nonlinear case, but we prefer to put it in a more general framework. Note that due to the convexity of LP problems, strong duality holds. Let us start with an LP problem (P_1) in the following canonical form:

$$(P_1) \qquad \min \quad \mathbf{c}^T \mathbf{x}$$
$$\text{s.t.} \quad \mathbf{Ax} \geq \mathbf{b}.$$

If we dualize the inequality constraints with a vector $\boldsymbol{\mu} \in \mathbb{R}_+^m$ of dual variables, we get the dual problem

$$\max_{\boldsymbol{\mu} \geq 0} \min_{\mathbf{x}} \left\{ \mathbf{c}^T \mathbf{x} + \boldsymbol{\mu}^T (\mathbf{b} - \mathbf{Ax}) \right\} = \max_{\boldsymbol{\mu} \geq 0} \left\{ \boldsymbol{\mu}^T \mathbf{b} + \min_{\mathbf{x}} \left(\mathbf{c}^T - \boldsymbol{\mu}^T \mathbf{A} \right) \mathbf{x} \right\}.$$

Since $\mathbf{x}$ is unrestricted in sign, the inner minimization problem has a finite value if and only if

$$\mathbf{c}^T - \boldsymbol{\mu}^T \mathbf{A} = \mathbf{0};$$

otherwise, each component of $\mathbf{x}$ is set to $\pm\infty$, depending on the sign of the corresponding cost coefficient, and this results in a value $-\infty$ for the dual function. Since we want to maximize the dual function, we may enforce the condition above, and the dual problem (D_1) turns out to be

$$(D_1) \qquad \max \quad \boldsymbol{\mu}^T \mathbf{b}$$
$$\text{s.t.} \quad \mathbf{A}^T \boldsymbol{\mu} = \mathbf{c}$$
$$\boldsymbol{\mu} \geq 0.$$

The dual problem is still an LP problem, resulting from exchanging $\mathbf{b}$ with $\mathbf{c}$ and by transposing $\mathbf{A}$. The duality relationship between (P_1) and (D_1) can be interpreted the other way round, too:

$$\left(\begin{array}{cc} \max & \mathbf{x}^T \mathbf{c} \\ \text{s.t.} & \mathbf{A}\mathbf{x} = \mathbf{b} \\ & \mathbf{x} \geq 0 \end{array} \right) \iff \left(\begin{array}{cc} \min & \mathbf{b}^T \boldsymbol{\nu} \\ \text{s.t.} & \mathbf{A}^T \boldsymbol{\nu} \geq \mathbf{c} \end{array} \right).$$

Given an LP problem (P_2) in standard form,

$$(P_2) \qquad \min \quad \mathbf{c}^T \mathbf{x}$$
$$\text{s.t.} \quad \mathbf{A}\mathbf{x} = \mathbf{b}$$
$$\mathbf{x} \geq 0,$$

we can use the relationship above to find its dual:

$$\left(\begin{array}{cc} \min & \mathbf{c}^T \mathbf{x} \\ \text{s.t.} & \mathbf{A}\mathbf{x} = \mathbf{b} \\ & \mathbf{x} \geq 0 \end{array} \right) \iff \left(\begin{array}{cc} \max & \mathbf{x}^T(-\mathbf{c}) \\ \text{s.t.} & (\mathbf{A}^T)^T \mathbf{x} = \mathbf{b} \\ & \mathbf{x} \geq 0 \end{array} \right)$$

$$\iff \left(\begin{array}{cc} \min & \mathbf{b}^T \boldsymbol{\nu} \\ \text{s.t.} & \mathbf{A}^T \boldsymbol{\nu} \geq -\mathbf{c} \end{array} \right) \iff \left(\begin{array}{cc} \min & -\mathbf{b}^T \boldsymbol{\mu} \\ \text{s.t.} & -\mathbf{A}^T \boldsymbol{\mu} \geq -\mathbf{c} \end{array} \right)$$

$$\iff \left(\begin{array}{cc} \max & \mathbf{b}^T \boldsymbol{\mu} \\ \text{s.t.} & \mathbf{A}^T \boldsymbol{\mu} \leq \mathbf{c} \end{array} \right),$$

where we have introduced $\boldsymbol{\mu} = -\boldsymbol{\nu}$; we obtain the dual (D_2) of (P_2).

Note the similarities and the differences between the two dual pairs. The dual variables are restricted in sign when the constraints of the primal problem are inequalities, and are unrestricted in sign in the other case (this is coherent with the Kuhn-Tucker conditions). When the variables are restricted in sign in the primal, we have inequality constraints in the dual, whereas in the case of unrestricted variables we have equality constraints in the dual. In table 3.1 we summarize the "recipe" for building the dual of a generic LP.

Given a primal-dual pair of LP problems, the following cases may occur:

- Both problems have a finite optimal solution, in which case the two objectives have the same value at the optimum.

- Both problems are infeasible.

- One problem is unbounded, in which case the other one is infeasible.

Table 3.1   Duality Relationships

Primal	Dual
$\min \mathbf{c}^T \mathbf{x}$	$\max \boldsymbol{\mu}^T \mathbf{b}$
$\mathbf{a}_i^T \mathbf{x} = b_i$	μ_i unrestricted
$\mathbf{a}_i^T \mathbf{x} \geq b_i$	$\mu_i \geq 0$
$x_j \geq 0$	$\boldsymbol{\mu}^T \mathbf{A}^j \leq c_j$
x_j unrestricted	$\boldsymbol{\mu}^T \mathbf{A}^j = c_j$

As a final remark, it is important to note that the dual feasibility constraint $\mathbf{A}^T \boldsymbol{\mu} \leq \mathbf{c}$ for the dual of the problem in standard form can be read as the nonnegativity condition on the reduced costs by equating $\boldsymbol{\mu}^T = \mathbf{c}^T \mathbf{A}_B^{-1}$. Recall the sufficient conditions (3.7) for global optimality. They correspond to:

- Primal feasibility

- Dual feasibility

- Complementary slackness

In fact, the simplex method works by maintaining primal feasibility and complementary slackness, and it iterates until dual feasibility is obtained. Switching roles between primal and dual problems, it is possible to devise a dual simplex method which works toward primal feasibility. This is sometimes advantageous over the primal simplex approach. However, there is still a third possibility: we can keep a pair of primal and dual feasible solutions and work to obtain complementary slackness. This approach leads to primal-dual algorithms, and it is exploited in the interior point method described in the next section.

3.4.4   Interior point methods

The simplex method works only on the extreme points of the feasible set. As the name suggests, interior point methods move on a path that lies within the feasible set. There are several variants of interior point algorithms; we describe just the basics of a rather simple approach, which may be called the *primal-dual barrier method*, as it exploits the correspondence between a primal and dual problems, and an interior penalty function. It is convenient to start with the LP problem written in canonical form for a maximization problem:[6]

$$\max \quad \mathbf{c}^T\mathbf{x}$$
$$\text{s.t.} \quad \mathbf{A}\mathbf{x} \leq \mathbf{b}$$
$$\mathbf{x} \geq \mathbf{0},$$

which may be converted to the standard form by adding slack variables $\mathbf{w}$:

$$\max \quad \mathbf{c}^T\mathbf{x}$$
$$\text{s.t.} \quad \mathbf{A}\mathbf{x} + \mathbf{w} = \mathbf{b}$$
$$\mathbf{x}, \mathbf{w} \geq \mathbf{0}.$$

Now suppose that we do not know anything about the simplex method. We could try applying what we know from the general theory of constrained optimization; one idea would be getting rid of the nonnegativity restriction by a suitable penalty function and then apply the method of Lagrangian multipliers. Using an interior penalty function based on a logarithmic barrier:

$$(PP) \qquad \max \quad \mathbf{c}^T\mathbf{x} + \sigma \sum_j \log x_j + \sigma \sum_i \log w_i$$
$$\text{s.t.} \quad \mathbf{A}\mathbf{x} + \mathbf{w} = \mathbf{b}.$$

Since this problem has only equality constraints, we may dualize them with Lagrangian multipliers $\mathbf{y}$, yielding the Lagrangian function

$$\mathcal{L}(\mathbf{x}, \mathbf{w}, \mathbf{y}) = \mathbf{c}^T\mathbf{x} + \sigma \sum_j \log x_j + \sigma \sum_i \log w_i + \mathbf{y}^T(\mathbf{b} - \mathbf{A}\mathbf{x} - \mathbf{w}).$$

The first-order stationarity conditions are then

$$\frac{\partial \mathcal{L}}{\partial x_j} = c_j + \sigma \frac{1}{x_j} - \sum_i y_i a_{ij} = 0 \qquad \forall j$$

$$\frac{\partial \mathcal{L}}{\partial w_i} = \sigma \frac{1}{w_i} - y_i = 0 \qquad \forall i$$

$$\frac{\partial \mathcal{L}}{\partial y_i} = b_i - \sum_j a_{ij} x_j - w_i = 0 \qquad \forall i.$$

Using the notation

$$\mathbf{X} = \begin{bmatrix} x_1 & & & \\ & x_2 & & \\ & & \ddots & \\ & & & x_n \end{bmatrix} \qquad \mathbf{e} = \begin{bmatrix} 1 \\ 1 \\ \vdots \\ 1 \end{bmatrix},$$

---

[6]The exposition here is based on [20].

the optimality equations may be rewritten in matrix form:

$$A^T y - \sigma X^{-1} e = c$$
$$y = \sigma W^{-1} e$$
$$Ax + w = b.$$

The addition of the auxiliary vector

$$z = \sigma X^{-1} e,$$

and a slight rearrangement yields the following set of optimality equations:

$$\begin{aligned} Ax + w &= b \\ A^T y - z &= c \\ XZe &= \sigma e \\ YWe &= \sigma e. \end{aligned}$$

These equations have a nice interpretation. We have just to recall that the starting problem has an LP dual:

$$\begin{aligned} \min \quad & b^T y \\ \text{s.t.} \quad & A^T y \geq c \\ & y \geq 0, \end{aligned}$$

or, adding slack variables z,

$$\begin{aligned} \min \quad & b^T y \\ \text{s.t.} \quad & A^T y - z = c \\ & y, z \geq 0. \end{aligned}$$

Hence, the equations we arrived at are simply the conditions of primal feasibility, dual feasibility, and (if $\sigma = 0$) complementary slackness (see theorem 3.7). For $\sigma > 0$, they are a set of nonlinear equations:

$$F(\xi) = 0,$$

where

$$\xi = \begin{bmatrix} x \\ y \\ w \\ z \end{bmatrix},$$

which may be tackled by Newton's method (section 2.4.2).

In principle, by solving this system of nonlinear equations for different values of σ we get a path $(x_\sigma, y_\sigma, w_\sigma, z_\sigma)$. This path is called *central path* and for $\sigma \to 0$, it leads to the optimal solution of the original LP. From a computational point of view, it is not convenient to start with a too small σ, nor to solve the nonlinear equations exactly for each σ. One idea is to reduce the value of the penalty parameter within the iterations of Newton's method, so that the central path is only a reference path leading to solution through the interior of the feasible set.

Interior point methods have a polynomial computational complexity which is, theoretically, better than the complexity of the simplex method, which is exponential in pathological cases. It should be stressed that many computational tricks are needed to implement both the simplex and interior point method in a very efficient way. These are beyond the scope of this book, but it should be clear that the two approaches may lead to qualitatively different, though cost-equivalent solutions, as illustrated in the next section.

3.4.5 Linear programming in MATLAB

The Optimization toolbox includes a function, linprog, which solves LP problems of the form

$$\begin{aligned} \min \quad & \mathbf{c}^T \mathbf{x} \\ \text{s.t.} \quad & \mathbf{Ax} \le \mathbf{b} \\ & \mathbf{A}_{eq}\mathbf{x} = \mathbf{b}_{eq} \\ & \mathbf{l} \le \mathbf{x} \le \mathbf{u}. \end{aligned}$$

We have seen that alternative algorithms are available for linear programming. What happens in MATLAB, then? Consider the following rather trivial LP problem:

$$\begin{aligned} \max \quad & x_1 + x_2 \\ \text{s.t.} \quad & x_1 + x_2 \le 1 \\ & x_1, x_2 \ge 0. \end{aligned}$$

It is easy to see that two basic optimal solutions are $(1,0)$ and $(0,1)$. All the solutions between these two extreme points are equivalent and optimal. We expect that the simplex algorithm should report one of the two extreme points. To use linprog, we have to change the sign of the coefficients in the objective function and to pass as null vectors the parameters we do not need:

```
>> x=linprog([-1 -1],[1 1],1,[],[],[0 0])
Optimization terminated successfully.
x =
    0.5000
    0.5000
```

We see that the reported solution is *on the center* of the face of equivalent solutions, and it is not basic. This happens since the default LP option in

MATLAB is an interior point algorithm. We may use an alternative algorithm by switching the LargeScale option off. In this case, an active set method is used (see section 3.3.5). With this method, an initial solution may be passed to the algorithm; the solution we obtain may depend on this initial solution:

```
>> options = optimset('LargeScale','off');
>> x=linprog([-1 -1],[1 1],1,[],[],[0 0],[],[],options)
Optimization terminated successfully.
x =
    0.5000
    0.5000
>> x=linprog([-1 -1],[1 1],1,[],[],[0 0],[],[0 0.5],options)
Optimization terminated successfully.
x =
    0.2500
    0.7500
```

We see that, starting from the initial solution, the search moves along the gradient until the constraint is reached, which is turned active and the process is stopped. So it is important to bear in mind that MATLAB does not implement a simplex method and that in some cases, an optimal but nonbasic solution is obtained. This may have consequences if linprog is embedded within an algorithm that requires basic optimal solutions. For instance, in some problems with special structure, the simplex method yields an integer solution; this is the case when the feasible set is a polyhedron whose extreme points have integer coordinates. But in the case of multiple optimal solutions, it could well happen that the active set method outputs a fractional solution rather than an integer one. This is no surprise actually, as no Optimization toolbox function is aimed at integer programming, but it is worth keeping in mind anyway.

3.5 BRANCH AND BOUND METHODS FOR NONCONVEX OPTIMIZATION

Consider a generic optimization problem:

$$P(S) \qquad \min_{\mathbf{x} \in S} f(\mathbf{x}),$$

and assume that it is a difficult one, as either the objective function or the feasible set is nonconvex. Consider figure 3.13; in the first case, the objective function has local minima; in the second case, the feasible set is discrete, and hence nonconvex. While solving nonconvex problems is very difficult in general, in some cases it could be made a straightforward task if a suitable convexification were available. For instance, if S is convex but f is not, we

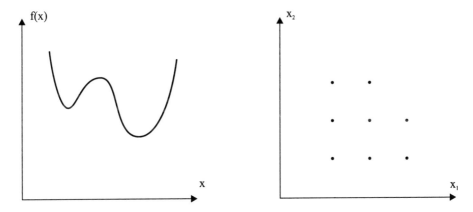

Fig. 3.13   Nonconvex objective function and discrete nonconvex feasible set.

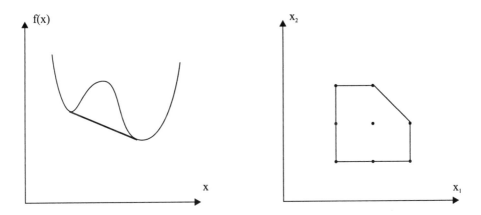

Fig. 3.14   Convexification of a nonconvex objective function and a discrete nonconvex feasible set.

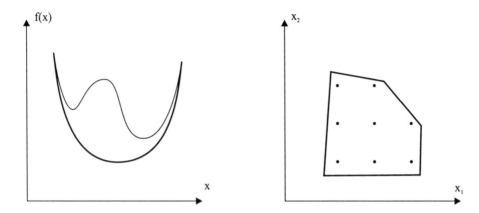

Fig. 3.15   Convex lower bounding function and a relaxation of a discrete feasible set.

could take the convex hull of the epigraph of f, as illustrated in figure 3.14. Taking the convex hull of the epigraph of f yields a function h such that:

- h is convex on S.

- $h(\mathbf{x}) \leq f(\mathbf{x})$ for any $\mathbf{x} \in S$.

- If g is a convex function such that $g(\mathbf{x}) \leq f(\mathbf{x})$ for any $\mathbf{x} \in S$, then $g(\mathbf{x}) \leq h(\mathbf{x})$ for any $\mathbf{x} \in S$.

In this case, we could think of replacing f by h and solve the problem by convex optimization techniques. By the same token, consider a linear integer programming problem:

$$\text{(PI)} \qquad \min \quad \mathbf{c}^T \mathbf{x}$$
$$\text{s.t.} \quad \mathbf{Ax} \leq \mathbf{b}$$
$$\mathbf{x} \in \mathbb{Z}_+^n.$$

The feasible set is a discrete set much like that in figure 3.13. If we knew its convex hull, illustrated in figure 3.14, we could simply tackle the problem as an ordinary LP problem by the simplex method. In fact, the convex hull of a discrete set of points is a polyhedron; if the points have integer coordinates, then the extreme points of the convex hull will be integer too, and one of them will turn out to be the optimal solution returned by the simplex method.

Unfortunately, we are rarely in the lucky position of being able to find such a convexification easily. However, we might be able to find weaker convex objects, as illustrated in figure 3.15. They are exploited to define a relaxation of the original problem.

DEFINITION 3.10 *An optimization problem,*

$$\text{RP}(T) \qquad \min_{\mathbf{x} \in T} h(\mathbf{x}),$$

is a relaxation of problem $P(S)$ if:

- $S \subseteq T$.

- $h(\mathbf{x}) \le f(\mathbf{x})$, *for any* $\mathbf{x} \in S$.

Solving a relaxation does not yield the optimal solution of the original problem in general, but it gives a lower bound for its optimal value.

Example 3.12 Consider a nonconvex function $f(\mathbf{x})$ on a hyperrectangle S defined by the bounds

$$l_j \le x_j \le u_j \qquad j = 1, \ldots, n.$$

Assume that f is twice continuously differentiable. In supplement S3.1.1 it is stated that a twice continuously differentiable function is convex if its Hessian matrix is positive semidefinite, which is equivalent to requiring that its eigenvalues are nonnegative. We may build a convex underestimating function for f by adding an additional term and considering

$$h(\mathbf{x}) = f(\mathbf{x}) + \alpha \sum_{i=1}^{n} (l_i - x_i)(u_i - x_i)$$

for some $\alpha > 0$. It is easy to see that the additional term is nonpositive on the region S and that it is zero on its boundary. Thus h is an underestimator for f. It will be convex if α is large enough. To see this, consider how the Hessian $\mathbf{H}$ of h is related to the Hessian $\mathbf{H}_f$ of the original objective f:

$$\frac{\partial^2 h}{\partial x_i^2} = \frac{\partial^2 f}{\partial x_i^2} + 2\alpha \qquad i = 1, \ldots, n$$

$$\frac{\partial^2 h}{\partial x_i \partial x_j} = \frac{\partial^2 f}{\partial x_i \partial x_j} \qquad i, j = 1, \ldots, n; i \ne j.$$

The eigenvalues of h are the solution of the following equation:

$$\det(\mathbf{H}_f + 2\alpha\mathbf{I} - \mu\mathbf{I}) = \det(\mathbf{H}_f - (\mu - 2\alpha)\mathbf{I}) = 0.$$

It is easy to see that if the eigenvalues of $\mathbf{H}_f$ are λ_i, $i = 1, \ldots, n$, then the eigenvalues of the Hessian of h are simply

$$\mu_i = \lambda_i + 2\alpha,$$

which may be made positive by choosing a suitably large value of α. We will see shortly that a relaxation should be as tight as possible. This means that the underestimating function should be as large as possible and that α should be as small as possible. Guidelines for the selection of α are given in the original reference [12]. ⬜

Example 3.13 Consider the integer programming problem (IP). A relaxation of the feasible set

$$S = \{\mathbf{x} : \mathbf{Ax} \le \mathbf{b}; \mathbf{x} \in \mathbb{Z}_+^n\}$$

can be obtained by dropping the integrality requirement:

$$T = \{\mathbf{x} : \mathbf{Ax} \le \mathbf{b}; \mathbf{x} \in \mathbb{R}_+^n\}.$$

This yields an LP problem which is readily solved by the simplex method. In general, some components of the solution of the relaxed problem will be fractional; this implies that the solution we obtain is not feasible, but we get a lower bound on the optimal value of the objective function.   ☐

We have seen in examples 3.12 and 3.13 that when the relaxed problem is convex, it is easily solved, but it will only yield a lower bound on the optimal value of the objective function.

A possible solution strategy is to decompose the original problem $P(S)$ by splitting the feasible set S into a collection of subsets $S_1, \ldots, S_q$ such that

$$S = S_1 \cup S_2 \cup \cdots \cup S_q;$$

then we have

$$\min_{\mathbf{x} \in S} f(\mathbf{x}) = \min_{i=1,\ldots,q} \left\{ \min_{\mathbf{x} \in S_i} f(\mathbf{x}) \right\}.$$

The rationale behind this decomposition of the feasible set is that we may expect that solving the problems over smaller sets is easier; or, at least, the lower bounds obtained by solving the relaxed problems will be tighter. For efficiency reasons it is advisable, but not strictly necessary, to partition the set S in such a way that

$$S_i \cap S_j = \emptyset \qquad i \ne j.$$

This type of decomposition is called *branching*.

Example 3.14 Consider the binary programming problem:

$$\begin{aligned} \min \quad & \mathbf{c}^T \mathbf{x} \\ \text{s.t.} \quad & \mathbf{x} \in S = \{\mathbf{x} \mid \mathbf{Ax} \ge \mathbf{b}; x_j \in \{0,1\}\}. \end{aligned}$$

The problem may be decomposed in two subproblems by picking a variable x_p and fixing it to 1 and 0:

$$\begin{aligned} S_1 &= \{\mathbf{x} \in S; \ x_p = 0\} \\ S_2 &= \{\mathbf{x} \in S; \ x_p = 1\}. \end{aligned}$$

The resulting problems $P(S_1)$ and $P(S_2)$ can be decomposed in turn, until eventually all the variables have been fixed. The branching process can be pictorially represented as a *search tree*, as shown in figure 3.16.   ☐

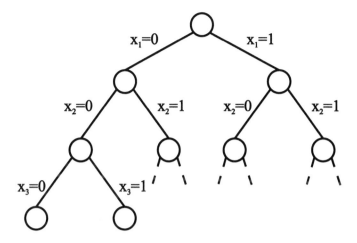

Fig. 3.16 Search tree for a binary programming problem.

The branching process leads to easier problems. In the example, the leaves of the search tree are trivial problems, since all the variables are fixed to a value; actually, the search tree is, in this case, just a way to enumerate the possible solutions. Unfortunately, there are a large number of leaves; if $\mathbf{x} \in \{0,1\}^N$, there are 2^N possible solutions. Actually, the constraints $\mathbf{Ax} \geq \mathbf{b}$ rule out many of them, but a brute-force enumeration is not feasible except for the smallest problems.

To reduce the computational burden, one can try to eliminate a subproblem $P(S_k)$, or equivalently, a node of the tree, by showing that it cannot lead to the optimal solution of $P(S)$. This can be accomplished if it is possible to compute a lower bound for each subproblem by a convex relaxation or by whatever method. Let $\nu[P(S_k)]$ denote the optimal value of problem $P(S_k)$. The lower bound $\beta[P(S_k)]$ is such that

$$\beta[P(S_k)] \leq \nu[P(S_k)].$$

Now assume that we know a feasible, but not necessarily optimal solution $\hat{\mathbf{x}}$ of $P(S)$. Such a solution, if any exists, is eventually found while searching the tree (with the exception of pathological cases). The value $f(\hat{\mathbf{x}})$ is an upper bound on the optimal value $\nu^* = \nu[P(S)]$. Clearly, there is no point in solving a subproblem $P(S_k)$ if

$$\beta[P(S_k)] \geq f(\hat{\mathbf{x}}). \tag{3.28}$$

In fact, solving this subproblem cannot yield an improvement with respect to feasible solution $\hat{\mathbf{x}}$ that we know already. In this case we can eliminate $P(S_k)$ from further consideration; this elimination, called *fathoming*, corresponds to pruning a branch of the search tree. Note that $P(S_k)$ can be fathomed only by comparing the lower bound $\beta[P(S_k)]$ with an upper bound on $\nu[P(S)]$. It is *not* correct to fathom $P(S_k)$ on the basis of a comparison with a subproblem $P(S_i)$ such that

$$\beta[P(S_i)] < \beta[P(S_k)].$$

The branching and fathoming mechanism is the foundation of a wide class of algorithms known as *branch and bound methods*. In the next subsection we outline the basic structure of branch and bound methods for mixed-integer linear programming (MILP) problems. These methods are widely available in commercial optimization software libraries. On the contrary, branch and bound methods for nonconvex continuous problems require ad hoc coding in practice.

3.5.1 LP-based branch and bound for MILP models

The fundamental branch and bound algorithm can be outlined as follows. At each step we work on a list of open subproblems, corresponding to nodes of a search tree, and we try to generate a sequence of improving incumbent solutions until we can prove that an incumbent solution is the optimal one. At intermediate steps, the incumbent solution is the best feasible (integer) solution found so far; the incumbent solution provides us with an upper bound on the value of the optimal solution. We give the algorithm for a minimization problem; it is easy to adapt the algorithm to a maximization problem.

Fundamental branch and bound algorithm

1. *Initialization.* The list of open subproblems is initialized to $P(S)$; the value of the incumbent solution ν^* is set to $+\infty$.

2. *Selecting a candidate subproblem.* If the list of open subproblems is empty, stop: the incumbent solution $\mathbf{x}^*$, if any has been found, is optimal; if $\nu^* = +\infty$, the original problem was infeasible. Otherwise, select a subproblem $P(S_k)$ from the list.

3. *Bounding.* Compute a lower bound $\beta(S_k)$ on $\nu[P(S_k)]$ by solving a relaxed problem $P(\overline{S}_k)$. Let $\overline{\mathbf{x}}_k$ be the optimal solution of the relaxed subproblem.

4. *Prune by optimality.* If $\overline{\mathbf{x}}_k$ is feasible, prune subproblem $P(S_k)$. Furthermore, if $f(\overline{\mathbf{x}}_k) < \nu^*$, update the incumbent solution $\mathbf{x}^*$ and its value ν^*. Go to step 2.

5. *Prune by infeasibility.* If the relaxed subproblem $P(\overline{S}_k)$ is infeasible, eliminate $P(S_k)$ from further consideration. Go to step 2.

6. *Prune by bound.* If $\beta(S_k) \geq \nu^*$, eliminate subproblem $P(S_k)$ and go to step 2.

7. *Branching.* Replace $P(S_k)$ in the list of open subproblems with a list of child subproblems $P(S_{k1})$, $P(S_{k2})$,..., $P(S_{kq})$, obtained by partitioning S_k; go to step 2.

To apply this algorithm successfully, we must cope with the following issues:

- How to compute a strong lower bound efficiently

- How to branch to generate subproblems

- How to select the right candidate from the list of open subproblems

Commercial branch and bound procedures compute bounds by the following LP-based (continuous) relaxation. Given a MILP problem

$$P(S) \quad \min \quad \mathbf{c}^T\mathbf{x} + \mathbf{d}^T\mathbf{y}$$
$$\text{s.t.} \quad \mathbf{Ax} + \mathbf{Ey} \leq \mathbf{b}$$
$$\mathbf{x} \in \mathbb{R}_+^{n_1} \quad \mathbf{y} \in \mathbb{Z}_+^{n_2},$$

the continuous relaxation is obtained by relaxing the integrality constraints:

$$P(\overline{S}) \quad \min \quad \mathbf{c}^T\mathbf{x} + \mathbf{d}^T\mathbf{y}$$
$$\text{s.t.} \quad \mathbf{Ax} + \mathbf{Ey} \leq \mathbf{b}$$
$$\begin{bmatrix} \mathbf{x} \\ \mathbf{y} \end{bmatrix} \in \mathbb{R}_+^{n_1+n_2}.$$

Ideally, the relaxed region $\overline{S}$ should be as close as possible to the convex hull of S; the smaller $\overline{S}$, the larger the lower bound. Tighter lower bounds make pruning by bound easier. To this end, careful model formulation may help.

Example 3.15 Consider a fixed-charge model in which the level of activity i is measured by the continuous decision variable x_i and the decision of starting that activity is modeled by the binary decision variable $\delta_i \in \{0, 1\}$. To relate the two decision variables, we may write the constraint

$$x_i \leq M_i \delta_i,$$

where M_i is an upper bound on the level x_i. When we solve the continuous relaxation, we drop the integrality constraint on δ_i, and we replace it by $\delta_i \in [0, 1]$. In principle, M_i may be a very large number, but to get a tight relaxation, we should select M_i as small as possible. ▯

Example 3.16 In section 1.8 we have considered a trivial model of a capital budgeting problem. We have a set of competing projects and we have to select the best subset of them, considering that our budget does not allow us to invest in all of them. In section 2.1.1 we have considered an extension of that basic model in which activity 0 may be started only if all the activities within a certain subset may be started. A possible constraint to model this requirement is

$$Nx_0 \leq \sum_{i=1}^{N} x_i,$$

where $x_0 \in \{0, 1\}$ models the decision of starting activity 0, and $x_i \in \{0, 1\}$ is related to the N activities in the subset conditioning activity 0. An alternative and equivalent formulation is

$$x_0 \leq x_i \qquad i = 1, \ldots, N.$$

On the one hand, this disaggregated form entails more constraints and probably require more work in solving the continuous relaxation. However, when we consider the continuous relaxation, all the points that are feasible for the disaggregate formulation are feasible for the aggregate constraint, but not vice versa. Hence, the feasible set for the relaxation of the disaggregate formulation is smaller, and the lower bound is tighter. Such a reformulation, as well as others, is carried out automatically by some packages (e.g., CPLEX) and may cut the computational effort of a branch and bound algorithm considerably.

$\Box$

As to branching, the following strategy is commonly applied. Assume that an integer variable y_j takes a noninteger value $\bar{y}_j$ in the optimal solution of the relaxed subproblem (one must exist; otherwise, we would prune by feasibility). Then two subproblems are generated; in the *down-child* we add the constraint

$$y_j \leq \lfloor \bar{y}_j \rfloor$$

to the formulation; in the *up-child* we add

$$y_j \geq \lfloor \bar{y}_j \rfloor + 1.$$

For instance, if $\bar{y}_j = 4.2$, we generate two subproblems with the addition of constraints $y_j \leq 4$ (for the down-child) and $y_j \geq 5$ (for the up-child).

A thorny issue is which variable we should branch on. Similarly, we should decide which subproblem we select from the list at step 2 of the branch and bound algorithm. As is often the case, there is no general answer; software packages offer different options to the user, and some experimentation may be required to come up with the best strategy.

Quite impressive improvements have been made in commercial branch and bound packages. Despite this, some large-scale problems cannot be solved to optimality within a reasonable amount of time. If this is the case, one possibility is to run branch and bound with a suboptimality tolerance. Instead of pruning a subproblem $P(S_k)$ only if the lower bound is larger than or equal to the incumbent, $\beta(S_k) \geq \nu^*$, we may introduce a tolerance parameter ϵ and eliminate a node in the tree whenever

$$\beta(S_k) \geq (1 - \epsilon)\nu^*.$$

Doing so, we have only the guarantee of finding a near-optimal solution, but we have a bound on the level of suboptimality. In exchange, we may considerably reduce the computational effort. An alternative approach is to resort to heuristic methods.

3.6 HEURISTIC METHODS FOR NONCONVEX OPTIMIZATION

When a branch and bound method is not able to yield an optimal or near-optimal solution with a reasonable effort, we may settle for a quick heuristic method able to provide us with a good solution. For any specific problem it is possible to devise an ad hoc method. However, it is interesting to consider relatively general principles which, with some adaptation, may yield good heuristics for a wide class of problems. Local search metaheuristics[7] are quite popular and have also been proposed for financial problems. They were originally developed for discrete optimization problems; however, they may also be applied to continuous nonlinear programming when the objective is nonconvex.

Local search algorithms are similar to the gradient method for nonlinear programming. The basic idea is to improve a known solution by applying a set of local perturbations. Consider a generic optimization problem

$$\min_{x \in S} f(x),$$

defined over a discrete set S. Given a feasible solution x, a neighborhood $\mathcal{N}(x)$ is defined as the set of solutions obtained by applying a set of simple perturbations to x. Different perturbations yield different *neighborhood structures*.

The simplest local search algorithm is *local improvement*. Given a current solution $\bar{x}$, an alternative (candidate) solution x° is searched for, such that

$$f(x^\circ) = \min_{x \in \mathcal{N}(\bar{x})} f(x).$$

If the neighborhood structure $\mathcal{N}(\cdot)$ is simple enough, the minimization above can be performed by an exhaustive search; we speak of a *best-improving method* since we try to find the best solution in the neighborhood. Clearly, there is a trade-off between the effectiveness of the neighborhood structure (the larger the better) and the efficiency of the algorithm. If $f(x^\circ) < f(\bar{x})$, then x° is set as the new current solution and the process is iterated. If $f(x^\circ) \geq f(\bar{x})$, the algorithm is stopped. A possible variation is to *partially* explore the neighborhood of the current solution until an improving solution

---

[7]This name reflects the relatively general nature of the principle. In practice, a good deal of customization is needed to come up with a truly effective method for a specific problem.

is found; this approach is known as *first-improving*, since we do not explore the entire neighborhood before committing to a new current solution.

The neighborhood structure is problem dependent. In the case of discrete optimization problems, devising a neighborhood structure may be relatively straightforward. For instance, in a capital budgeting problem, the solution is represented by the subset of selected projects. The neighborhood might be generated by exchanging a project within the current subset with a project not included in it. In a general programming problem with binary variables one might consider complementing each variable in turn. Actually, devising a clever and effective neighborhood is not trivial, as it might seem, as due attention must be paid to constraints. In the case of continuous variables, a further complication arises; we may generate neighboring points by moving along a set of directions, but we must find a way to select the step size. To this aim, dynamic strategies have been devised (see, e.g., [6] for a financial application).

This basic idea is generally easy to apply, but it has one major drawback; the algorithm usually stops in a *locally* (with respect to the neighborhood structure) optimal solution. This is the difficulty we face when applying the gradient method to a nonconvex objective function; the reason behind the trouble is that only improving perturbations [i.e., such that $\Delta f = f(x^\circ) - f(\bar{x}) < 0$] are accepted. To avoid getting stuck in a local optimum, we must relax this assumption.

In the following we describe three local search approaches that have been proposed to overcome the limitations of local improvement: simulated annealing, tabu search, and genetic algorithms.

Simulated annealing   It has been pointed out that to overcome the problem of local minima, we have to accept, in some disciplined way, nonimproving perturbations, i.e., perturbations for which $\Delta f > 0$. Simulated annealing is based on an analogy between cost minimization in discrete optimization and energy minimization in physical systems. The local improvement strategy behaves much like physical systems do, according to classical mechanics; it is impossible for a system to have a certain energy at a certain time and to increase it without external input. This is not true in thermodynamics and statistical mechanics; according to these physical models, at a temperature above absolute zero, thermal noise makes an increase in the energy of a system possible. An increase in energy is more likely to occur at high temperatures. The probability P of this upward jump depends on the amount of energy ΔE acquired and the temperature T, according to the Boltzmann distribution

$$P(\Delta E, T) = \exp\left(-\frac{\Delta E}{KT}\right)$$

where K is the Boltzmann constant.

Annealing is a metallurgical process by which a melted material is slowly cooled in order to obtain good (low-energy) solid-state configurations. If the

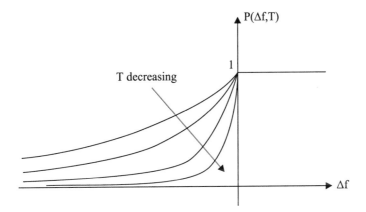

Fig. 3.17 Acceptance probabilities as a function of cost increase for different temperatures.

temperature is decreased too fast, the system gets trapped in a local energy minimum, and a glass is produced. But if the process is slow enough, random kinetic fluctuations due to thermal noise allow the system to escape from local minima, reaching a point very close to the global optimum.

In strict analogy with statistical mechanics, in the simulated annealing method a perturbation of the current solution yielding $\Delta f < 0$ is always accepted; a perturbation with $\Delta f > 0$ is accepted with a probability given by a Boltzmann-like probability distribution

$$P(\Delta f, T) = \exp\left(-\frac{\Delta f}{T}\right).$$

This probability distribution is a decreasing exponential in Δf, whose shape depends on the parameter T, acting as a temperature (see figure 3.17). The probability of accepting a nonimproving perturbation decreases as the deterioration of the solution increases. For a given Δf, the acceptance probability is higher at high temperatures. For $T \to 0$ the probability collapses into a step function, and the method behaves like local improvement. For $T \to +\infty$ the probability is 1 everywhere, and we have a random exploration of the solutions space. The parameter T allows balancing the need to *exploit* the solution at hand by improving it and the need to *explore* the solution space.

The simulated annealing method simply substitutes the deterministic acceptance rule of local improvement with a probabilistic rule. The temperature is set to a relatively high initial value T_1, and the algorithm is iterated using at step k a temperature T_k until some termination criterion is satisfied. The strategy by which the temperature is decreased is called the *cooling schedule*. The simplest cooling schedule is

$$T_k = \alpha T_{k-1} \qquad 0 < \alpha < 1.$$

In practice, it is advisable to keep the temperature constant for a certain number of steps, in order to reach a thermodynamic equilibrium before changing the control parameter. More sophisticated adaptive cooling strategies have been proposed, but the increase in complexity does not always seem justified. A very simple implementation of the annealing algorithm could be the following:

Step 1. Choose an initial solution x_{old}, an initial temperature T_1, and a decrease parameter α; let $k = 1$, $f_{old} = f(x_{old})$; let $\hat{f} = f_{old}$ and $\hat{x} = x_{old}$ be the current optimal value and optimal solution, respectively.

Step 2. Randomly choose a candidate solution x_{new} from the neighborhood of x_{old}, and compute its value f_{new}.

Step 3. Set the acceptance probability

$$P = \min\left\{1, \exp\left(-\frac{f_{new} - f_{old}}{T_k}\right)\right\}.$$

Step 4. Accept the new solution with probability P; if accepted, set $x_{old} = x_{new}$ and $f_{old} = f_{new}$; if necessary, update $\hat{f}$ and $\hat{x}$.

Step 5. If some termination condition is met, stop; otherwise, set $k = k + 1$, set the new temperature according to the cooling schedule, and go to step 2.

The probabilistic acceptance is easy to implement. P is evaluated according to the Boltzmann distribution; then a pseudorandom number r, uniformly distributed between 0 and 1, is computed and the move is accepted if $r \leq P$ (pseudorandom number generation is dealt with in section 4.2).

The termination condition could be related to a maximum iteration number, to a minimum temperature, or to a maximum number of steps in which the current solution remains unchanged. Note that we do not explore the entire neighborhood of the current solution; the method is of the first-improving type. If a candidate solution is rejected, we select another candidate in the neighborhood of the current solution. In principle, it is possible to visit the same solution twice; if the neighborhood structure is rich enough, this is unlikely. It is necessary to save the best solution found, since the freezing point (the last current solution) need not be the best solution visited.

An implementation of the annealing algorithm is therefore characterized by the solution space, the neighborhood structure, the rule by which the neighborhood is explored, and the cooling schedule. It can be shown that under some conditions, the method asymptotically converges (in a probabilistic sense) to the global optimum. The convergence property is a reassuring one, but it is usually considered of little practical value, since its conditions would require impractical running times. However, the experience suggests that in many practical settings, very good solutions (often optimal) are actually found. The running time of the algorithm to obtain high-quality solutions, however, is problem dependent.

Tabu search   Like simulated annealing, tabu search is a neighborhood search-based metaheuristic aimed at escaping local minima. Unlike simulated annealing, tabu search tries to keep the search biased toward good solutions.

The basic idea of tabu search is that the best solution in the neighborhood $\mathcal{N}$ of the current solution should be chosen as the new current solution, even if this implies increasing the cost. If we are in a local minimum, this means accepting a nonimproving perturbation. The problem with this basic idea is that the possibility of cycling arises. If we try to escape from a local minimum by choosing the best solution in its neighborhood, it might well be the case that at the next iteration, we fall back into the local minimum, since this could be the best solution in the new neighborhood.

To prevent cycling, we must prevent revisiting solutions. One way would be to keep a record of the already visited solutions; however, this would be both memory- and time-consuming, since checking a candidate solution against the list of visited ones would require a substantial effort. A better idea could be to record only the most recent solutions. A practical alternative is to keep in memory only some *attributes* of the solutions or of the applied perturbations; such attributes are called *tabu*. For instance, the reverse of the selected perturbation at each step can be marked as tabu, restricting the neighborhood to be considered. Consider a pure integer program involving only binary variables; if we complement variable x_i, in the next few iteration we might forbid any perturbation complementing this variable again. As an alternative, a tabu attribute of a solution could be the value of the objective function. In practice, it is necessary to keep only a record of the most recent tabu attributes to avoid cycling; the data structure implementing this function is the *tabu list*.

The basic tabu navigation algorithm can be described as follows:

Step 1. Choose an initial current solution x_{cur}, a tabu list size; let $k = 1$, $\hat{f} = f(x_{\text{cur}})$, $\hat{x} = x_{\text{cur}}$.

Step 2. Evaluate the neighborhood $\mathcal{N}(x_{\text{cur}})$; update the current solution with the best nontabu solution in the neighborhood; if necessary, update the current optimal solution $\hat{x}$ and the current optimal value $\hat{f}$.

Step 3. Add some attribute of the new solution or of the applied perturbation to the tabu list.

Step 4. If the maximum iteration number has been reached, stop; otherwise, set $k = k + 1$, and go to step 2.

Note that unlike simulated annealing, this version of tabu search explores the *entire* neighborhood of the current solution; basic tabu search is a strategy of the *best-improving* rather than first-improving type. However, it is possible to restrict the neighborhood to reduce the computational burden.

There are several issues and refinements to consider in order to implement an effective and efficient algorithm. They are rather problem specific; this shows that although local search metaheuristics are general-purpose, a certain degree of "customization" is necessary.

Genetic algorithms Unlike simulated annealing and tabu search, genetic algorithms work on a set of solutions rather than a single point. In this sense they are similar to the simplex search method of section 3.2.4. The idea is based on the survival-of-the-fittest mechanism of biological evolution. Each solution is represented by a string of numbers or symbols; strings are subject to random evolution mechanisms which change the current population. One evolution mechanism is mutation; an attribute of a string is randomly selected and modified using a neighborhood structure. Mutation is very similar to the usual local search mechanism, but there is another mechanism which is peculiar to genetic algorithms: crossover. In the crossover mechanism, two elements of the current set of solutions are selected and merged in some way. Given two strings, we select a "breakpoint" position k and merge the strings as follows:

$$\left\{ \begin{array}{c} x_1, x_2, \ldots, x_k, x_{k+1}, \ldots, x_n \\ y_1, y_2, \ldots, y_k, y_{k+1}, \ldots, y_n \end{array} \right\} \Rightarrow \left\{ \begin{array}{c} x_1, x_2, \ldots, x_k, y_{k+1}, \ldots, y_n \\ y_1, y_2, \ldots, y_k, x_{k+1}, \ldots, x_n \end{array} \right\}.$$

Different variations are possible; for instance, a double crossover may be exploited, in which two breakpoints are selected for the crossover.

The set of solutions is updated at each iteration, selecting the "best" individuals for mutation and crossover and/or letting only the best individuals survive. Rather than selecting the best individuals deterministically, based on the value of the objective function, random selection mechanisms are employed to avoid freezing the population to a locally optimal solution. Genetic algorithms may be integrated with local search strategies; one idea is to use genetic mechanisms to find a set of initial points from which a local improvement search is carried out.

The idea of genetic algorithms certainly has a good potential for solving quite complex problems; the evident downside is that considerable experimentation may be needed to come up with the best strategy and the best setting of numerical parameters regulating the evolution mechanisms.

3.7 L-SHAPED METHOD FOR TWO-STAGE LINEAR STOCHASTIC PROGRAMMING

In principle, stochastic programs with a discrete set of scenarios can be tackled by ordinary LP techniques such as the simplex algorithm or the active set method. However, their large-scale nature, and the numerical difficulties of their solution even when thy are not so large, often make this approach not practical. Interior point methods are worth trying, but alternative methods may be developed by exploiting the peculiar structure of a stochastic program. Consider a two-stage problem with a fixed recourse matrix $\mathbf{W}$:

$$\min \quad \mathbf{c}^T\mathbf{x} + \sum_{s\in S} p_s\mathbf{q}_s^T\mathbf{y}^s$$
$$\text{s.t.} \quad \mathbf{A}\mathbf{x} = \mathbf{b}$$
$$\mathbf{W}\mathbf{y}_s + \mathbf{T}_s\mathbf{x} = \mathbf{r}_s \quad \forall s \in S$$
$$\mathbf{x}, \mathbf{y}_s \geq \mathbf{0},$$

where p_s is the probability of scenario s. It may be seen that the problem lends itself to a decomposition approach: in fact, once the first-stage decisions $\mathbf{x}$ are fixed, the problem is decomposed into a set of small subproblems, one for each scenario s. This point may be appreciated by looking at the sparse structure of the overall technological matrix for this problem:

$$\begin{bmatrix} \mathbf{A} & \mathbf{0} & \mathbf{0} & \cdots & \mathbf{0} \\ \mathbf{T}_1 & \mathbf{W} & \mathbf{0} & \cdots & \mathbf{0} \\ \mathbf{T}_2 & \mathbf{0} & \mathbf{W} & \cdots & \mathbf{0} \\ \vdots & \vdots & \vdots & \ddots & \vdots \\ \mathbf{T}_S & \mathbf{0} & \mathbf{0} & \cdots & \mathbf{W} \end{bmatrix}.$$

This matrix is almost block-diagonal. The recourse function is

$$H(\mathbf{x}) = \sum_{s\in S} p_s h_s(\mathbf{x}),$$

where

$$h_s(\mathbf{x}) \equiv \quad \min \quad \mathbf{q}_s^T\mathbf{y}^s \qquad (3.29)$$
$$\text{s.t.} \quad \mathbf{W}\mathbf{y}_s = \mathbf{r}_s - \mathbf{T}_s\mathbf{x}$$
$$\mathbf{y}_s \geq \mathbf{0}.$$

Evaluating the recourse function for a given first-stage decision $\hat{\mathbf{x}}$ entails solving a set of independent LP problems. For simplicity, we assume here that all these problems are solvable, i.e., $h_s(\hat{\mathbf{x}}) < +\infty$ for any scenario s, for any $\hat{\mathbf{x}}$ that is feasible with respect to the first-stage constraints. We say in this case that the problem has relatively complete recourse. This may be a reasonable assumption in financial problems. Consider, for instance, an asset and liability problem; you want to manage a portfolio of assets so as to meet a stream of future, possibly stochastic liabilities. If we include extreme and pessimistic financial scenarios in our model, it might be the case that some liabilities are not met; in such a case, we may relax the constraints by suitable penalties (see section 6.2). These penalties make the recourse complete. If the recourse is not complete, the approach we describe here may easily be extended.

It can be shown that the recourse function $H(\mathbf{x})$ is convex; hence we may consider the application of Kelley's cutting plane algorithm, illustrated in section 3.3.4. To this end, let us rewrite the two-stage problem as

$$
\begin{aligned}
\min \quad & \mathbf{c}^T\mathbf{x} + \theta \\
\text{s.t.} \quad & \mathbf{Ax} = \mathbf{b} \\
& \theta \geq H(\mathbf{x}) \\
& \mathbf{x} \geq \mathbf{0}.
\end{aligned}
\tag{3.30}
$$

We may relax the constraint (3.30), obtaining a relaxed master problem, and then add cutting planes of the form

$$
\theta \geq \boldsymbol{\alpha}^T\mathbf{x} + \beta.
$$

The coefficients of each cut are obtained by solving the scenario subproblems for given first-stage decisions. To see how, let $\hat{\mathbf{x}}$ be the optimal solution of the initial master problem. Consider the dual of problem (3.29):

$$
\begin{aligned}
h_s(\hat{\mathbf{x}}) \equiv \quad \max \quad & (\mathbf{r}_s - \mathbf{T}_s\hat{\mathbf{x}})^T \boldsymbol{\pi}_s \\
\text{s.t.} \quad & \mathbf{W}^T\boldsymbol{\pi}_s \leq \mathbf{q}_s.
\end{aligned}
$$

Given an optimal dual solution $\hat{\boldsymbol{\pi}}_s$, it is easy to see that the following relationships hold:

$$
\begin{aligned}
h_s(\hat{\mathbf{x}}) \quad &= \quad (\mathbf{r}_s - \mathbf{T}_s\hat{\mathbf{x}})^T \hat{\boldsymbol{\pi}}_s \\
h_s(\mathbf{x}) \quad &\geq \quad (\mathbf{r}_s - \mathbf{T}_s\mathbf{x})^T \hat{\boldsymbol{\pi}}_s \qquad \forall \mathbf{x}.
\end{aligned}
$$

The inequality derives from the fact that $\hat{\boldsymbol{\pi}}_s$ is the optimal dual solution for $\hat{\mathbf{x}}$ but not for a generic $\mathbf{x}$. Summing over the scenarios, we get the cutting plane:

$$
\theta \geq \sum_{s \in S} p_s (\mathbf{r}_s - \mathbf{T}_s\mathbf{x})^T \hat{\boldsymbol{\pi}}_s.
$$

The L-shaped decomposition algorithm is obtained by iterating the solution of the relaxed master problem, which yields $\hat{\theta}$ and $\hat{\mathbf{x}}$, and of the corresponding scenario subproblem. At each iteration, cuts are added to the master problem. The algorithm stops when the optimal solution of the master problem satisfies

$$
\hat{\theta} \leq H(\hat{\mathbf{x}}).
$$

This condition may be relaxed if a near-optimal solution is enough for our purposes.

If the recourse is not complete, some of scenario subproblems may be infeasible for certain first-stage decisions. In this case we may again exploit the dual of the scenario subproblem. Note that the feasibility region of this dual does not depend on the first-stage decisions. Thus, if a dual problem is infeasible, it means that the second-stage problem for the corresponding scenario will be infeasible for any first-stage decision. Ruling out this case, which is likely to be due to a modeling error, when the primal problem is infeasible, the dual will be unbounded. Hence, there is an extreme ray of the dual feasible set along which the optimal solution goes to infinity. In this case we may easily add an infeasibility cut to the master, cutting the first-stage decisions which lead to an infeasible second-stage problem. Thus, at any iteration, we discover either an extreme point or an extreme ray of the dual feasible sets of each second-stage subproblem. The finite convergence of the method derives from the fact that any polyhedron has a finite number of extreme points and extreme rays (see supplement S3.1.2).

S3.1 ELEMENTS OF CONVEX ANALYSIS

Convexity is arguably the most important concept in optimization theory. In the next two sections we want first to recall the related concepts of convex set and convex function, and then to outline a few concepts in polyhedral theory which are important for linear and mixed-integer programming.

S3.1.1 Convexity in optimization

Convexity is a possible attribute of the feasible set S of an optimization problem.

Definition. A set $S \subseteq \mathbb{R}^n$ is a *convex set* if

$$\mathbf{x}, \mathbf{y} \in S \Rightarrow \lambda \mathbf{x} + (1 - \lambda)\mathbf{y} \in S \quad \forall \lambda \in [0, 1].$$

Example 3.17 The concept of convexity can be grasped intuitively by considering that the points of the form $\lambda \mathbf{x} + (1 - \lambda)\mathbf{y}$, where $0 \leq \lambda \leq 1$, are simply the points on the straight line joining $\mathbf{x}$ and $\mathbf{y}$. A set S is convex if the line joining any pair of points $\mathbf{x}, \mathbf{y} \in S$ is contained in S. Consider the three subsets of $\mathbb{R}^2$ depicted in Figure 3.18. S_1 is convex, but S_2 is not. S_3 is a discrete set and it is not convex; this fact has important consequences for discrete optimization problems.   ⬜

The following property is easy to verify.

PROPERTY 3.11 *The intersection of convex sets is a convex set.*

Note that the union of convex sets need not be convex. The *convex combination* of p points $\mathbf{x}_1, \mathbf{x}_2, \ldots, \mathbf{x}_p \in \mathbb{R}^n$ is defined as

$$\mathbf{x} = \sum_{i=1}^{p} \mu_i \mathbf{x}_i \qquad \mu_1, \ldots, \mu_p \geq 0, \quad \sum_{i=1}^{p} \mu_i = 1.$$

Given a set $S \subset \mathbb{R}^n$, the set of points which are the convex combinations of points in S is the *convex hull* of S (denoted by $[S]$). If S is a convex set,

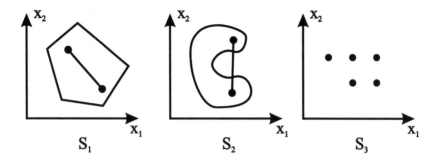

Fig. 3.18   Convex and nonconvex sets.

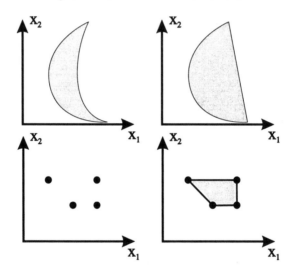

Fig. 3.19   Nonconvex sets and their convex hulls.

then $S \equiv [S]$. The convex hull of a generic set S is the smallest convex set containing S; it can also be regarded as the intersection of all the convex sets containing S. Two nonconvex sets and their convex hulls are shown in figure 3.19.

Definition. A scalar function $f: \mathbb{R}^n \to \mathbb{R}$ defined over a convex set $S \subseteq \mathbb{R}^n$ is a *convex function* if

$$\mathbf{x}, \mathbf{y} \in S, \ \lambda \in [0, 1] \Rightarrow f(\lambda \mathbf{x} + (1 - \lambda)\mathbf{y}) \leq \lambda f(\mathbf{x}) + (1 - \lambda)f(\mathbf{y}).$$

If this condition is met with strict inequality for all $\mathbf{x} \neq \mathbf{y}$, the function is *strictly* convex.

Definition. A function f is *concave* if $(-f)$ is convex.

The concept of convex function is illustrated in figure 3.20. The first func-

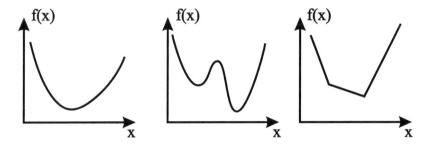

Fig. 3.20   Convex and nonconvex functions.

tion is convex, whereas the second is not. Also, the third function is convex; a convex function need not be differentiable everywhere. The definition can be interpreted as follows. Given any two points $\mathbf{x}$ and $\mathbf{y}$, consider another point which is a convex combination of $\mathbf{x}$ and $\mathbf{y}$; then the function value in this point is overestimated by the convex combination of the function values $f(\mathbf{x})$ and $f(\mathbf{y})$, since the line segment joining $(\mathbf{x}, f(\mathbf{x}))$ and $(\mathbf{y}, f(\mathbf{y}))$ lies above the graph of the function between $\mathbf{x}$ and $\mathbf{y}$. In other words, a function is convex if its epigraph is a convex set. A further link between convex sets and convex functions is that the set $S = \{\mathbf{x} \in \mathbb{R}^n \mid g(\mathbf{x}) \leq 0\}$ is convex if g is a convex function. Convexity of functions is preserved by some operations; in particular, a linear combination of convex functions f_i,

$$f(\mathbf{x}) = \sum_{i=1}^{n} \lambda_i f_i(\mathbf{x}),$$

is a convex function if $\lambda_i \geq 0$, for any i.

There are alternative characterizations of a convex function. For our purposes the most important is the following. If f is a differentiable function, it is convex (over S) if and only if

$$f(\mathbf{x}) \geq f(\mathbf{x}_0) + \nabla f^T(\mathbf{x}_0)(\mathbf{x} - \mathbf{x}_0) \qquad \forall \mathbf{x}, \mathbf{x}_0 \in S. \tag{3.31}$$

Note that the hyperplane

$$z = f(\mathbf{x}_0) + \nabla f^T(\mathbf{x}_0)(\mathbf{x} - \mathbf{x}_0)$$

is the usual tangent hyperplane, i.e., the first-order Taylor expansion of f at $\mathbf{x}_0$. For a differentiable function, convexity implies that the first-order approximation at a certain point $\mathbf{x}_0$ consistently underestimates the true value of the function at all the other points $\mathbf{x} \in S$. The concept of a tangent hyperplane applies only to differentiable convex functions, but it can be generalized by the concept of a support hyperplane.

Definition. Given a convex function f and a point $\mathbf{x}^0$, the hyperplane (in $\mathbb{R}^{n+1}$) given by $z = f(\mathbf{x}^0) + \boldsymbol{\gamma}^T(\mathbf{x} - \mathbf{x}^0)$, which meets the epigraph of f in $(\mathbf{x}^0, f(\mathbf{x}^0))$ and lies below it is called the *support hyperplane* of f at $\mathbf{x}^0$.

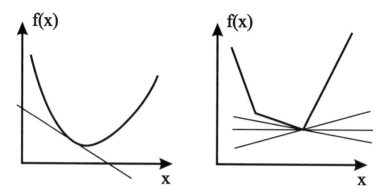

Fig. 3.21 Illustration of the support hyperplane

The concept of a support hyperplane is depicted in figure 3.21. A support hyperplane at $\mathbf{x}_0$ is essentially defined by a vector $\boldsymbol{\gamma}$ such that

$$f(\mathbf{x}) \geq f(\mathbf{x}_0) + \boldsymbol{\gamma}^T(\mathbf{x} - \mathbf{x}_0) \qquad \forall \mathbf{x} \in S. \tag{3.32}$$

The vector $\boldsymbol{\gamma}$ in inequality (3.32) plays the same role as the gradient does in inequality (3.31). If f is differentiable in $\mathbf{x}_0$, the support hyperplane is the usual tangent hyperplane and $\boldsymbol{\gamma} = \nabla f(\mathbf{x}_0)$. This is why a vector $\boldsymbol{\gamma}$ such that inequality (3.32) holds is called a *subgradient* of f at $\mathbf{x}_0$. If f is nondifferentiable, the support hyperplane need not be unique and there is a set of subgradients. The set of subgradients is called the *subdifferential* of f at $\mathbf{x}_0$ and is denoted by $\partial f(\mathbf{x}_0)$. It can be shown that a convex function on a set S is subdifferentiable on the interior of S, i.e., we can always find a subgradient (on the boundary of the set S some difficulties may occur, but we need not be concerned with this technicality in the following).

A further characterization of convex functions can be given for twice-differentiable functions.

THEOREM 3.12 *If f is a twice-differentiable function, defined on a non-empty and open convex set S, then f is convex if and only if its Hessian matrix is positive semidefinite at any point in S.*

We recall that the Hessian matrix $\mathbf{H}(\mathbf{x})$ is the (symmetric) matrix of second-order derivatives of $f(\mathbf{x})$:

$$\mathbf{H}_{ij} = \frac{\partial^2 f}{\partial x_i \, \partial x_j}.$$

We also recall that a symmetric (hence square) matrix $\mathbf{A}(\mathbf{x})$ is positive semi-definite on S if

$$\mathbf{x}^T \mathbf{A}(\mathbf{x})\mathbf{x} \geq 0 \qquad \forall \mathbf{x} \in S.$$

The matrix is positive definite if the inequality above is strict for all $\mathbf{x} \neq \mathbf{0}$. If the Hessian matrix is positive definite, the function is strictly convex; however, the converse is not necessarily true. The definiteness of a matrix may be investigated by checking the sign of its eigenvalues; the matrix is positive semidefinite if all of its eigenvalues are nonnegative, and it is positive definite if all of its eigenvalues are positive.

S3.1.2   Convex polyhedra and polytopes

Consider the *hyperplane* (in $\mathbb{R}^n$) $\mathbf{a}_i^T \mathbf{x} = b_i$, where $b_i \in \mathbb{R}$ and $\mathbf{a}_i, \mathbf{x} \in \mathbb{R}^n$ are column vectors.[8] A hyperplane divides $\mathbb{R}^n$ into two *half-spaces* expressed by the linear inequalities $\mathbf{a}_i^T \mathbf{x} \leq b_i$ and $\mathbf{a}_i^T \mathbf{x} \geq b_i$.

Definition. A *polyhedron* $P \subseteq \mathbb{R}^n$ is a set of points satisfying a finite collection of linear inequalities, i.e.,

$$P = \{\mathbf{x} \in \mathbb{R}^n \mid \mathbf{A}\mathbf{x} \geq \mathbf{b}\}.$$

A polyhedron is therefore the intersection of a finite collection of half-spaces.

PROPERTY 3.13 *A polyhedron is a convex set (it is the intersection of convex sets).*

Definition. A polyhedron is *bounded* if there exists a positive number M such that

$$P \subseteq \{\mathbf{x} \in \mathbb{R}^n \mid -M \leq x_j \leq M \ \ j = 1, \ldots, n\}.$$

A bounded polyhedron is called a *polytope*. A polytope and an unbounded polyhedron are shown in figure 3.22.

Definition. A point $\mathbf{x}$ is an *extreme point* of a polyhedron P if $\mathbf{x} \in P$ and it is not possible to express $\mathbf{x}$ as $\mathbf{x} = \frac{1}{2}\mathbf{x}' + \frac{1}{2}\mathbf{x}''$ with $\mathbf{x}', \mathbf{x}'' \in P$ and $\mathbf{x}' \neq \mathbf{x}''$.

A polytope P has a finite number of extreme points $\mathbf{x}^1, \ldots, \mathbf{x}^{\mathcal{J}}$. Any point $\mathbf{x}$ in a polytope P can be expressed as a convex combination of its extreme points:

$$\mathbf{x} = \sum_{j=1}^{\mathcal{J}} \lambda_j \mathbf{x}^j \qquad \sum_{j=1}^{\mathcal{J}} \lambda_j = 1, \ \lambda_j \geq 0;$$

in other words, a polytope is the convex hull of its extreme points. In the case of an unbounded polyhedron, this is not true and we must introduce another concept.

Definition. A vector $\mathbf{r} \in \mathbb{R}^n$ is called a *ray* of the polyhedron

$$P = \{\mathbf{x} \in \mathbb{R}^n \mid \mathbf{A}\mathbf{x} \geq \mathbf{b}\}$$

---

[8] Unless the contrary is stated, we assume that all vectors are columns.

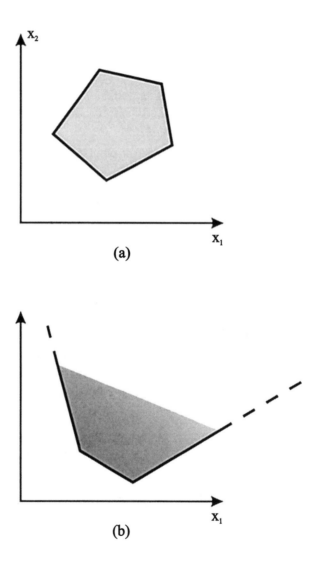

Fig. 3.22   Two-dimensional polytope (a) and unbounded polyhedron (b).

if $\mathbf{Ar} \geq 0$.

If $\mathbf{x}_0$ is a point in a polyhedron P and $\mathbf{r}$ is a ray of P, then

$$\mathbf{y} = \mathbf{x}_0 + \lambda \mathbf{r} \in P \qquad \forall \lambda \geq 0.$$

Clearly, only unbounded polyhedra have rays.

Definition. A ray $\mathbf{r}$ of a polyhedron P is called an *extreme ray* if it cannot be expressed as $\mathbf{r} = \frac{1}{2}\mathbf{r}_1 + \frac{1}{2}\mathbf{r}_2$ where $\mathbf{r}_1, \mathbf{r}_2$ are rays of P such that $\mathbf{r}_1 \neq \lambda \mathbf{r}_2$ for any number $\lambda > 0$.

A polyhedron P can be described in terms of its extreme rays and points, in the sense that any point $\mathbf{x} \in P$ can be expressed combining extreme rays and points:

$$\mathbf{x} = \sum_{j=1}^{\mathcal{J}} \lambda_j \mathbf{x}^j + \sum_{k=1}^{\mathcal{K}} \mu_k \mathbf{r}^k \qquad \sum_{j=1}^{\mathcal{J}} \lambda_j = 1, \ \lambda_j, \mu_k \geq 0.$$

For further reading

In the literature

- A general and introductory book on optimization theory is [19].

- See, e.g., [2] for nonlinear programming and [20] for linear programming.

- Interior point methods are dealt with in [22].

- If you are interested in the theory behind convex optimization, you should check [8] or [18]. If you are more interested in the numerical aspects of optimization, [5] is for you.

- As to more specific references, good sources are [4] and [10] for stochastic programming, [13] and [17] for integer programming, and [9] for global optimization.

- The recent book [21] is also a good source of information about commercial branch and bound codes for MILP problems.

- We have not dealt with dynamic programming. A comprehensive reference is [3]; [14] includes optimal control models in finance; [11] is a more recent reference.

- Good general sources on tabu search and genetic algorithms are [7] and [16], respecively.

- The use of metaheuristics for continuous global optimization is proposed, among others, in [1] and [15].

On the Web

- A good source for information on the practical application of optimization models and methods to a variety of problems is

  `http://e-OPTIMIZATION.COM`.

- Relevant academic societies in the field are:

  - `http://www.informs.org` (INFORMS: Institute for Operations Research and the Management Sciences)

  - `http://www.siam.org` (SIAM: Society for Industrial and Applied Mathematics)

  - `http://www.caam.rice.edu/~mathprog` (MPS: Mathematical Programming Society)

- A good pointer for interior point methods is

  `http://www-unix.mcs.anl.gov/otc/InteriorPoint`.

- Michael Trick's Web page lists several useful links to journals, societies, people, etc.; see `http://mat.gsia.cmu.edu`.

REFERENCES

1. R. Battiti and G. Tecchiolli. The Continuous Reactive Tabu Search: Blending Combinatorial Optimization and Stochastic Search for Global Optimization. *Annals of Operations Research*, 63:153–188, 1996.

2. M.S. Bazaraa, H.D. Sherali, and C.M. Shetty. *Nonlinear Programming. Theory and Algorithms (2nd ed.)*. Wiley, Chichester, West Sussex, England, 1993.

3. D. Bertsekas. *Dynamic Programming and Optimal Control (vols. 1 and 2)*. Athena Scientific, Belmont, MA, 1995.

4. J.R. Birge and F. Louveaux. *Introduction to Stochastic Programming*. Springer-Verlag, New York, 1997.

5. R. Fletcher. *Practical Methods of Optimization (2nd ed.)*. Wiley, Chichester, West Sussex, England, 1987.

6. F. Glover, J.M. Mulvey, and K. Hoyland. Solving Dynamic Stochastic Control Problems in Finance Using Tabu Search with Variable Scaling. In I.H. Osman and J.P. Kelly, editors, *Meta-Heuristics: Theory and Applications*, pages 429–448. Kluwer Academic, Dordrecht, The Netherlands, 1996.

7. F.W. Glover and M. Laguna. *Tabu Search.* Kluwer Academic, Dordrecht, The Netherlands, 1998.

8. J.-B. Hiriart-Urruty and Claude Lemaréchal. *Convex Analysis and Minimization Algorithms (vols. 1 and 2).* Springer-Verlag, Berlin, 1993.

9. R. Horst, P.M. Pardalos, and N.V. Thoai. *Introduction to Global Optimization.* Kluwer Academic, Dordrecht, The Netherlands, 1995.

10. P. Kall and S.W. Wallace. *Stochastic Programming.* Wiley, Chichester, West Sussex, England, 1994.

11. R. Korn. *Optimal Portfolios: Stochastic Models for Optimal Investment and Risk Management in Continuous Time.* World Scientific Publishing, Singapore, 1997.

12. C.D. Maranas and C.A. Floudas. Global Minimum Potential Energy Conformations of Small Molecules. *Journal of Global Optimization,* 4:135–170, 1994.

13. R.K. Martin. *Large Scale Linear and Integer Optimization: A Unified Approach.* Kluwer Academic, Dordrecht, The Netherlands, 1999.

14. R.C. Merton. *Continuous-Time Finance.* Blackwell Publishers, Malden, MA, 1990.

15. Z. Michalewicz. Evolutionary Computation Techniques for Nonlinear Programming Problems. *International Transactions of Operations Research,* 1:223–140, 1994.

16. Z. Michalewicz. *Genetic Algorithms + Data Structures = Evolution Programs.* Springer-Verlag, Berlin, 1996.

17. G.L. Nemhauser and L.A. Wolsey. *Integer Programming and Combinatorial Optimization.* Wiley, Chichester, West Sussex, England, 1998.

18. R.T. Rockafellar. *Convex Analysis.* Princeton University Press, Princeton, NJ, 1970.

19. R.K. Sundaram. *A First Course in Optimization Theory.* Cambdridge University Press, Cambridge, 1996.

20. R.J. Vanderbei. *Linear Programming: Foundations and Extensions.* Kluwer Academic, Dordrecht, The Netherlands, 1996.

21. L.A. Wolsey. *Integer Programming.* Wiley, New York, 1998.

22. S.J. Wright. *Primal-Dual Interior-Point Methods.* Society for Industrial and Applied Mathematics, Philadelphia, 1997.

4

Principles of Monte Carlo simulation

Simulation is widely used to solve problems that are intractable from an analytical point of view. This may be due to the inherent complexity of a problem, to the presence of uncertainty, or both.

Basically, Monte Carlo simulation is based on statistical sampling, and it may be visualized as a black box in which a stream of psuedorandom numbers enter; an estimate of a quantity of interest is obtained by analyzing the output. Typically, we want to estimate an expected value with respect to an underlying probability distribution; for instance, an option price may be evaluated by computing the expected value of the payoff with respect to a risk-neutral probability measure. In other cases we want to evaluate a portfolio policy by simulating a suitably large number of scenarios. We may be interested not only in average values, but also in what happens on the tail of probability distributions, as in the case of value-at-risk; however, even this case boils down to computing expected values, possibly conditioned ones. In all these examples, the underlying issue is actually computing an integral in a possibly high-dimensional space. This is why we start this chapter by considering Monte Carlo integration in section 4.1. Monte Carlo simulation is based on random number generation; actually, we must speak of pseudorandom numbers, since nothing is random on a computer. How this is accomplished is described in section 4.2.

If we fed random numbers into a simulation procedure, the output will be a sequence of random numbers. Given this output, we use statistical techniques to build an estimate of a quantity of interest. We would like to evaluate the reliability of this estimate in some way, e.g., by a confidence interval, or the other way around, we would like to carry out the simulation experiments in such a way that the estimation error is controlled. Section 4.3 deals with the issue of setting the number of simulation experiments (replications) properly. Intuitively, the more replications we run, the more reliable our estimates will be. Unfortunately, reaching a suitable precision might require a prohibitive number of experiments. Improving the quality of the estimates without incurring huge CPU times calls for proper variance reduction techniques, which are the subject of section 4.4.

Using pseudorandom numbers on a computer and then applying statistical techniques may raise some philosophical issues; after all, the sequences of numbers we use are deterministic. It can be argued that the success of Monte Carlo simulation simply shows that there are some deterministic sequences that work well and that there could be others that work even better. Pursuing this idea leads to quasi-Monte Carlo simulation, which is dealt with in section 4.5.

A final consideration is that simulation may be used to evaluate the consequences of a certain policy, but it cannot generate the policy itself. To this end, we should use the optimization methods described in chapter 3. Unfortunately, most of those techniques require an analytical model that may be too complex or not available at all, which is the very reason we so often resort to simulation. Possible ways to couple simulation and optimization techniques are described in section 4.6.

4.1   MONTE CARLO INTEGRATION

In section 2.5 we have recalled the basic numerical approaches to computing integrals. While in principle those approaches can be extended to multi-dimensional spaces, when we integrate in several dimensions the use of quadrature formulas becomes computationally quite expensive. A possible alternative is resorting to random sampling-based Monte Carlo integration.

Consider the problem of computing a multidimensional integral of the form

$$I = \int_{\mathcal{A}} \phi(\mathbf{x})\, d\mathbf{x} \qquad (4.1)$$

where $\mathcal{A} \subset \mathbb{R}^n$. We may estimate I by randomly sampling a sequence of points $\mathbf{x}^i \in \mathcal{A}$, $i = 1, \ldots, m$, and building the estimator

$$\hat{I}_m = \frac{\mathrm{vol}(\mathcal{A})}{m} \sum_{i=1}^{m} \phi(\mathbf{x}^i), \qquad (4.2)$$

where $\mathrm{vol}(\mathcal{A})$ denotes the volume of the region $\mathcal{A}$. Quite often we consider the unit hyperrectangle, i.e.,

$$\mathcal{A} = [0, 1] \times [0, 1] \times \cdots \times [0, 1],$$

hence $\mathrm{vol}(\mathcal{A}) = 1$; in the following we assume that we integrate over such a region. We may interpret equation (4.2) as a way to estimate the average value of the function ϕ, which is then multiplied by the volume of the region to yield the value of the integral. The strong law of large numbers implies that with probability 1

$$\lim_{m \to \infty} \hat{I}_m = I.$$

We see that a random sampling mechanism may be used to build an estimate of a deterministic quantity. If we have a vector random variable

$$\mathbf{X} = \begin{bmatrix} X_1 \\ X_2 \\ \vdots \\ X_n \end{bmatrix}$$

with joint density function $f(x_1, \ldots, x_n)$, we may use Monte Carlo integration to estimate the expected value of an arbitrary function of $\mathbf{X}$:

$$E[g(\mathbf{X})] = \int \int \cdots \int g(x_1, \ldots, x_n) f(x_1, \ldots, x_n) \, dx_1 \cdots dx_n.$$

But how is random sampling accomplished? Consider the simple case

$$I = \int_0^1 g(x) \, dx.$$

We may think of this integral as the expected value $E[g(U)]$, where U is a uniform random variable on the interval $(0, 1)$, i.e., $U \sim (0, 1)$. So what we have to do is to generate a sequence $\{U_i\}$ of *independent* random numbers from the uniform distribution and then evaluate the sample mean:

$$\frac{1}{m} \sum_{i=1}^{m} g(U_i).$$

Example 4.1 Consider the trivial case

$$I = \int_0^1 e^x \, dx = e - 1 \approx 1.7183.$$

To generate uniformly distributed random numbers, we may use the MATLAB rand function; a call like rand(m,n) yields a $m \times n$ matrix of uniform random numbers.

```
>> rand('seed',0)
>> mean(exp(rand(1,10)))
ans =
     1.6318
>> mean(exp(rand(1,100)))
ans =
```

```
     1.7744
>> mean(exp(rand(1,1000)))
ans =
     1.7051
>> mean(exp(rand(1,10000)))
ans =
     1.7195
```

Apparently, the estimate tends to the correct value as the number of samples increases. Actually, we should be very careful in qualifying our estimate with some measure of its reliability, such as a confidence interval. Furthermore, we should understand what "random" means on a computer; in particular, we should understand the role of the command rand('seed',0) in generating pseudorandom numbers.   ⬛

4.2 GENERATING PSEUDORANDOM VARIATES

The usual way to generate pseudorandom variates, i.e., samples from a given probability distribution, starts from the generation of pseudorandom numbers, which are simply variates from the uniform distribution on the interval (0,1). Then suitable transformations are applied in order to obtain the desired distribution. We discuss briefly the most common transformations: the inverse transform method, the acceptance-rejection approach, and ad hoc strategies such as those used to generate standard normal variates. The MATLAB Statistics toolbox provides the user with a rich library of random generators, so the user need not herself program the procedures we describe in the following. Nevertheless, we believe it is important to have at least a grasp of what is done, in order to properly apply variance reduction procedures to improve the estimates.

4.2.1 Generating pseudorandom numbers

To generate $U(0,1)$ variables, the standard method is based on linear congruential generators (LCGs). A LCG generates a sequence of nonnegative integer numbers Z_i as follows; given an integer number Z_{i-1}, we generate the next number in the sequence by computing

$$Z_i = (aZ_{i-1} + c)(\text{mod } m),$$

where a, c, and m are properly chosen parameters and mod denotes the remainder of integer division(e.g., $15 \text{mod} 6 = 3$). Then, to generate a $U(0,1)$ variable, we return the number (Z_i/m). It is clear that there is nothing random in this sequence. To begin with, it must start from an initial number Z_0; this is called the seed of the sequence. Starting the sequence from the same *seed* will always yield the same sequence. Indeed, any time you start MAT-LAB and type rand, you get the same number; if you keep typing rand, you see a sequence of numbers that look to be random and uniformly distributed. However, this sequence is always the same, since starting MATLAB sets the seed to a precise value. This may seem rather dull, and using a command like

```
rand('seed',sum(100*clock)),
```

which sets the seed of the random generator to a number depending on the current clock value, may seem a brilliant idea. In practice this is not a good idea at all; on the one hand, it makes debugging difficult; on the other the variance reduction techniques we describe in the following call for the ability to control the seeds. The command rand('seed',0) we used in example 4.1 may be used to reset the seed to a specific value, in order to reinitialize the random sequence.

A few remarks are in order. A first observation is that with a LCG we actually generate rational numbers rather than real ones; this is not a serious problem, provided that m is large enough. But there is another reason to choose a large value for m; the generator is periodic. In fact, we may generate at most m distinct integer numbers Z_i, and whenever we repeat a number previously generated, the sequence repeats itself (which is not very random at all). Since the maximum possible period is m, we should make it very large in order to have a large period. The proper choice of a and c ensures that the period is maximized and that the sequence looks random. A sequence like

$$U_i = \frac{i}{m} \qquad i = 0, 1, \ldots, m,$$

has a maximum period and is, in some sense, uniformly distributed on the interval (0,1), but it is far from satisfactory. The point is that the samples should also look independent; to be more precise, they should be able to trick statistical testing procedures into "believing" that they are a sequence of independent samples from the uniform distribution. This is why designing a good random number generator is not easy; luckily, when you purchase good numerical software, someone has already solved the issue for you.

We close this section with a couple of remarks concerning MATLAB. By issuing a command like rand('seed',0), we tell MATLAB to use its older random number generator. Actually, there is a more recent one, which uses a "state" consisting of a vector of 35 numbers; this state is controlled by commands like rand('state',0). This new generator has been introduced as an improvement over the existing one, and it allows a much larger period, among other things. For our didactic purposes, using the older LCG may be enough, and we refer the reader to [12] for more information. Another important point is that when generating normal variates, MATLAB uses the randn function; this function generates standard normal variates, and it has a separate seed from the uniform generator. The seed mechanism for randn is similar to that of **rand**; the important point to keep in mind is that they are separate, and resetting the seed for the uniform generator is no use when you are generating normal variates (which is a common task when pricing options). The reader is urged to explore this issue with online help.

4.2.2   Inverse transform method

Suppose we are given the distribution function $F(x) = P\{X \le x\}$ and that we want to generate random variates according to F. If we are able to invert F easily, we may apply the following inverse transform method:

1. We draw a random number $U \sim U(0, 1)$.

2. We return $X = F^{-1}(U)$.

It is easy to see that the random variate X generated by this method is actually characterized by the distribution function F:

$$P\{X \le x\} = P\{F^{-1}(U) \le x\} = P\{U \le F(x)\} = F(x),$$

where we have used the monotonicity of F and the fact that U is uniformly distributed.

Example 4.2 A typical distribution which can be simulated easily by the inverse transform method is the exponential distribution. If $X \sim \exp(\mu)$, where $1/\mu$ is the expected value of X, its distribution function is

$$F(x) = 1 - e^{-\mu x}.$$

Direct application of the inverse transform yields

$$x = -\frac{1}{\mu} \ln(1 - U).$$

Since the distributions of U and $(1 - U)$ are actually the same, it is customary to generate exponential variates by drawing a random number U and by returning $-\ln(U)/\mu$. We may check that this is indeed the method used in the Statistics toolbox to simulate exponential random variables through the exprnd function:

```
>> rand('seed',0)
>> exprnd(1)
ans =
    1.5189
>> rand('seed',0)
>> -log(rand)
ans =
    1.5189
```

Generating exponential random variables is useful when you have to simulate a Poisson process, which is a possible model for shocks in asset prices or credit rating.   ▯

The inverse transform method is quite simple, and it may also be applied when no theoretical distribution model is available and all you have is a set of empirical data. You just have to build a sensible distribution function based on your data set (see, e.g., [11]); one way to build a distribution function in this case is linear interpolation, and inverting a piecewise linear function is easily accomplished. However, we may not apply the inverse transform method when F is not invertible, which is the case with discrete distributions (in this case the distribution function is piecewise constant). Nevertheless, we may still adapt the method. Consider a discrete empirical distribution with a finite support:

$$P\{X = x_j\} = p_j \qquad j = 1, 2, \ldots, n.$$

Then we must generate a uniform random variate U and return X as

$$X = \begin{cases} x_1 & \text{if } U < p_1 \\ x_2 & \text{if } p_1 \le U < p_1 + p_2 \\ \vdots \\ x_j & \text{if } \sum_{k=1}^{j-1} p_k \le U < \sum_{k=1}^{j} p_k \\ \vdots \end{cases}$$

For many relevant distributions, the distribution function is invertible, but this is not easily accomplished. In such a case, one possibility is to resort to the acceptance-rejection method.

4.2.3 Acceptance-rejection method

Suppose we must generate random variates according to a probability density $f(x)$, and that the difficulty in inverting the corresponding distribution function makes the inverse transform method unattractive. Assume that we know a function $t(x)$ such that

$$t(x) \ge f(x) \qquad \forall x \in I,$$

where I is the support of f. The function $t(x)$ is not a probability density, but the related function $r(x) = t(x)/c$ is, provided that we select

$$c = \int_I t(x) \, dx.$$

If the distribution $r(x)$ is easy to simulate, it can be shown that the following acceptance-rejection method generates a random variate X distributed according to the density f:

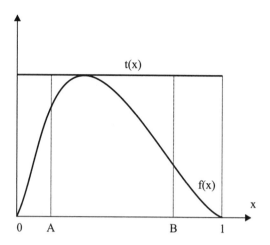

Fig. 4.1 Graphical example of the acceptance-rejection method.

1. Generate $Y \sim r$.

2. Generate $U \sim U(0, 1)$, independent of Y.

3. If $U \leq f(Y)/t(Y)$, return $X = Y$; otherwise, repeat the procedure.

If the support I is bounded, a natural choice for $r(x)$ is simply the uniform distribution on I, and we may choose

$$t(x) = \max_{x \in I} f(x).$$

We will not prove the correctness of the method, but an intuitive grasp can be gained from figure 4.1. In the figure, the support of $f(x)$ is the unit interval. A typical distribution that looks like f is the *beta distribution*:

$$f(x) = \frac{x^{\alpha_1 - 1}(1 - x)^{\alpha_2 - 1}}{\mathrm{B}(\alpha_1, \alpha_2)} \qquad x \in [0, 1],$$

provided that the parameters satisfy $\alpha_1, \alpha_2 > 1$ (the beta distribution does not require this condition, but its appearance would be different from figure 4.1). The *beta function* is defined as

$$\mathrm{B}(\alpha_1, \alpha_2) = \int_0^1 x^{\alpha_1 - 1}(1 - x)^{\alpha_2 - 1} \, dx.$$

The Y variables are generated according to the uniform distribution and will spread evenly over the unit interval. Consider point A; since $f(A)$ is close to $t(A)$, A is likely to be accepted, as the ratio $f(A)/t(A)$ is close to 1. When we consider point B, where the value of the density f is small, we see that the ratio $f(B)/t(B)$ is small; hence, B is unlikely to be accepted, which is what we would expect. It can also be shown that the average number of iterations to terminate the procedure with an accepted value is c.

Example 4.3 Consider the density

$$f(x) = 30(x^2 - 2x^3 + x^4) \qquad x \in [0,1].$$

The reader is urged to verify that this is indeed a density (actually, it is the beta density with $\alpha_1 = \alpha_2 = 3$). If we apply the inverse transform method, we have to invert a fifth-degree polynomial at each generation, which suggests use of the acceptance-rejection method. By ordinary calculus we see that

$$\max_{x \in [0,1]} f(x) = 30/16$$

for $x^* = 0.5$. Using the uniform density as the easy density r, we get the following algorithm:

1. Draw two independent and uniformly distributed random variables U_1 and U_2.

2. If $U_2 \le 16(U_1^2 - 2U_1^3 + U_1^4)$, accept $X = U_1$; otherwise, reject and go back to step 1.

The average number of iterations to generate one random variate is $30/16$.

$\square$

4.2.4 Generating normal variates by the polar approach

The inverse transform and acceptance-rejection methods are general purpose, but they are not always applicable. In the case of normal variables we cannot invert the distribution function, since no analytical form for it is known (actually, numerical approaches could be applied, but this is not necessarily efficient), nor may we easily find a majorant function for the normal density, since its support is not finite. A way to overcome the difficulty is resorting to ad hoc methods. Recall first that if $X \sim N(0,1)$, then $\mu + \sigma X \sim N(\mu, \sigma^2)$; hence we just need a method for generating standard normal variables.

One old-fashioned possibility, which is still suggested in some textbooks, is to exploit the central limit theorem and to generate and sum a suitable number of uniform variates. Although this approach would work in the limit, computational efficiency would restrict the number of uniform variates that we use. The result is that we obtain a variate which could be of sufficient quality in noncritical simulations in which we are interested in average values, but is of debatable quality when we are interested in critical behavior in the tail of the distribution (as is the case in value-at-risk computations).

An alternative method is the Box-Muller approach. Consider two independent variables $X, Y \sim N(0,1)$, and let (R, θ) be the polar coordinates of the point of Cartesian coordinates (X, Y) in the plane, so that

$$d = R^2 = X^2 + Y^2 \qquad \theta = \tan^{-1} Y/X$$

The joint density of X and Y is

$$f(x,y) = \frac{1}{\sqrt{2\pi}}e^{-x^2/2}\frac{1}{\sqrt{2\pi}}e^{-y^2/2} = \frac{1}{2\pi}e^{-(x^2+y^2)/2} = \frac{1}{2\pi}e^{-d/2}.$$

The last expression looks like a product of an exponential density for d and a uniform distribution; the term $1/2\pi$ may be interpreted as the uniform distribution for the angle $\theta \in (0, 2\pi)$. However, we are missing some constant term in order to obtain the exponential density. To express the density in terms of (d, θ), we should properly take the Jacobian of the transformation from (x, y) to (d, θ) into account.[1] Some calculations yield

$$J = \begin{vmatrix} \dfrac{\partial d}{\partial x} & \dfrac{\partial d}{\partial y} \\[2ex] \dfrac{\partial \theta}{\partial x} & \dfrac{\partial \theta}{\partial y} \end{vmatrix} = 2,$$

and the correct density in the alternative coordinates is

$$f(d, \theta) = \frac{1}{2}\frac{1}{2\pi}e^{-d/2}.$$

Hence, we may generate R^2 as an exponential variable with mean 2 and θ as a uniformly distributed angle, and then transform back into Cartesian coordinates in order to obtain two independent standard normal variates. The Box-Muller algorithm may be implemented as follows:

1. Generate two independent uniform variates $U_1, U_2 \sim U(0, 1)$.

2. Set $R^2 = -2\log U_1$ and $\theta = 2\pi U_2$.

3. Set $X = R\cos\theta$, $Y = R\sin\theta$.

In practice, this algorithm may be improved by avoiding the costly evaluation of trigonometric functions and integrating the Box-Muller approach with the rejection approach. The idea results in the following polar rejection method:

1. Generate two independent uniform variates $U_1, U_2 \sim U(0, 1)$.

2. Set $V_1 = 2U_1 - 1$, $V_2 = 2U_2 - 1$, $S = V_1^2 + V_2^2$.

3. If $S > 1$, return to step 1; otherwise, return the independent standard normal variates:

$$X = \sqrt{\frac{-2\ln S}{S}}V_1 \qquad Y = \sqrt{\frac{-2\ln S}{S}}V_2.$$

---

[1]See, e.g., [16] for details.

```
% MultiNormrnd.m
function Z = MultiNormrnd(mu,sigma,howmany)
n = length(mu);
Z = zeros(howmany,n);
mu = mu(:); % make sure it's a column vector
L = chol(sigma);
for i=1:howmany
   Z(i,:) = mu' + randn(1,n) * L;
end
```

Fig. 4.2 Code to simulate multivariate normal variables.

We refer the reader to [17, section 5.3] for a justification of the polar rejection method.

In many financial applications one has to generate variates according to a multivariate normal distribution with expected value μ and covariance matrix Σ. This task may be accomplished by obtaining the Cholesky factor for Σ, i.e., an upper triangular matrix $\mathbf{L}$ such that $\Sigma = \mathbf{L}^T\mathbf{L}$. Then we may apply the following algorithm:

1. Generate n independent standard normal variates $Z_1, \ldots, Z_n \sim N(0,1)$.

2. Return $X = \mu + \mathbf{L}^T\mathbf{Z}$, where $\mathbf{Z} = [Z_1, \ldots, Z_n]^T$.

Example 4.4 A rough code to simulate multivariate normal variables is illustrated in figure 4.2. The code builds a matrix whose columns correspond to the different variables, and the rows correspond to the different realizations of them. Assume that we have the following parameters:

```
>> Sigma = [4 1 -2 ; 1 3 1 ; -2 1 5];
>> mu = [ 8 ; 6 ; 10];
>> eig(Sigma)
ans =
    4.1433
    6.5712
    1.2855
```

Note that we make sure that the matrix Σ is positive definite, as it should be. Now we may generate a few samples and verify the results.

```
>> rand('seed',0);
>> Z = MultiNormrnd(mu,Sigma,10000);
>> mean(Z)
ans =
```

```
      8.0363       6.0108       9.9747
>> cov(Z)
ans =
      3.9066       1.0178      -1.8950
      1.0178       3.0848       1.0319
     -1.8950       1.0319       4.8936
```

We leave to the reader the exercise of improving the code, by checking that the vector and matrix sizes of the input arguments agree, by checking that the matrix Sigma is a positive definite symmetric matrix, and by avoiding the for loop. Then have a look at the function mvnrnd, included in the Statistics toolbox, which does just this job.   ⏏

4.3   SETTING THE NUMBER OF REPLICATIONS

Carrying out a Monte Carlo simulation entails the generation of samples of the quantity of interest and then an estimation of the relevant parameters. One would expect that the larger the number of samples, or replications, the better the quality of the estimates will be. From appendix B we recall that given a sequence of *independent* (and we stress the independence) samples X_i, drawn from the same underlying distribution, we may build the sample mean:

$$\bar{X}(n) = \frac{1}{n} \sum_{i=1}^{n} X_i,$$

which is an unbiased estimator of the parameter $\mu = \mathrm{E}[X_i]$ that we are trying to estimate, and the sample variance:

$$S^2(n) = \frac{1}{n-1} \sum_{i=1}^{n} \left[X_i - \bar{X}(n) \right]^2.$$

We may try to quantify the quality of our estimator by considering the expected value of square error:

$$\mathrm{E}[(\bar{X}(n) - \mu)^2] = \mathrm{Var}[\bar{X}(n)] = \frac{\sigma^2}{n},$$

where σ^2 may be estimated by the sample variance. Clearly, increasing the number n of replications improves the estimate; but how can we reasonably set the value of n?

Recall that the confidence interval at level $(1 - \alpha)$ may be computed as

$$\bar{X}(n) \pm z_{1-\alpha/2} \sqrt{S^2(n)/n}, \tag{4.3}$$

where $z_{1-\alpha/2}$ is a critical number from the standard normal distribution. Strictly speaking, this is just an approximation, which will be a good one provided that n is large enough, so that by virtue of the central limit theorem $\bar{X}(n)$ is approximately normally distributed.

Suppose you are interested in controlling the *absolute* error in such a way that with some confidence level $(1 - \alpha)$,

$$| \bar{X}(n) - \mu | \leq \beta,$$

where β is the maximum acceptable tolerance. From equation (4.3) we see that the absolute error is actually the half-length

$$H = z_{1-\alpha/2} \sqrt{S^2(n)/n}$$

of the confidence interval. So we should simply run replications until H is less than or equal to the tolerance β, and the number n must satisfy

$$z_{1-\alpha/2} \sqrt{S^2(n)/n} \leq \beta. \tag{4.4}$$

Actually, we are chasing our tail a bit here, since we cannot estimate the sample variance $S^2(n)$ until the number n has been set. The way out is to run a suitable number, say $k = 30$, of pilot replications, in order to come up with an estimate $S^2(k)$. Then we may apply (4.4) using $S^2(k)$ to determine n. After running the n replications, it is advisable to check that equation (4.4) holds with the new estimate $S^2(n)$. Alternatively, we may simply add replications, updating the sample variance, until the criterion is met.

If you are interested in controlling the relative error, so that

$$\frac{| \bar{X}(n) - \mu |}{| \mu |} \leq \gamma$$

holds with probability $(1-\alpha)$, things are a little more involved. The difficulty is that we may run replications until the half-length H satisfies

$$\frac{H}{| \bar{X}(n) |} \leq \gamma,$$

but in this inequality we are using the known quantity $\bar{X}(n)$ rather than the unknown parameter μ. Nevertheless, if the inequality above holds, we may write

$$
\begin{aligned}
1 - \alpha &\approx P\left\{ \frac{| \bar{X}(n) - \mu |}{| \bar{X}(n) |} \leq \frac{H}{| \bar{X}(n) |} \right\} \\
&= P\left\{ | \bar{X}(n) - \mu | \leq \gamma | \bar{X}(n) | \right\} \\
&= P\left\{ | \bar{X}(n) - \mu | \leq \gamma | \bar{X}(n) - \mu + \mu | \right\} \\
&\leq P\left\{ | \bar{X}(n) - \mu | \leq \gamma | \bar{X}(n) - \mu | + | \mu | \right\} \tag{4.5} \\
&= P\left\{ \frac{| \bar{X}(n) - \mu |}{| \mu |} \leq \frac{\gamma}{1 - \gamma} \right\},
\end{aligned}
$$

where inequality (4.5) follows from the triangle inequality and the last equation is obtained by a slight rearrangement. Therefore, we see that if we proceed without care, the actual relative error we get is bounded by $\gamma/(1 - \gamma)$, which is larger than the desired bound γ; so we should choose n such that the following criterion is met:

$$\frac{z_{1-\alpha/2}\sqrt{S^2(n)/n}}{|\bar{X}(n)|} \leq \gamma', \tag{4.6}$$

where

$$\gamma' = \frac{\gamma}{1+\gamma} < \gamma.$$

Again, we should run some pilot replications in order to get a first estimate of the sample variance $S^2(n)$.

From equation (4.3) we see that the rate of improvement of the quality of our estimate, i.e., the rate of decrease of the error, is something like $O(1/\sqrt{n})$. In practice, this means that the more samples we get the better, but the rate of improvement is slower and slower as we keep adding samples. Thus a brute-force Monte Carlo simulation may take quite some amount of computation to yield an acceptable estimate. One way to overcome this issue is to adopt a clever sampling strategy in order to reduce the variance σ^2 of our samples; the other one is to adopt a quasi-Monte Carlo approach.

4.4 VARIANCE REDUCTION TECHNIQUES

We have seen in section 4.3 that one way to improve the accuracy of an estimate is to increase the number of replications n, since $\mathrm{Var}(\bar{X}(n)) = \mathrm{Var}(X_i)/n$. However, this brute-force approach may require an excessive computational effort. An alternative is to work on the numerator of this fraction and to reduce the variance of the samples X_i directly. This may be accomplished in different ways, more or less complicated, and more or less rewarding as well.

4.4.1 Antithetic variates

A first approach that is easy to apply and does not require deep knowledge of what we are simulating is antithetic sampling. In the plain Monte Carlo approach we generate a sequence of independent samples. However, inducing some correlation in a clever way may be helpful. Consider the idea of generating a sequence of paired replications (X_i^1, X_i^2), $i = 1, \ldots, n$:

$$\begin{matrix} X_1^1 & X_2^1 & \cdots & X_n^1 \\ X_1^2 & X_2^2 & \cdots & X_n^2. \end{matrix}$$

These samples are "horizontally" independent, in the sense that $X_{i_1}^j$ and $X_{i_2}^k$ are independent if $i_1 \neq i_2$, however we choose j and k. Thus the pair-averaged samples $X_i = (X_i^1 + X_i^2)/2$ are independent, and we may build a confidence interval based on them. However, we do not require "vertical" independence, since for a fixed i, X_i^1 and X_i^2 may be dependent. If we build the sample mean $\bar{X}(n)$ based on the samples X_i,

$$
\begin{aligned}
\mathrm{Var}[\bar{X}(n)] &= \frac{\mathrm{Var}(X_i)}{n} \\
&= \frac{\mathrm{Var}(X_i^1) + \mathrm{Var}(X_i^2) + 2\,\mathrm{Cov}(X_i^1, X_i^2)}{4n}.
\end{aligned}
$$

We see that in order to reduce the sample mean variance, we should take negatively correlated replications within each pair, so that $\mathrm{Cov}(X_i^1, X_i^2) < 0$. Each sample x_i^k is obtained by generating random variates according to one of the methods we have described before; but all of these methods exploit a stream of uniformly distributed random numbers. Hence, to induce a negative correlation, we may use a random number sequence $\{U_k\}$ for the first replication in each pair, and then $\{1 - U_k\}$ in the second one. Since the input streams are negatively correlated, we hope that the output streams will, too.

Example 4.5 Let us repeat example 4.1, where we wanted to use Monte Carlo integration to estimate

$$
I = \int_0^1 e^x \, dx = e - 1 \approx 1.7183.
$$

We want to draw 100 samples and find a 90% confidence interval for I. First we run a naive Monte Carlo experiment:

```
>> rand('seed',0)
>> U = rand(1,100);
>> X=exp(U);
>> I=mean(X)
I =
    1.7461
>> S2 = var(X)
S2 =
    0.2499
>> z=norminv(0.95)
z =
    1.6449
>> H=z*sqrt(S2 / 100)
H =
    0.0822
>> I-H
ans =
    1.6638
```

```
>> I+H
ans =
    1.8283
```

In this MATLAB snapshot we first draw a sequence U of 100 random numbers; then we compute the sample points X, the sample mean I, and the sample variance S; then, using the critical number z for the selected confidence interval, we compute the half-length H, and finally, the confidence interval (1.7050, 1.7872). The confidence interval does bracket the correct value, but it is not very small; increasing the confidence level to 95% would make the confidence interval even wider. It is also worth noting that using the normfit function provided by the Statistic toolbox, we get a slightly wider confidence interval:

```
>> [mu,s,ci] = normfit(X,0.1)
mu =
    1.7461
s =
    0.4999
ci =
    1.6631
    1.8291
```

This is due to the fact that normfit does not use the critical numbers from the normal distribution (which indeed is not quite correct for a small number of samples) but the critical numbers from the t distribution. Do not get fooled by the fact that the estimate is close to the actual value; if you keep repeating the experiment (without resetting the seed), you will see that the estimates do swing around the correct value. Let us try now with 50 pairs of antithetic samples:

```
>> rand('seed',0)
>> U1 = rand(1,50);
>> U2 = 1-U1;
>> X1 = exp(U1);
>> X2 = exp(U2);
>> X = 0.5 * (X1+X2);
>> I=mean(X)
I =
    1.7175
>> S2 = var(X)
S2 =
    0.0044
>> H=z*sqrt(S2 / 50)
H =
    0.0154
>> I-H
```

```
ans =
    1.7021
>> I+H
ans =
    1.7329
```

Now the confidence interval is much smaller and, despite the limited number of samples, the estimate is fairly reliable.   ▯

The antithetic sampling method looks quite easy to apply, but does it always work? The following counterexample shows that this method may actually backfire, resulting in an increase in the variance.

Example 4.6 Consider the function $h(x)$, defined as

$$h(x) = \begin{cases} 0 & x < 0 \\ 2x & 0 \le x \le 0.5 \\ 2 - 2x & 0.5 \le x \le 1 \\ 0 & x > 1 \end{cases}$$

and suppose that we want to take a Monte Carlo approach to estimate

$$\int_0^1 h(x)\, dx.$$

The function we want to integrate is obviously a triangle with both basis and height equal to 1; note that unlike the exponential function of example 4.5, this is not a monotone function with respect to x. It is easy to compute the integral as the area of a triangle:

$$\int_0^1 h(x)\, dx \Rightarrow \mathrm{E}[h(U)] = \int_0^1 h(u) \cdot 1\, du = 1/2.$$

Now let

$$X_I = \frac{h(U_1) + h(U_2)}{2},$$

where U_1 and U_2 independent uniform variates, be the usual sample based on independent sampling, and

$$X_A = \frac{h(U) + h(1 - U)}{2}$$

be the pair-averaged sample built by antithetic sampling. We may compare the two variances:

$$\mathrm{Var}(X_I) = \frac{\mathrm{Var}[h(U)]}{2}$$

$$\mathrm{Var}(X_A) = \frac{\mathrm{Var}[h(U)]}{2} + \frac{\mathrm{Cov}[h(U), h(1 - U)]}{2}.$$

The difference between the two variances is

$$\Delta \ = \ \text{Var}(X_A) - \text{Var}(X_I) = \frac{\text{Cov}[h(U), h(1-U)]}{2}$$

$$= \ \frac{1}{2}\left\{E[h(U)h(1-U)] - E[h(U)]E[h(1-U)]\right\}.$$

But in this case, due to the shape of h, we have

$$E[h(U)] = E[h(1-U)] = 1/2$$

and

$$E[h(U)h(1-U)] \ = \ \int_0^{1/2} 2u \cdot (2 - 2(1-u)) \, du + \int_{1/2}^1 2(1-u) \cdot (2-2u) \, du$$

$$= \ \int_0^{1/2} 4u^2 \, du + \int_{1/2}^1 (2-2u)^2 \, du = 1/3.$$

Therefore, $\text{Cov}[h(U), h(1-U)] = 1/3 - 1/4 = 1/12$ and $\Delta = 1/24 > 0$, and antithetic sampling actually increases variance in this case.

Indeed, there is a trivial explanation. The two antithetic samples have the same value $h(U) = h(1-U)$, so that $\text{Cov}[h(U), h(1-U)] = \text{Cov}[h(U), h(U)] = \text{Var}[h(U)]$. In this (pathological) case, the variance of the single sample is doubled by applying antithetic sampling. □

What is wrong with the example 4.6? The variance of the antithetic pair is actually increased due to the nonmonotonicity of $h(x)$. In fact, while it is true that the random number sequences $\{U_i\}$ and $\{1 - U_i\}$ are negatively correlated, there is no guarantee that the same holds for the sequences X_i^1 and X_i^2 in general. To be sure that the negative correlation in the input random numbers yields a negative correlation in the output samples, we must require a monotonic relationship between them. The exponential function is a monotonic function, but the triangle function of the second example is not. We should also pay attention to the way that random variates are generated. The inverse transform method is based on the distribution function, which is a monotonic function; hence, there is a monotonic relationship between the input random numbers and the random variates generated. This is not necessarily the case with the acceptance-rejection method or the Box-Muller method. Luckily, when we need normal variates, we may simply generate a sequence Z_i, where $Z_i \sim N(0,1)$, and use the sequence $-Z_i$ for the antithetic samples.

4.4.2   Common random numbers

The common random numbers (CRN) technique is very similar to antithetic sampling, but it is applied in a different situation. Suppose that we use Monte

Carlo simulation to estimate a value depending on a parameter α. In formulas, we are trying to estimate something like

$$h(\alpha) = E_\omega[f(\alpha; \omega)],$$

where we have emphasized randomness through the variable ω. We could also be interested in evaluating the sensitivity of this value on the parameter α:

$$\frac{dh(\alpha)}{d\alpha}.$$

This would be of interest when dealing with option sensitivities beyond the Black-Scholes model. Clearly, we cannot compute the derivative analytically; otherwise, we wouldn't use simulation to evaluate h in the first place. So the simplest idea would be using simulation to estimate the value of the finite difference,

$$\frac{h(\alpha + \delta\alpha) - h(\alpha)}{\delta\alpha},$$

for a small value of the increment $\delta\alpha$. However, what we can really do is to generate samples of the difference:

$$\frac{f(\alpha + \delta\alpha; \omega) - f(\alpha; \omega)}{\delta\alpha}$$

and to estimate its expected value. Unfortunately, when the increment $\delta\alpha$ is small, it is difficult to tell if the difference we obtain from the simulation is due to random noise or to variation in the parameter. A similar problem arises when we want to compare two portfolio management policies on a set of scenarios; in this case, too, what we need is an estimate of the expected value of the difference between two random variables.

Let us abstract a little and consider the difference of two random variables

$$Z = X_1 - X_2,$$

where, in general, $E[X_1] \neq E[X_2]$, since they come from simulating two different systems, possibly differing only in the value of a single parameter. By Monte Carlo simulation we get a sequence of independent samples

$$Z_j = X_{1,j} - X_{2,j}$$

and use statistical techniques to build a confidence interval for $E[X_1 - X_2]$. To improve our estimate, it would be useful to reduce the variance of the samples Z_j:

$$\text{Var}(X_{1j} - X_{2j}) = \text{Var}(X_{1j}) + \text{Var}(X_{2j}) - 2\,\text{Cov}(X_{1j}, X_{2j}).$$

To achieve this, we may try inducing some positive correlation between X_{1j} and X_{2j}. This can be obtained by using the same stream of random numbers in simulating both X_1 and X_2. The technique works much like antithetic sampling, and the same monotonicity assumption is required to ensure that the technique does not backfire.

4.4.3   Control variates

Antithetic sampling and common random numbers are two almost foolproof techniques that, provided the monotonicity assumption is valid, do not require much knowledge about the systems we are simulating. Better results might be obtained by exploiting some more knowledge. Suppose that we want to estimate $\theta = \mathrm{E}[X]$ and there is another random variable Y, with a *known* expected value ν, which is somehow correlated with X. Such a case occurs when we use Monte Carlo simulation to price an option for which an analytical formula is not known: θ is the unknown price of the option, and ν is the price of a corresponding vanilla option.

The variable Y is called the *control variate*. Additional knowledge about Y may be exploited by adopting the controlled estimator

$$X_C = X + c(Y - \nu),$$

where c is a parameter we must choose. Intuitively, when we run a simulation and we observe that our estimates are such that

$$\hat{\mathrm{E}}[Y] > \nu,$$

we may argue that the estimate $\hat{\mathrm{E}}[X]$ should be increased or reduced accordingly, depending on the sign of the correlation between X and Y. Indeed, we may see that

$$\mathrm{E}[X_C] = \theta$$
$$\mathrm{Var}(X_C) = \mathrm{Var}(X) + c^2 \mathrm{Var}(Y) + 2c\,\mathrm{Cov}(X, Y).$$

The first formula says that the controlled simulator is, for any choice of the control parameter c, an unbiased estimator of θ. The second formula suggests that by a suitable choice of c, we could reduce the variance of the estimator. We could even minimize the variance by choosing the optimal value for c:

$$c^* = -\frac{\mathrm{Cov}(X, Y)}{\mathrm{Var}(Y)},$$

in which case we get

$$\frac{\mathrm{Var}(X_C^*)}{\mathrm{Var}(X)} = 1 - \rho_{XY}^2,$$

where ρ_{XY} is the correlation between X and Y. Note that the sign of c depends on the sign of this correlation. For instance, if $\mathrm{Cov}(X, Y) > 0$, then $c < 0$. This implies that if $\hat{\mathrm{E}}[Y] > \nu$, we should reduce $\hat{\mathrm{E}}[X]$, which does make sense, because if our sample values for Y are larger than usual, the sample values for X probably are too.

In practice, the optimal value of c must be estimated since $\mathrm{Cov}(X, Y)$, and possibly $\mathrm{Var}(Y)$ are not known. This may be accomplished by a set of pilot replications. It would be tempting to use these replications both for selecting c^* and to estimate θ; however, in doing so you induce some bias in the estimate of θ, since in this case c^* is a random variable depending on X itself. So, unless suitable statistical techniques are used, which are beyond the scope of this book, the pilot replications should be discarded.

The control variates approach may be generalized to as many control variates as we want, with a possible improvement in the quality of the estimates. Of course, this requires more knowledge about the system we are simulating and more effort in setting the control parameters.

4.4.4 Variance reduction by conditioning

Computing expected values by conditioning is a common tool in probability theory. When we want to compute (or estimate) $E[X]$, it is sometimes useful to condition with respect to another variable Y, as the following formula holds:

$$E[X] = E[E[X \mid Y]]. \qquad (4.7)$$

Variances may be computed by conditioning, too. We recall the conditional variance formula [see also equation (B.2) in appendix B]

$$\text{Var}(X) = E[\text{Var}(X \mid Y)] + \text{Var}(E[X \mid Y]).$$

We do not use the conditional variance formula directly in this book. However, since all the involved quantities are nonnegative, we immediately see that the formula implies two consequences:

1. $\text{Var}(X) \geq E[\text{Var}(X \mid Y)]$.

2. $\text{Var}(X) \geq \text{Var}(E[X \mid Y])$.

Using the first inequality to reduce the variance of an estimator leads to variance reduction by stratification, which is discussed in the next section. The second one leads to variance reduction by conditioning.

Using conditioning is useful when our aim is to estimate $\theta = E[X]$ and there is another random variable Y such that the value of $E[X \mid Y = y]$ is known. From equation (4.7) we see that $E[X \mid Y]$ is also an unbiased estimator for θ, and the conditional variance formula implies that it may be a better one. In practice, to apply variance reduction by conditioning we simulate Y rather than X. Unlike antithetic sampling, variance reduction by conditioning requires some careful thinking and is strongly problem dependent. We see an application in section 7.4.

4.4.5 Stratified sampling

Suppose, as usual, that we are interested in estimating $E[X]$ and that X is somehow dependent on the value of another variable random Y, which may take a finite set of values y_j with known probability. Thus, Y has a discrete probability distribution with a known probability mass function:

$$P\{Y = y_j\} = p_j \qquad j = 1, \ldots, m.$$

Using conditioning, we see that

$$E[X] = \sum_{j=1}^{m} E[X \mid Y = y_j] p_j.$$

So we may use simulation to estimate the values $E[X \mid Y = y_j]$, for $j = 1, \ldots, m$, and use the formula above to put the results together. The conditional variance formula implies that this may yield a variance reduction with respect to crude sampling. The approach may look like variance reduction by conditioning. The key difference is that here we select a value for Y and then we sample X, conditioned on the event $Y = y_j$; this event is a *stratum*. In variance reduction by conditioning, you actually sample Y, not X. The following example justifies why such sampling is called *stratified*.

Example 4.7 As a simple example of stratification, consider using simulation to compute

$$\theta = \int_0^1 h(x) \, dx = E[h(U)].$$

In crude Monte Carlo simulation you would simply draw n uniform random numbers $U_i \sim U(0, 1)$ and compute the sample mean

$$\frac{1}{n} \sum_{i=1}^{n} h(U_i).$$

An improved estimator over crude sampling may be obtained by partitioning the integration interval $(0, 1)$ into m subintervals $((j - 1)/m, j/m)$, $j = 1, \ldots, m$. Each event $Y = y_j$ corresponds to a random number falling in the jth subinterval; in this case we have $p_j = 1/m$. For each stratum $j = 1, \ldots, m$ we may generate n_j random numbers $U_k \sim U(0, 1)$ to estimate

$$\hat{\theta}_j = \frac{1}{n_j} \sum_{k=1}^{n_j} h\left(\frac{U_k + j - 1}{m}\right).$$

Then we build the overall estimator:

$$\hat{\theta} = \sum_{j=1}^{m} \hat{\theta}_j p_j. \qquad \qquad \square$$

How should we determine the number of samples n_j to be allocated to each stratum? A uniform allocation in example 4.7 makes sure that we sample uniformly over the integration interval $(0, 1)$, but this need not be the optimal solution. Consider the variance of the estimator $\hat{\theta}$, and denote by X_j the random variable sampled in each stratum. If the strata are independently sampled, we have

$$\text{Var}(\hat{\theta}) = \sum_{j=1}^{m} p_j^2 \, \text{Var}(\hat{\theta}_j) = \sum_{j=1}^{m} \frac{p_j^2}{n_j} \, \text{Var}(X_j).$$

To minimize the overall variance, we should allocate more samples to the strata where $\text{Var}(X_j)$ is larger. So we could run a set of pilot replications to estimate $\text{Var}(X_j)$ by sample variances S_j^2 and then obtain the fraction of samples to be allocated to each stratum by solving a nonlinear programming problem:

$$\min \quad \sum_{j=1}^{m} \frac{p_j^2 S_j^2}{n_j}$$

$$\text{s.t.} \quad \sum_{j=1}^{m} n_j = n$$

$$n_j \geq 0.$$

4.4.6 Importance sampling

Unlike the other variance reduction methods, importance sampling is based on the idea of "distorting" the underlying probability measure. It may be particularly useful when simulating rare events or sampling from the tails of a distribution. Consider the problem of estimating

$$\theta = \text{E}[h(\mathbf{X})] = \int h(\mathbf{x}) f(\mathbf{x}) \, d\mathbf{x}$$

where $\mathbf{X}$ is a random vector with joint density $f(\mathbf{x})$. If we know another density g such that $f(\mathbf{x}) = 0$ whenever $g(\mathbf{x}) = 0$, we may write

$$\theta = \int \frac{h(\mathbf{x}) f(\mathbf{x})}{g(\mathbf{x})} g(\mathbf{x}) \, d\mathbf{x} = \text{E}_g \left[\frac{h(\mathbf{X}) f(\mathbf{X})}{g(\mathbf{X})} \right], \qquad (4.8)$$

where the notation E_g is used to stress the fact that the last expected value is taken with respect to another density. That changing the underlying probability measure may be useful should not be a surprise for people interested in finance; risk-neutral valuation does just that. However, it is not so obvious why this should be helpful in reducing variance. Indeed, the method may backfire if g is not chosen with care. Intuitively, we may argue that when looking for rare but important events, as is the case in VaR calculations, we should distort the probability measure in order to sample from the critical region, provided that we compensate for this bias. This is exactly what is done in equation (4.8).

```
function out=estpi(m)
z=sqrt(1-rand(1,m).^2);
out = 4*sum(z)/m;
```

Fig. 4.3 Trivial code to estimate π.

To get a more precise idea of why importance sampling may work, it is useful to consider a discrete version of the integration problem. Say that we want to estimate the following sum:

$$F(N) = \sum_{i=1}^{N} h(x_i),$$

where x_i belongs to a set of N discrete points. If N is very large, we might sample a subset of M points ($M \ll N$) and compute the estimate

$$N \frac{\sum_{j=1}^{M} h(x_j)}{M}. \tag{4.9}$$

Let p_i be the probability of sampling a certain point x_i; in crude sampling, we obviously have $p_i = 1/N$ for all points x_i, and we may read equation (4.9) as

$$\frac{1}{M} \frac{\sum_{j=1}^{M} h(x_j)}{1/N} = \frac{1}{M} \sum_{j=1}^{M} \frac{h(x_j)}{p_j}. \tag{4.10}$$

Now, importance sampling is based on the idea of using some clever probability distribution p_i rather than the uniform distribution. An ideal choice for this distribution would be

$$p_i = \frac{h(x_i)}{F(N)} \qquad i = 1, \ldots, N. \tag{4.11}$$

This would be a particularly nice choice since, applying equation (4.10), we have

$$\frac{1}{M} \sum_{j=1}^{M} \frac{h(x_j)}{p_j} = \frac{1}{M} \sum_{j=1}^{M} \frac{h(x_j)F(N)}{h(x_j)} = F(N).$$

In other words, however we sample, we always get the right answer, and the variance is reduced to zero. Of course, there is a little fly in the ointment: we may use this probability distribution only if we already know the answer we are looking for, i.e., the value of $F(N)$. Still, the ideal probability distribution gives us a hint about how to build a good approximate distribution.

Example 4.8 We may use a trivial integration example to illustrate the idea. Let us consider a way to compute π. We know that[2]

$$\int_0^1 \sqrt{1 - x^2} \, dx = \frac{\pi}{4},$$

since this is simply the area of a quarter of a unit circle; so estimating the integral is a possible way to obtain an estimate of π. A trivial code to do this is shown in figure 4.3, where the input parameter m is the number of points we want to sample. From the snapshot below we see that with 1000 samples, the estimates are not so reliable.

```
>> pi
ans =
    3.1416
>> rand('seed',0)
>> estpi(1000)
ans =
    3.1659
>> estpi(1000)
ans =
    3.1263
>> estpi(1000)
ans =
    3.1907
>> estpi(1000)
ans =
    3.1049
```

So let us try to improve our estimates by using importance sampling. A possible idea to approximate the ideal probability distribution (4.11) is to divide the integration interval $[0, 1]$ into L equally spaced subintervals of width $1/L$. The extreme points of the kth subinterval $(k = 1, \ldots, L)$ are $(k - 1)/L$ and k/L, and the midpoint of this subinterval is $s_k = (k - 1)/L + 1/(2L)$. A rough estimate of the integral is obtained by computing

$$\frac{\sum_{k=1}^{L} h(s_k)}{L}.$$

Now consider the quantities

$$q_k = \frac{h(s_k)}{\sum_{j=1}^{L} h(s_j)} \qquad k = 1, \ldots, L.$$

---

[2]This example is based on [3].

```
function z=estpiIS(m,L)
s= (0:(1/L):(1-1/L)) + 1/(2*L);
hvals = sqrt(1 - s.^2);
cs=cumsum(hvals);
for j=1:m
    loc=sum(rand*cs(L) > cs) +1;
    x=(loc-1)/L + rand/L;
    p=hvals(loc)/cs(L);
    est(j) = sqrt(1 - x.^2)/(p*L);
end
z = 4*sum(est)/m;
```

Fig. 4.4   Importance sampling-based code to estimate π.

Clearly, $\sum_k q_k = 1$ and $q_k \geq 0$, since our function h is nonnegative; hence, the numbers q_k may be interpreted as probabilities. In our case, they may be used as the probabilities of selecting a sample point from the kth subinterval. When sampling, we may select a subinterval according to the probabilities q_k, and then sample a point uniformly within the interval. To summarize, and to cast the problem within the general framework, we have

$$
\begin{aligned}
h(x) &= \sqrt{1-x^2} \\
f(x) &= 1 \\
g(x) &= Lq_k \qquad (k-1)/L \leq x < k/L.
\end{aligned}
$$

So $g(x)$ is a piecewise constant density; the L factor multiplying the q_k in $g(x)$ is just needed to obtain the uniform density over an interval of length $1/L$. The resulting code is illustrated in figure 4.4, where m is the number of sampled points and L is the number of subintervals.

```
>> rand('seed',0)
>> estpiIS(1000,10)
ans =
      3.1306
>> estpiIS(1000,10)
ans =
      3.1454
>> estpiIS(1000,10)
ans =
      3.1347
>> estpiIS(1000,100)
ans =
```

```
      3.1436
>> estpiIS(1000,100)
ans =
      3.1407
>> estpiIS(1000,100)
ans =
      3.1413
```

We see that the improved code, although not a very sensible way to compute π, yields a remarkable reduction in variance.

The approach we have just taken looks suspiciously like stratified sampling. Actually, there is a subtle difference. In stratified sampling we define a set of strata, which correspond to events of known probability; here we have not used strata with known probability, as we have used sampling to estimate the probabilities q_k. ◻

Importance sampling is often used when small probabilities are involved. Consider, for instance, a random vector $\mathbf{X}$ with joint density f, and suppose that we want to estimate:

$$\theta = \mathrm{E}[h(\mathbf{X}) \mid \mathbf{X} \in \mathcal{A}],$$

where $\{\mathbf{X} \in \mathcal{A}\}$ is a rare event with a small, but unknown probability $P\{\mathbf{X} \in \mathcal{A}\}$. Such an event could be the occurrence of a loss larger than the value-at-risk. The conditional density is

$$f(\mathbf{x}|\mathbf{X} \in \mathcal{A}) = \frac{f(\mathbf{x})}{P\{\mathbf{X} \in \mathcal{A}\}}$$

for $\mathbf{x} \in \mathcal{A}$. Defining the indicator function $I_{\mathcal{A}}(\mathbf{X})$ as

$$I_{\mathcal{A}}(\mathbf{X}) = \begin{cases} 1 & \text{if } \mathbf{X} \in \mathcal{A} \\ 0 & \text{if } \mathbf{X} \notin \mathcal{A}, \end{cases}$$

we may rewrite θ as

$$\theta = \frac{\int_{\mathbf{x} \in \mathcal{A}} h(\mathbf{x})f(\mathbf{x})\, d\mathbf{x}}{P\{\mathbf{X} \in \mathcal{A}\}} = \frac{\mathrm{E}[h(\mathbf{X})I_{\mathcal{A}}(\mathbf{X})]}{\mathrm{E}[I_{\mathcal{A}}(\mathbf{X})]}.$$

If we use crude Monte Carlo simulation, many samples will be wasted, as the event $\{\mathbf{X} \in \mathcal{A}\}$ will rarely occur. Now assume that there is a density g such that this event is more likely under the corresponding probability measure. Then, we may generate the samples $\mathbf{X}_i$ according to g, and estimate

$$\hat{\theta} = \frac{\sum_{i=1}^{k} h(\mathbf{X}_i)I_{\mathcal{A}}(\mathbf{X}_i)f(\mathbf{X}_i)/g(\mathbf{X}_i)}{\sum_{i=1}^{k} I_{\mathcal{A}}(\mathbf{X}_i)f(\mathbf{X}_i)/g(\mathbf{X}_i)}.$$

Importance sampling is certainly more difficult to apply than antithetic sampling or control variates, as it requires more knowledge about what we are simulating, as we must be able to figure out a suitably distorted probability measure. See section 7.4 for an example in which we integrate importance sampling with conditioning.

4.5   QUASI-MONTE CARLO SIMULATION

In the preceding sections we have considered the use of variance reduction techniques, which are based on the idea that random sampling is *really* random. However, the random numbers produced by a LCG or by more sophisticated algorithms are not random at all. Hence, one could take a philosophical view and wonder about the very validity of variance reduction methods, and even the Monte Carlo approach itself. Taking a more pragmatic view, and considering the fact that Monte Carlo methods have proven their value over the years, we should conclude that this shows that there are some deterministic number sequences that work well in generating samples. So one could try to devise alternative deterministic sequences of numbers which are in some sense evenly distributed. This idea may be made more precise by defining the *discrepancy* of a sequence of numbers.

Assume that we want to generate a sequence of N "random" vectors $\mathbf{X}^1, \mathbf{X}^2, \ldots, \mathbf{X}^N$ in the m-dimensional hypercube $I^m = [0,1]^m \subset \mathbb{R}^m$. Now, given a sequence of such vectors, if they are well distributed, the number of points included in any subset G of I^m should be roughly proportional to its volume $\mathrm{vol}(G)$. Given a vector $\mathbf{X} = (x_1, x_2, \ldots, x_m)$, consider the rectangular subset $G_\mathbf{X}$ defined as

$$G_\mathbf{X} = [0, x_1) \times [0, x_2) \times \cdots \times [0, x_m),$$

which has a volume $x_1 x_2 \cdots x_m$. If we denote by $S_N(G)$ the function counting the number of points in the sequence, which are contained in a subset $G \subset I^m$, a possible definition of discrepancy is

$$D(\mathbf{x}^1, \ldots, \mathbf{x}^N) = \sup_{\mathbf{X} \in I^m} \mid S_N(G_\mathbf{X}) - N x_1 x_2 \cdots x_m \mid .$$

When computing a multidimensional integral on the unit hypercube, it is natural to look for low-discrepancy sequences; an alternative name for a low-discrepancy sequence is *quasirandom sequence*, which is why the term *quasi-Monte Carlo* is used. Actually, the quasirandom term is a bit misleading, as there is no randomness at all. Some theoretical results suggest that low-discrepancy sequences may perform better than pseudorandom sequences obtained through a LCG or its variations. The point is that from section 4.3 we know that the estimation error with Monte Carlo simulation is something like $O(1/\sqrt{N})$, where N is the number of samples. With certain low-discrepancy sequences, it can be shown that the error is something like $O(\ln N)^m/N$, where m is the dimension of the space in which we are integrating. We refer the reader to the comprehensive book [13] for a detailed and rigorous account on this subject. Different sequences have been proposed in the literature. In the following we simply illustrate the basic ideas behind two low-discrepancy sequences, Halton's and Sobol's sequences, and their implementation. Low-discrepancy sequences are sequences in the unit interval $(0, 1)$; from what we know about the generation of generally distributed random variates, we see that this is what we need to simulate according to any distribution we need.

4.5.1   Generating Halton's low-discrepancy sequences

Halton's low-discrepancy sequences are based on a simple recipe:

- Representing an integer number n in a base b, where b is a prime number:

$$n = (\cdots d_4 d_3 d_2 d_1 d_0)_b$$

- Reflecting the digits and adding a radix point to obtain a number within the unit interval:
$$h = (0.d_0 d_1 d_2 d_3 d_4 \cdots)_b$$

More formally, if we represent an integer number n as

$$n = \sum_{k=0}^{m} d_k b^k,$$

the nth number in the Halton's sequence with base b is

$$h(n, b) = \sum_{k=0}^{m} d_k b^{-(k+1)}.$$

Using the principles illustrated in section 2.1.2 on the binary representation of numbers on a computer, it is easy to generate the nth number in a Halton's sequence with base b. The code is illustrated in figure 4.5. Let us generate the first 10 numbers in the sequence with base 2:

```
>> seq = zeros(10,1);
>> for i=1:10, seq(i) = Halton(i,2);, end
>> seq
seq =
    0.5000
    0.2500
    0.7500
    0.1250
    0.6250
```

```
function h=Halton(n,b)
n0 = n;
h = 0;
f = 1/b;
while (n0 > 0)
   n1 = floor(n0/b);
   r = n0 - n1*b;
   h = h+f*r;
   f = f/b;
   n0=n1;
end
```

Fig. 4.5 MATLAB code to generate the nth element of a Halton sequence with a given base.

```
0.3750
0.8750
0.0625
0.5625
0.3125
```

We see how Halton's sequences work; by reflecting and adding more bits, we fill the space between 0 and 1 with finer and finer intervals. A code to obtain a whole sequence is illustrated in figure 4.6; the input parameters are HowMany, i.e., how long the sequence should be, and the base Base. Rather than generating each number in the sequence one at a time, we generate the sequence $1, \ldots, n$ by incrementing the bit representation in base b, which is immediately converted into $H(n, b)$.

Example 4.9 It is instructive to compare how a pseudorandom sample covers the square $(0, 1) \times (0, 1)$ in two dimensions. Using the MATLAB 4 random generator, we get the plot of figure 4.7:

```
>> rand('seed',0);
>> plot(rand(100,1),rand(100,1),'o')
>> grid on
```

To do the same with Halton sequences we must use different bases, which should be prime numbers. Let us try with 2 and 7:

```
>> plot(GetHalton(100,2),GetHalton(100,7),'o')
>> grid on
```

The result is shown in figure 4.8. The judgment is a bit subjective here, but it could be argued that the covering of the Halton sequence is more even. On the other hand, using a nonprime number as the base, as in

```
function Seq = GetHalton(HowMany, Base)
Seq = zeros(HowMany,1);
NumBits = 1+ceil(log(HowMany)/log(Base));
VetBase = Base.^(-(1:NumBits));
WorkVet = zeros(1,NumBits);
for i=1:HowMany
   % increment last bit and carry over if necessary
   j=1;
   ok = 0;
   while ok == 0
      WorkVet(j) = WorkVet(j)+1;
      if WorkVet(j) < Base
         ok = 1;
      else
         WorkVet(j) = 0;
         j = j+1;
      end
   end
   Seq(i) = dot(WorkVet,VetBase);
end
```

Fig. 4.6 MATLAB code to generate a Halton low-discrepancy sequence with a given base.

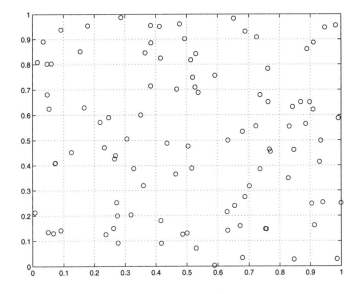

Fig. 4.7 Random sample in two dimensions.

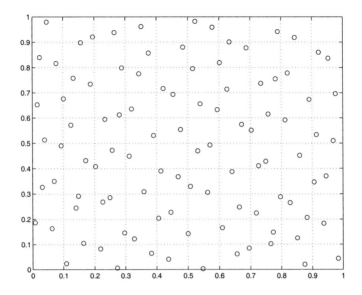

Fig. 4.8   Covering the bidimensional unit square with Halton sequences.

```
>> plot(GetHalton(100,2), GetHalton(100,4), 'o')
>> grid on
```

may result in quite unsatisfactory patterns, such as the one shown in figure 4.9.   □

Example 4.10 Let us explore the use of Halton low-discrepancy sequences in a bidimensional integration context. Suppose that we want to compute

$$\int_0^1 \int_0^1 e^{-xy} \left(\sin 6\pi x + \cos 8\pi y \right) \ dx \ dy.$$

To begin with, let us set up an inline function in order to plot the integrand and to use the `dblquad` MATLAB function to get an estimate by traditional quadrature formulas.

```
>> f=inline('exp(-x.*y).*(sin(6*pi*x)+cos(8*pi*y))');
>> dblquad(f,0,1,0,1)
ans =
    0.0199
>> [X,Y] = meshgrid(0:0.01:1 , 0:0.01:1);
>> Z = f(X,Y);
>> surf(X,Y,Z)
```

Please note how the inline function is defined using the dot operator, in order to receive vector or matrix arguments and to compute the vector or matrix

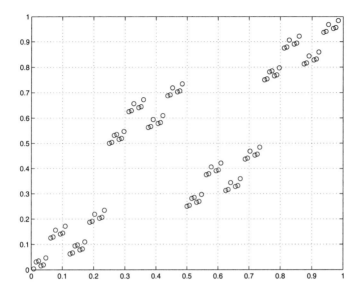

Fig. 4.9 Bad choice of bases in Halton sequences.

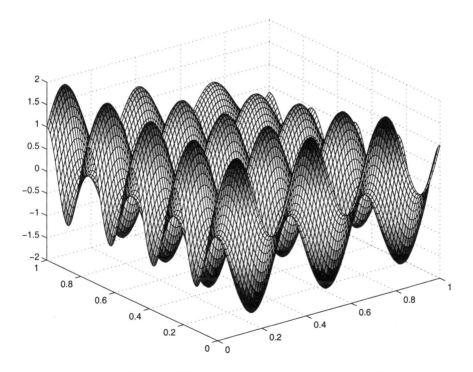

Fig. 4.10 Plot of the integrand function in example 4.10.

of the corresponding function values. The resulting surface is illustrated in figure 4.10. It is easy to see that Monte Carlo estimates with 10000 sampled points are not reliable.

```
>> rand('seed',0);
>> mean(f(rand(1,10000),rand(1,10000)))
ans =
    0.0086
>> mean(f(rand(1,10000),rand(1,10000)))
ans =
    0.0244
>> mean(f(rand(1,10000),rand(1,10000)))
ans =
    0.0194
```

So we may try with Halton's sequences, changing the bases and keeping the same number of samples:

```
>> seq2 = GetHalton(10000,2);
>> seq4 = GetHalton(10000,4);
>> seq5 = GetHalton(10000,5);
>> seq7 = GetHalton(10000,7);
>> mean(f(seq2,seq5))
ans =
    0.0200
>> mean(f(seq2,seq4))
ans =
    0.0224
>> mean(f(seq2,seq7))
ans =
    0.0199
>> mean(f(seq5,seq7))
ans =
    0.0198
```

We see that provided that we use prime numbers as the bases, the results are much more accurate. It is also instructive to compare the results for a small number of samples.

```
>> rand('seed',0)
>> mean(f(rand(1,100),rand(1,100)))
ans =
    -0.0330
>> mean(f(rand(1,500),rand(1,500)))
ans =
    -0.0412
```

```
>> mean(f(rand(1,1000),rand(1,1000)))
ans =
    0.0220
>> mean(f(rand(1,1500),rand(1,1500)))
ans =
    0.0338
>> mean(f(rand(1,2000),rand(1,2000)))
ans =
    0.0021

>> mean(f(seq2(1:100),seq7(1:100)))
ans =
    0.0267
>> mean(f(seq2(1:500),seq7(1:500)))
ans =
    0.0197
>> mean(f(seq2(1:1000),seq7(1:1000)))
ans =
    0.0210
>> mean(f(seq2(1:1500),seq7(1:1500)))
ans =
    0.0190
>> mean(f(seq2(1:2000),seq7(1:2000)))
ans =
    0.0197
```

The potential advantage of low-discrepancy sequences is evident even if the optimal choice of bases is an issue. ▯

4.5.2 Generating Sobol's low-discrepancy sequences

Halton's low-discrepancy sequences are arguably the simplest, but not necessarily the best. Choosing the best sequence in practice is still an open problem, but we would like at least to take a look at a more sophisticated alternative, i.e., Sobol's sequences. For the sake of clarity, it is better to consider the generation of a one-dimensional sequence x^n in the $[0,1]$ interval. A Sobol's sequence is generated on the basis of a set of "direction numbers" $v_1, v_2, \ldots$; we sill see shortly how direction numbers are selected, but for now just think of them as numbers which are less than 1. To get the nth number in the sequence, consider the binary representation of the integer n:

$$n = (\ldots b_3 b_2 b_1)_2.$$

The result is obtained by computing the bitwise exclusive or of the direction numbers v_i for which $b_i \neq 0$:

$$x^n = b_1 v_1 \oplus b_2 v_2 \oplus \cdots \tag{4.12}$$

If direction numbers are chosen properly, a low-discrepancy sequence will be generated [18]. A direction number may be thought as a binary fraction:

$$v_i = (0.v_{i1}v_{i2}v_{i3}\ldots)_2,$$

or as

$$v_i = \frac{m_i}{2^i},$$

where $m_i < 2^i$ is an odd integer. To generate direction numbers, we exploit primitive polynomials over the field $\mathbb{Z}_2$, i.e., polynomials with binary coefficients:

$$P = x^d + a_1 x^{d-1} + \cdots + a_{d-1}x + 1 \qquad a_k \in \{0,1\}.$$

Irreducible polynomials are those polynomials which cannot be factored; primitive polynomials are a subset of the irreducible polynomials and are strongly linked to the theory of error-correcting codes, which is beyond the scope of the book. Some irreducible polynomials over the field $\mathbb{Z}_2$ are listed, e.g., in [15, chapter 7], to which the reader is referred for further information. Given a primitive polynomial of degree d, the procedure for generating direction numbers is based on the recurrence formula

$$v_i = a_1 v_{i-1} \oplus a_2 v_{i-2} \oplus \cdots \oplus a_{d-1}v_{i-d+1} \oplus v_{i-d} \oplus \left[v_{i-d}/2^d\right] \qquad i > d.$$

This is better implemented in integer arithmetic as

$$m_i = 2a_1 m_{i-1} \oplus 2^2 a_2 m_{i-2} \oplus \cdots \oplus 2^{d-1}a_{d-1}m_{i-d+1} \oplus 2^d m_{i-d} \oplus m_{i-d}.$$

Some numbers $m_1,\ldots,m_d$ are needed to initialize the recursion. They may be chosen arbitrarily, provided that each m_i is odd and $m_i < 2^i$.

Example 4.11 As an example, let us build the set of direction numbers on the basis of the primitive polynomial

$$x^3 + x + 1.$$

The recursive scheme runs as follows:

$$m_1 = 4m_{i-2} \oplus 8m_{i-3} \oplus m_{i-3},$$

which may be initialized with $m_1 = 1$, $m_2 = 3$, $m_3 = 7$.[3] We may carry out the necessary computations step by step in MATLAB, using the `bitxor` function.

```
>> m = [1 3 7];
```

[3]The reasons why this may be a good choice are given in [4].

```
function [v, m] = GetDirNumbers(p,m0,n)
degree = length(p)-1;
p = p(2:degree);
m = [ m0 , zeros(1,n-degree) ];
for i= (degree+1):n
    m(i) = bitxor(m(i-degree), 2^degree * m(i-degree));
    for j=1:(degree-1)
        m(i) = bitxor(m(i), 2^j * p(j) * m(i-j));
    end
end
v=m./(2.^(1:length(m)));
```

Fig. 4.11   MATLAB code to generate direction numbers for Sobol's sequences.

```
>> i=4;
>> m(i) = bitxor( 4 * m(i-2) , bitxor(8*m(i-3) , m(i-3)));
>> i=5;
>> m(i) = bitxor( 4 * m(i-2) , bitxor(8*m(i-3) , m(i-3)));
>> i=6;
>> m(i) = bitxor( 4 * m(i-2) , bitxor(8*m(i-3) , m(i-3)));
>> m
m =
      1     3     7     5     7     43
```

Given the integer numbers m_i, we may build the direction numbers v_i. To implement the generation of direction numbers, we may use a function like GetDirNumbers, which is given in figure 4.11. The function requires a primitive polynomial p, a vector of initial numbers m, and the number n of direction numbers we want to generate. On exit we obtain the direction numbers v and the integer numbers m.

```
>> p = [1 0 1 1];
>> m0 = [1 3 7];
>> [v,m]=GetDirNumbers(p,m0,6)
v =
    0.5000    0.7500    0.8750    0.3125    0.2188    0.6719
m =
      1     3     7     5     7     43
```

The code is not optimized; for instance, the first and last coefficients of the polynomial should be 1 by default, and no check is done on the congruence in size of the input vectors.   ⎕

After computing the direction numbers, we could generate a Sobol's sequence according to equation (4.12). However, an improved method was proposed by Antonov and Saleev [1], who proved that the discrepancy is not changed by using the Gray code representation of n. Gray codes are discussed, e.g., in [15, chapter 20]; all we need to know is the following:

1. A Gray code is a function mapping an integer i to a corresponding binary representation $G(i)$; the function, for a given integer N, is one-to-one for $0 \leq i \leq 2^N - 1$.

2. A Gray code representation for the integer n is obtained from its binary representation by computing

$$\ldots g_3 g_2 g_1 = (\ldots b_3 b_2 b_1)_2 \oplus (\ldots b_4 b_3 b_2)_2.$$

3. The main feature of such a code is that the codes for consecutive numbers n and $n + 1$ differ only in one position.

Example 4.12 Computing a Gray code is easily accomplished in MATLAB. For instance, we may define an inline function and compute the Gray codes for the numbers $i = 0, 1, \ldots, 15$ as follows:

```
>> gray = inline('bitxor(x,bitshift(x,-1))');
>> codes = zeros(16,4);
>> for i=1:16, codes(i,:)=bitget(gray(i-1), [4 3 2 1]);, end
>> codes
codes =
     0     0     0     0
     0     0     0     1
     0     0     1     1
     0     0     1     0
     0     1     1     0
     0     1     1     1
     0     1     0     1
     0     1     0     0
     1     1     0     0
     1     1     0     1
     1     1     1     1
     1     1     1     0
     1     0     1     0
     1     0     1     1
     1     0     0     1
     1     0     0     0
```

We have used the function `bitshift` to shift the binary representation of x one position to the right and the function `bitget` to get specific bits of the binary representation of a number. We see that indeed the Gray codes for consecutive numbers i and $i + 1$ differ in one position; that position corresponds to the

```
function SobSeq = GetSobol(GenNumbers, x0, HowMany)
Nbits = 20;
factor = 2^Nbits;
BitNumbers = GenNumbers * factor;
SobSeq = zeros(HowMany + 1, 1);
SobSeq(1) = fix(x0*factor);
for i=1:HowMany
   c = min(find( bitget(i-1,1:16) == 0));
   SobSeq(i+1) = bitxor(SobSeq(i), BitNumbers(c));
end
SobSeq = SobSeq / factor;
```

Fig. 4.12 MATLAB code to generate a Sobol's sequence by the Antonov and Saleev approach.

rightmost zero bit in the binary representation of i (adding leading zeros if necessary).   ⎕

Using the feature of Gray codes, we may streamline generation of a Sobol sequence. Given x^n, we have

$$x^{n+1} = x^n \oplus v_c,$$

where c is the index of the rightmost zero bit b_c in the binary representation of n.

Example 4.13 To implement the mechanism in MATLAB, we need a way to find the rightmost zero bit in the binary representation of a number. A function like the following one will do (provided that at most eight bits are used to represent x):

```
rightbit = inline('min(find( bitget(x,1:8) == 0))')
```

Now we may put it all together. First, we generate the direction numbers. Then we initialize the sequence in some way, e.g., $x^0 = 0$, and apply the code of figure 4.12. The code is straightforward; the only point is that in theory we should compute the exclusive or on bits of a binary fraction; however, `bitxor` works on integer numbers only. This is why we shift everything to the left by `Nbits` position, which is accomplished multiplying by `factor` and dividing on exit from the function. Also, we truncate the initial number in order to make sure that we are "xoring" integer numbers.

```
>> p = [1 0 1 1];
>> m0 = [1 3 7];
>> [v,m]=GetDirNumbers(p,m0,6);
```

```
>> GetSobol(v,0,10)
ans =
           0
      0.5000
      0.2500
      0.7500
      0.1250
      0.6250
      0.3750
      0.8750
      0.6875
      0.1875
      0.9375
```

Using a different set of generating numbers and a different starting point, we generate different sequences.

```
>> p = [1 0 1 1 1 1];
>> m0 = [1 3 5 9 11];
>> [v,m]=GetDirNumbers(p,m0,8);
>> GetSobol(v,0.124,10)
ans =
      0.1240
      0.6240
      0.3740
      0.8740
      0.4990
      0.9990
      0.2490
      0.7490
      0.1865
      0.6865
      0.4365
```

Note that to generate longer sequences, more generating numbers are needed.

□

4.6 INTEGRATING SIMULATION AND OPTIMIZATION

Simulation models are a convenient way to evaluate the performance of complex and stochastic systems for which analytical models may be very hard or even impossible to come up with. However, they are just able to evaluate a performance measure given a set of input parameters. In option pricing, this may be just what we need, but we could also be interested in finding the optimal set of parameters; in other words, in many settings, such as portfolio optimization, we would like to integrate simulation and optimization (see, e.g., [6]). Such an integration may certainly be worthwhile, as it provides us with a way to optimize complex and stochastic systems which cannot be dealt with by deterministic and even stochastic programming. However, we may have to face at least some of the following issues:

- The objective function may be nonconvex.

- Some of the input parameters may be discrete rather than continuous.

- The evaluation of the objective function may be affected by noise.

- Using gradient-based methods may be difficult, as gradients must be estimated.

Let us start from the last point and assume for simplicity that we want to solve an unconstrained optimization whereby the objective function is the expected value of some random performance measure depending on a vector of parameters $\mathbf{x} \in \mathbb{R}^n$:

$$\min f(\mathbf{x}) = E_\omega[h(\mathbf{x}, \omega)].$$

For optimization purposes it would be useful to have a way to compute the gradient $\nabla f(\mathbf{x})$ at any point. As pointed out in section 4.4.2, a gradient could be estimated by finite differences, but this is made difficult by the noise in the estimates. Using common random numbers to reduce variance is the least we should do; an alternative is represented by using some form of regression. The idea is to use a simulation model to build a sort of empirical metamodel, the response surface, which yields an analytical approximation $g(\mathbf{x})$ of the performance measure $f(\mathbf{x})$ with respect to the input parameters. If we want to estimate the gradient at a certain point $\mathbf{x}$, we may consider a linear approximation, such as

$$g(\mathbf{x}) = \alpha + \sum_{i=1}^{n} \beta_i x_i = \alpha + \boldsymbol{\beta}^T \mathbf{x}.$$

We may estimate α and $\boldsymbol{\beta}$ by evaluating f for a set of test values $\mathbf{x}^j$ and by minimizing a function of the regression errors. Let $\hat{f}_j$ be the estimate of f corresponding to the point $\mathbf{x}^j$ $(j = 1, \ldots, m)$. We have

$$\hat{f}_j = \alpha + \boldsymbol{\beta}^T \mathbf{x}^j + \epsilon_j,$$

where ϵ_j is an error term. Using least squares approximation, we may find α and $\boldsymbol{\beta}$ in such a way that $\sum_j \epsilon_j^2$ is minimized. Let us define the matrix

$$\mathbf{X} = \begin{bmatrix} 1 & x_1^1 & x_2^1 & \cdots & x_n^1 \\ 1 & x_1^2 & x_2^2 & \cdots & x_n^2 \\ \vdots & \vdots & \vdots & \ddots & \vdots \\ 1 & x_1^m & x_2^m & \cdots & x_n^m \end{bmatrix},$$

where x_i^j is the jth setting of the parameter x_i. It can be shown that the sum of the squared errors is minimized by

$$\begin{bmatrix} \hat{\alpha} \\ \hat{\beta} \end{bmatrix} = (\mathbf{X}^T\mathbf{X})^{-1}\mathbf{X}^T\hat{\mathbf{f}},$$

where $\hat{\mathbf{f}}$ is the vector of the m estimates of f. Then we may set $\hat{\nabla}f^{(k)} = \beta$ and use it within a gradient optimization method. A first-order fit is suitable when we are not close to the optimum. When we are approaching the minimizer, a quadratic polynomial can be fitted:

$$f(\mathbf{x}) = \alpha + \beta^T\mathbf{x} + \mathbf{x}^T\mathbf{\Gamma}\mathbf{x},$$

where $\mathbf{\Gamma}$ is a square matrix, and quadratic programming may be used to find the optimal set of parameters for the metamodel, which is successively updated until some convergence criterion is met [8]. This results in a method resembling the quasi-Newton methods for nonlinear programming.

An obvious disadvantage of an approach based on the response surface methodology is that it is likely to be quite expensive in computational terms. Alternative methods, such as perturbation analysis, have been proposed to estimate sensitivities with a single simulation runs. An example of an application to estimate option sensitivities can be found in [5]. A treatment of these methods require deep mathematical knowledge, so we refer, e.g., to [14] for a thorough treatment of these topics. We would only like to point out a subtle issue of using gradient-based methods for simulation optimization. In principle, we should evaluate

$$\nabla f(\mathbf{x}) = \nabla \mathrm{E}_\omega[h(\mathbf{x},\omega)],$$

but simulation actually yields something like

$$\mathrm{E}_\omega[\nabla h(\mathbf{x},\omega)].$$

That expectation can be commuted with differentiation is not granted at all. This issue is well explored in [9].

Given all of the considerations above, it's no surprise that nonlinear programming methods that do not exploit derivatives in any way are of interest for simulation optimization. One such method is the simplex search procedure we have outlined in section 3.2.4; see [10] for a recent paper on this topic.[4]

Although using a simplex search procedure has its merit, it does not overcome the possible difficulties due to the nonconvexity of the objective function or the discrete character of some decision parameters. For such cases the integration of simulation with metaheuristics such as tabu search or genetic algorithms, which we have described in section 3.6, is probably the only practical

---

[4]It may also be worth noting that MATLAB allows the integration of simplex search and other no-derivatives methods with the dynamic systems simulator SIMULINK.

solution approach. Indeed, this is the approach taken in some commercial stochastic simulation packages. The application of a population-based approach like genetic algorithms or their variants has the further advantage of making the noisy function evaluations less critical.

For further reading

In the literature

- For a general introduction to simulation, see [11] or [17], both of which have heavily influenced the presentation in this chapter.

- For a more theoretical treatment of Monte Carlo simulation and random number generation, see [7]. The random number generators used in MATLAB are described in [12].

- Low-discrepancy sequences are treated in [13].

- For a tutorial survey on the integration of simulation and optimization, see [8]. A deep mathematical treatment is given in [14].

- The integration of simulation and optimization is surveyed in [8]. The use of simplex search to drive a simulator is explored in [2] and [10].

On the Web

- For a list of resources on Monte Carlo and quasi-Monte Carlo simulation, see http://www.mcqmc.org.

- See also http://www.mat.sbg.ac.at/~schmidw/links.html.

REFERENCES

1. I.A. Antonov and V.M. Saleev. An Economic Method of Computing LP_τ Sequences. *USSR Computational Mathematics and Mathematical Physics*, 19:252–256, 1979.

2. R.R. Barton and Jr. J.S. Ivey. Nelder-Mead Simplex Modifications for Simulation Optimization. *Management Science*, 42:954–973, 1996.

3. I. Beichl and F. Sullivan. The Importance of Importance Sampling. *Computing in Science and Engineering*, 1:71–73, March-April 1999.

4. P. Bratley and B.L. Fox. Algorithm 659: Implementing Sobol's Quasirandom Sequence Generator. *ACM Transactions on Mathematical Software*, 14:88–100, 1988.

5. M. Broadie and P. Glasserman. Estimating Security Price Derivatives Using Simulation. *Management Science*, 42:269–285, 1996.

6. A. Consiglio and S.A. Zenios. Designing Portfolios of Financial Products via Integrated Simulation and Optimization Models. *Operations Research*, 47:195–208, 1999.

7. G.S. Fishman. *Monte Carlo: Concepts, Algorithms, and Applications*. Springer-Verlag, Berlin, 1996.

8. M.C. Fu. Optimization by Simulation: A Review. *Annals of Operations Research*, 53:199–247, 1994.

9. P. Glasserman. *Gradient Estimation via Perturbation Analysis*. Kluwer Academic, Boston, MA, 1991.

10. D.G. Humphrey and J.R. Wilson. A Revised Simplex Search Procedure for Stochastic Simulation Response Surface Optimization. *INFORMS Journal on Computing*, 12:272–283, 2000.

11. A.M. Law and W.D. Kelton. *Simulation Modeling and Analysis (2nd ed.)*. McGraw-Hill, New York, 1991.

12. C. Moler. Random Thoughts. *Matlab News & Notes*, pages 2–3, Fall 1995. This paper may be downloaded from The Mathworks' Web site at `http://www.mathworks.com/company/newsletter/pdf/Cleve.pdf`.

13. H. Niederreiter. *Random Number Generation and Quasi-Monte Carlo Methods*. Society for Industrial and Applied Mathematics, Philadelphia, PA, 1992.

14. G.C. Pflug. *Optimization of Stochastic Models: the Interface between Simulation and Optimization*. Kluwer Academic, Dordrecht, The Netherlands, 1996.

15. W.H. Press, S.A. Teukolsky, W.T. Vetterling, and B.P. Flannery. *Numerical Recipes in C*. Cambridge University Press, Cambridge, 1993.

16. S. Ross. *Introduction to Probability Models (6th ed.)*. Academic Press, San Diego, CA, 1997.

17. S. Ross. *Simulation*. Academic Press, San Diego, CA, 1997.

18. I.M. Sobol. On the Distribution of Points in a Cube and the Approximate Evaluation of Integrals. *USSR Computational Mathematics and Mathematical Physics*, 7:86–112, 1967.

5

Finite difference methods for partial differential equations

Partial differential equations (PDEs) play a major role in financial engineering. Since the seminal work leading to the the Black-Scholes equation, which we introduced in section 1.4.2, PDEs have become an important tool in option valuation. It turns out that PDEs provide a powerful and consistent framework for pricing rather complex derivatives. Unfortunately, as analytical solutions like the Black and Scholes formula are not available in general, one must often resort to numerical methods.

The numerical solution of PDEs is a common tool in mathematical physics and engineering, and quite sophisticated methods have been developed. The complexity of the methods also depends on the specific type of PDE at hand. As expected, nonlinear equations are generally more difficult than linear ones, but there is also a subtler dependence on numerical parameters, since a change in the sign of a coefficient may drastically change the characteristics of an equation. In the financial engineering case, it happens that most of the time rather simple methods are enough to obtain a reasonably accurate solution. Indeed, we deal here only with relatively straightforward finite difference methods, which are based on the natural idea of approximating partial derivatives with difference quotients. Even so, the topic is not as trivial as one may think, since careless use of finite difference schemes may lead to unreasonable results. In fact, while some authors suggest the use of PDEs as the single most useful tool in derivatives pricing [8, p. 615], others suggest that they are quite vulnerable to numerical issues and, while acknowledging the role of finite difference methods, suggest the use of lattice-based methods when possible (see, e.g., [2, p. 365]). Actually, this is a bit a matter of taste, and when confident with one method, one is able to squeeze the most out of it. Fortunately, when numerical difficulties occur in solving a PDE for a financial problem, usually the answers we get from the algorithm are so blatantly senseless that we may easily spot the trouble. In this chapter we introduce concepts related to convergence, consistency, and stability in order to understand the basic issues connected with the numerical solution of PDEs.

It should be stressed that PDEs are actually a difficult topic requiring advanced mathematical concepts for a detailed treatment. We limit ourselves to an indispensable background; a further limitation is that we do not consider systems of PDEs. We first classify PDEs in section 5.1. Then in section 5.2 we introduce different ways to approximate partial derivatives by finite differences, leading to different solution schemes which may turn out numerically stable or unstable. We devote a particular attention to the heat equation, which is the subject of section 5.3, since the Black-Scholes PDE is strongly linked to diffusion processes. In section 5.4 we briefly point out a few theoretical concepts concerning the convergence of finite difference methods. Finally, supplement S5.1 gives some information about the concept of characteristic curves, justifying the classification schemes for second-order quasilinear PDEs.

5.1 INTRODUCTION AND CLASSIFICATION OF PDEs

We introduced the Black-Scholes PDE in section 1.4.2 to find the theoretical price $f(S, t)$ of a derivative security depending on the price S of one underlying asset at time t. Using a stochastic differential equation to model the dynamics of the underlying asset price and using no arbitrage arguments, we have found that f must satisfy the equation

$$\frac{\partial f}{\partial t} + \frac{1}{2}\sigma^2 S^2 \frac{\partial^2 f}{\partial S^2} + rS\frac{\partial f}{\partial S} - rf = 0, \qquad (5.1)$$

where r is the risk-free interest rate and σ is the asset price volatility. Suitable boundary conditions must be added to find a specific solution depending on the option type we are considering. This equation has various features:

- It is second-order.

- It is linear.

- It is a parabolic equation.

All these features refer to how PDEs are classified; such a classification is relevant in that the choice of a numerical method to cope with a PDE generally depends on its characteristics.

In order to classify PDEs, let us abstract from the financial interpretation of the variables involved and refer to an unknown function $\phi(x, y)$, depending on variables x and y; for simplicity we deal with a function of two independent variables only, but the classification scheme may be applied in a more general setting. The *order* of a PDE is the highest order of the derivatives involved. For instance, a generic first-order equation has the form

$$a(x,y)\frac{\partial \phi}{\partial x} + b(x,y)\frac{\partial \phi}{\partial y} + c(x,y)\phi + d(x,y) = 0,$$

where a, b, c, d are given functions of the independent variables. This equation is first-order since only first-order derivatives are involved. Furthermore, it is linear, since the functions a, b, c, and d depend only on the independent variables x and y and not on ϕ itself. By the same token, the generic form of a linear second-order equation is

$$a\frac{\partial^2 \phi}{\partial x^2} + b\frac{\partial^2 \phi}{\partial x\, \partial y} + c\frac{\partial^2 \phi}{\partial y^2} + d\frac{\partial \phi}{\partial x} + e\frac{\partial \phi}{\partial y} + f\phi + g = 0,$$

where again all the given functions depend only on x and y. An example of a first-order nonlinear equation is

$$\left(\frac{\partial \phi}{\partial x}\right)^2 + \left(\frac{\partial \phi}{\partial y}\right)^2 = 1. \tag{5.2}$$

An example of a second-order nonlinear equation is

$$a\left(x, y, \frac{\partial \phi}{\partial y}\right)\frac{\partial^2 \phi}{\partial x^2} + d(x, y, \phi)\frac{\partial \phi}{\partial x} + e(x, y)\frac{\partial \phi}{\partial y} + f(x, y)\phi = 0. \tag{5.3}$$

Equation (5.3) is nonlinear but in a different way than (5.2). In this equation, the coefficient a of the highest-order derivative depends only on the first-order derivative. We have a *quasilinear equation* whenever the highest-order derivatives occur linearly, with coefficients depending only on the independent variables, the unknown function ϕ, and its lower-order derivatives. For the sake of simplicity, in this introductory book we deal only with linear equations. It should be noted that while most of the models you will see in finance are linear, nonlinear equations may be obtained when relaxing some of the assumptions behind the Black-Scholes model; for an example of a nonlinear equation that arises when introducing transaction costs, see [8, chapter 21].

It is customary to classify quasilinear second-order equations depending on the sign of the expression $b^2 - 4ac$:

- If $b^2 - 4ac > 0$, the equation is hyperbolic.

- If $b^2 - 4ac = 0$, the equation is parabolic.

- If $b^2 - 4ac < 0$, the equation is elliptic.

It is easy to see that the discriminant term $b^2 - 4ac$ is formally similar to the analogous term we have in second-degree algebraic equations. In the algebraic case, when it is positive we have two real roots, when it is null we have one, and when it is negative we have no real roots. The idea here is that with hyperbolic equations there are two distinct curves in the x, y plane, called *characteristics*, along which the PDE can be transformed to an ordinary differential equation. Parabolic equations have one characteristic, whereas elliptic equations have none. We should also note that depending on the functions a, b, and c, a PDE may be of one type in a region of the plane and of another type in another region. The concept of characteristic is better explained in supplement S5.1 at the end of the chapter.[1]

Elliptic equations may arise in equilibrium models (where time is not involved). A typical example is the Laplace equation

$$\frac{\partial^2 \phi}{\partial x^2} + \frac{\partial^2 \phi}{\partial y^2} = 0.$$

Here we have $a = c = 1$ and $b = 0$, so that $b^2 - 4ac = -4 < 0$. The wave equation

$$\frac{\partial^2 \phi}{\partial t^2} - \rho^2 \frac{\partial^2 \phi}{\partial x^2} = 0,$$

where t is time, is a typical example of a hyperbolic equation, since the discriminant term is $4\rho^2 > 0$. The prototype parabolic equation is the *heat* (or *diffusion*) *equation*:

$$\frac{\partial \phi}{\partial t} = k \frac{\partial^2 \phi}{\partial x^2},$$

where t is time and ϕ is the temperature of a point with coordinate x on a line. In this case, $b^2 - 4ac = 0$. By a change of variables, the equation may be cast into a dimensionless form:

$$\frac{\partial \phi}{\partial t} = \frac{\partial^2 \phi}{\partial x^2}. \tag{5.4}$$

Now consider the Black-Scholes equation; again $b = c = 0$, so the equation is parabolic. This does not happen by chance, since with some work it can be shown that the Black-Scholes equation actually boils down to the heat equation.

An equation like (5.4) must be integrated with suitable conditions in order to pinpoint a meaningful solution. For instance, assume that $\phi(x, t)$ is the "temperature" at point $x \in [0, 1]$ of a rod of length 1 at time t; the endpoints are kept at a constant temperature u_0, and the initial temperature of the rod is given over all of its length. Then we must add the initial condition

$$\phi(x, 0) = u(x) \qquad 0 \le x \le 1,$$

---

[1]This supplement is optional and may be skipped with no loss in continuity.

and the boundary conditions

$$\phi(0,t) = \phi(1,t) = u_0 \qquad t > 0.$$

Here the domain is bounded with respect to space and unbounded with respect to time. In financial problems, the initial condition is usually replaced by a terminal condition, as the option payoff is known at expiration; therefore, the time domain is bounded, whereas the domain with respect to the price of the underlying asset may be (in principle) unbounded. From a computational point of view, the domain must be limited in some sensible way. Boundary conditions are easy to spot for vanilla European options. With exotic options, enforcing boundary conditions may be more complicated, e.g., when the boundary conditions must themselves be approximated by some numerical scheme. In other cases, such as barrier options, the boundary conditions may actually result in a simplification of the problem. American options raise another issue; for each time before expiration, there is a critical value for the price of the underlying asset at which it is optimal to exercise the option (see figure 1.17 on page 57); depending on the option type (call or put), it will also be optimal to exercise the optimal for prices above and below the critical price.[2] So with American options we should cope with a *free boundary*, i.e., a boundary within the domain, which separates the exercise and no-exercise region. We deal with these issues in chapter 8.

A noteworthy feature of the heat equation is that any discontinuity in the initial conditions is somehow smoothed out, so that the solution for $t > 0$ is differentiable everywhere. On the contrary, in the wave equation, the irregularities are propagated along the characteristic lines. Another feature of parabolic equations is that they are relatively easy to work with from the numerical point of view.

A final remark is that the form of the equation and the boundary conditions determine if a given problem involving a PDE is *well posed*. A problem is well posed if:

- There exists a solution.

- The solution is unique (at least within a certain class of functions of interest).

- The solution depends in a nice way on the problem data (i.e., a small perturbation in the problem data results in a small perturbation of the solution).

We will trust our intuition that the equations we write make sense and will assume implicitly that all our problems are well posed.

---

[2]Recall that a vanilla American call should never be exercised unless the stock pays dividends.

5.2   NUMERICAL SOLUTION BY FINITE DIFFERENCE METHODS

Finite difference methods to solve PDEs are based on the simple idea of approximating each partial derivative by a difference quotient. This transforms the functional equation into a set of algebraic equations. As in many numerical algorithms, the starting point is a finite series approximation. Under suitable continuity and differentiability hypotheses, Taylor's theorem states that a function $f(x)$ may be represented as

$$f(x+h) = f(x) + hf'(x) + \frac{1}{2}h^2 f''(x) + \frac{1}{6}h^3 f'''(x) + \cdots. \qquad (5.5)$$

If we neglect the terms of order h^2 and higher, we get

$$f'(x) = \frac{f(x+h) - f(x)}{h} + O(h). \qquad (5.6)$$

This is the *forward* approximation for the derivative; indeed, the derivative is just defined as a limit of the difference quotient above as $h \to 0$. There are alternative ways to approximate first-order derivatives. By similar reasoning, we may write

$$f(x-h) = f(x) - hf'(x) + \frac{1}{2}h^2 f''(x) - \frac{1}{6}h^3 f'''(x) + \cdots, \qquad (5.7)$$

from which we obtain the *backward* approximation,

$$f'(x) = \frac{f(x) - f(x-h)}{h} + O(h). \qquad (5.8)$$

In both cases we get a truncation error of order $O(h)$. A better approximation can be obtained by subtracting equation (5.7) from equation (5.5) and rearranging:

$$f'(x) = \frac{f(x+h) - f(x-h)}{2h} + O(h^2). \qquad (5.9)$$

This is the *central* approximation, and for small h it is a better approximation, since the truncation error is $O(h^2)$. Why this is the case may also be seen from figure 5.1. However, this does not imply that forward and backward approximations must be disregarded; they may be useful to come up with efficient numerical schemes, depending on the type of boundary conditions.

The reasoning may be extended to higher-order derivatives. To cope with the Black-Scholes equation, we must approximate second-order derivatives, too. This is obtained by adding equations (5.5) and (5.7), which yields

$$f(x+h) + f(x-h) = 2f(x) + h^2 f''(x) + O(h^4),$$

and rearranging yields

$$f''(x) = \frac{f(x+h) - 2f(x) + f(x-h)}{h^2} + O(h^2). \qquad (5.10)$$

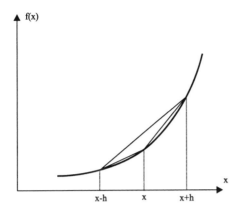

Fig. 5.1 Graphical illustration of forward, backward, and symmetric approximations of a derivative.

In order to apply the ideas above to a PDE involving a function $\phi(x, y)$, it is natural to set up a discrete grid of points of the form $(i\,\delta x, j\,\delta y)$, where δx and δy are discretization steps, and to look for the values of ϕ on this grid. It is customary to use the notation

$$\phi_{ij} = \phi(i\,\delta x, j\,\delta y).$$

$$\phi_{ij} = \phi(i\,\delta x, j\,\delta y).$$

Depending on the type of equation and on how the derivatives are approximated, we obtain a set of algebraic equations which may be more or less easily solved. A possible difficulty is represented by boundary conditions. If the equation is defined over a rectangular domain in the (x, y) space, it is easy to set up a grid such that the boundary points are on the grid. Other cases might not be so easy, and a sensible way to approximate the boundary conditions must be devised. Nevertheless, we would expect that for $\delta x, \delta y \to 0$ the solution of this set of equations converges (in some sense) to the solution of the PDE. Actually, this is not granted at all, as different complications may arise.

5.2.1 Bad example of a finite difference scheme

Consider the following example of a first-order linear equation:[3]

$$\frac{\partial \phi}{\partial t} + c \frac{\partial \phi}{\partial x} = 0, \tag{5.11}$$

where $\phi = \phi(x, t)$, $c > 0$, and the initial condition

$$\phi(x, 0) = f(x) \qquad \forall x$$

[3]The example is taken from [1, chapter 2].

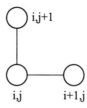

Fig. 5.2 Representing a finite difference scheme by a computational diagram.

is given. It is easy to verify that the solution is of the form

$$\phi(x, t) = f(x - ct);$$

in other words, the solution is simply a translation of $f(x)$ with velocity of propagation c. In fact, this type of equation is called the *transport equation* [and may involve a function $c(x)$ rather than a constant velocity c]. We take for granted that the problem is well posed, and we do not check the uniqueness of the solution (see [1, pp. 21-25] for a thorough discussion). Now let us ignore what we know about the solution and try a finite difference scheme based on forward approximations. Equation (5.11) may be approximated by

$$\frac{\phi(x, t + \delta t) - \phi(x, t)}{\delta t} + c \frac{\phi(x + \delta x, t) - \phi(x, t)}{\delta x} + O(\delta t) + O(\delta x) = 0,$$

which, neglecting the truncation error and using the grid notation $x = i \, \delta x$, $t = j \, \delta t$, yields

$$\frac{\phi_{i,j+1} - \phi_{ij}}{\delta t} + c \frac{\phi_{i+1,j} - \phi_{ij}}{\delta x} = 0, \tag{5.12}$$

with the initial condition

$$\phi_{i0} = f(i\delta x) = f_i \qquad \forall i. \tag{5.13}$$

In practice, in order to solve the problem on a computer, we should restrict the domain in some way, enforcing some limits on i and j. For now, we simply assume that we are interested in the solution for $t > 0$, hence $j = 1, 2, 3, \ldots$ Now, how can we solve equation (5.12) in a systematic way? If we look at time 0, i.e., when $j = 0$, we know the solution; if we step forward to $j = 1$, we see that there is an unknown value in equation (5.12), namely $\phi_{i,j+1}$, which may be obtained as an explicit function of known values. In fact, solving for the unknown value, we get

$$\phi_{i,j+1} = \left(1 + \frac{c}{\rho}\right) \phi_{ij} - \frac{c}{\rho} \phi_{i+1,j}, \tag{5.14}$$

where $\rho = \delta x / \delta t$. This computational scheme can be represented by the computational diagram depicted in figure 5.2, and it is easy to understand

and implement. Unfortunately, it need not converge to the solution of the equation. Consider the following initial condition:

$$f(x) = \begin{cases} 0 & x < -1, \\ x + 1 & -1 \le x \le 0, \\ 1 & x > 0, \end{cases} \tag{5.15}$$

which implies that

$$\phi_{i0} = f(i\delta x) = 1 \qquad \forall i \ge 0.$$

Now, using the computational scheme (5.14), for $j = 1$ we have

$$\phi_{i,1} = \left(1 + \frac{c}{\rho}\right)\phi_{i0} - \frac{c}{\rho}\phi_{i+1,0} = 1 \qquad \forall i \ge 0.$$

Repeating this argument for any time instant $(j = 2, 3, \ldots)$, it is easily seen that, however small we take the discretization steps,

$$\phi_{ij} = 1 \qquad i, j \ge 0,$$

which is certainly not the correct solution. Some readers might wonder if this is due to some irregularity in the initial values. In fact, the derivative of $f(x)$ is discontinuous at certain points,[4] but it is easy to see that using a regular function would not change the issue.

5.2.2 Instability in a finite difference scheme

The reason for the failure of the finite difference scheme in section 5.2.1 is that it does not reflect the physical propagation process, where the initial condition is translated "to the right" with respect to space. Intuitively, we could fix the problem by adopting the computational scheme represented in figure 5.3, which is obtained by using a backward difference for the partial derivative with respect to x. This yields

$$\frac{\phi_{i,j+1} - \phi_{ij}}{\delta t} + c\frac{\phi_{ij} - \phi_{i-1,j}}{\delta x} = 0, \tag{5.16}$$

and solving for $\phi_{i,j+1}$, we get the scheme

$$\phi_{i,j+1} = \left(1 - \frac{c}{\rho}\right)\phi_{ij} + \frac{c}{\rho}\phi_{i-1,j}. \tag{5.17}$$

Note that here $\phi_{i,j+1}$ still depends on the data at the previous time instant but "to the left" with respect to space. Let us try this scheme with MATLAB.

---

[4]The irregularity in the initial condition is propagated along a characteristic curve, as described in supplement S5.1. This example shows that nondifferentiable functions may look like acceptable solutions of a PDE, which is a bit odd since derivatives are not defined everywhere for such functions; a rigorous investigation of this question leads to the concept of weak solution of a PDE [1].

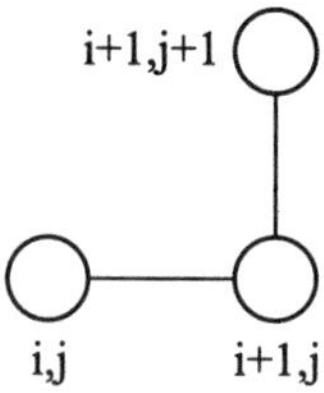

Fig. 5.3 Computational diagram of the modified scheme for the transport equation.

```
% f0transp.m
function y=f0transp(x)
if (x < -1)
    y=0;
elseif (x <= 0)
    y=x+1;
else
    y=1;
end
```

Fig. 5.4 Function to evaluate the initial values for the transport equation.

Example 5.1 In order to apply the computational scheme (5.17) with initial condition (5.15), we have to write a few M-files. In figure 5.4 we have the code to evaluate the initial value at a given point x at $t = 0$. In figure 5.5 we see the MATLAB code for solving the equation. Note that we must truncate the domain between minimum and maximum x values, and with respect to time as well. We use a fixed value for the leftmost value in space, assuming that for smaller x the initial value is constant. Finally, the function TransportPlot illustrated in figure 5.6 is simply used to plot the numerical solution at different times: four time subscripts are passed as an argument and the corresponding four plots are obtained. To begin with, we may solve the equation on the domain $-2 \le x \le 3$, $0 \le t \le 2$, with discretization steps $\delta x = 0.05$, $\delta t = 0.01$:

```
>> xmin = -2;
>> xmax = 3;
>> dx = 0.05;
>> tmax = 2;
>> dt = 0.01;
>> c = 1;
>> sol = transport(xmin, dx, xmax, dt, tmax, c, 'f0transp');
>> TransportPlot(xmin, dx, xmax, [1 50 100 200], sol)
```

```
% transport.m
function [solution, N, M] = transport(xmin, dx, xmax, dt, tmax, c, f0)
N = ceil((xmax - xmin) / dx);
xmax = xmin + N*dx;
M = ceil(tmax/dt);
k1 = 1 - dt*c/dx;
k2 = dt*c/dx;
solution = zeros(N+1,M+1);
vetx = xmin:dx:xmax;
for i=1:N+1
   solution(i,1) = feval(f0,vetx(i));
end
fixedvalue = solution(1,1);
% this is needed because of finite domain
for j=1:M
   solution(:,j+1) = k1*solution(:,j)+k2*[fixedvalue ; solution(1:N,j)];
end
```

Fig. 5.5 Code implementing the finite difference scheme for the transport equation.

```
% TransportPlot.m
function TransportPlot(xmin, dx, xmax, times, sol)
subplot(2,2,1)
plot(xmin:dx:xmax, sol(:,times(1)))
axis([xmin xmax -0.1 1.1])
subplot(2,2,2)
plot(xmin:dx:xmax, sol(:,times(2)))
axis([xmin xmax -0.1 1.1])
subplot(2,2,3)
plot(xmin:dx:xmax, sol(:,times(3)))
axis([xmin xmax -0.1 1.1])
subplot(2,2,4)
plot(xmin:dx:xmax, sol(:,times(4)))
axis([xmin xmax -0.1 1.1])
```

Fig. 5.6 Function for plotting the numerical solution of the transport equation.

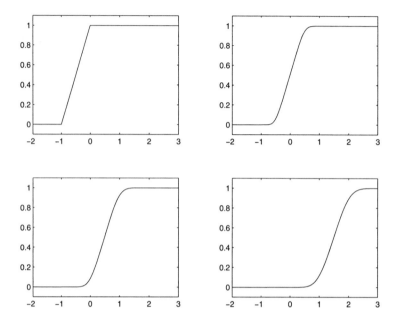

Fig. 5.7 Numerical solution of the transport equation for $\delta x = 0.05$, $\delta t = 0.01$; $t = 0.01$, $t = 0.5$, $t = 1$, and $t = 2$.

The solution, plotted in figure 5.7, gets progressively translated as we would expect, but it also looks progressively "smoothed." This could be due to a coarse discretization along the x axis. So we may try with $\delta x = 0.01$:

```
>> dx = 0.01;
>> sol = transport(xmin, dx, xmax, dt, tmax, c, 'f0transp');
>> TransportPlot(xmin, dx, xmax, [1 50 100 200], sol)
```

The solution is depicted in figure 5.8, and it looks much better. So, why don't we try a finer discretization, say $\delta x = 0.005$?

```
>> dx = 0.005;
>> sol = transport(xmin, dx, xmax, dt, tmax, c, 'f0transp');
>> TransportPlot(xmin, dx, xmax, [1 6 7 8], sol)
```

The solution we see in figure 5.9 is not really satisfactory. Something is definitely going wrong.   ⬚

As we may see, for certain settings of the discretization steps, the finite difference method is subject to numerical instability. By looking at equation (5.17), we may see that what we are doing is similar to a convex combination of two values; indeed, it will be a convex combination, provided that $c/\rho \geq 0$,

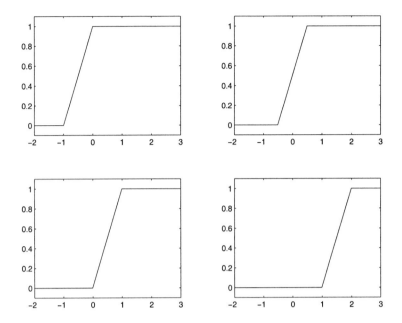

Fig. 5.8 Numerical solution of the transport equation for $\delta x = 0.01$, $\delta t = 0.01$; $t = 0.01$, $t = 0.5$, $t = 1$, and $t = 2$.

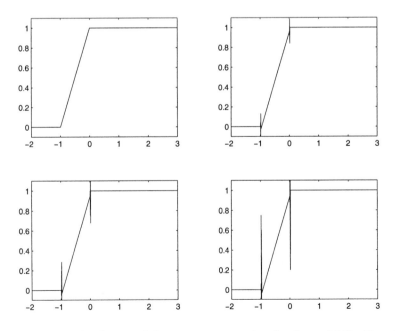

Fig. 5.9 Numerical solution of the transport equation for $\delta x = 0.005$, $\delta t = 0.01$; $t = 0.01$, $t = 0.06$, $t = 0.07$, and $t = 0.08$.

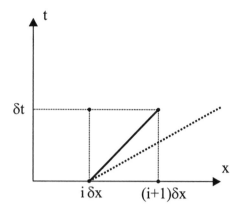

Fig. 5.10 Physical interpretation of the stability condition (5.18).

which is the case as we assumed that $c > 0$, and $c/\rho \leq 1$, i.e.,

$$c\,\delta t \leq \delta x. \tag{5.18}$$

If this condition is not met, we have a negative coefficient in the linear combination (5.17); but if the initial data are positive, we would not expect negative quantities.

It is also possible to give a more physical interpretation of the stability condition (5.18) in terms of a domain of influence. Consider figure 5.10. Due to the structure of the numerical scheme (5.17), the value $\phi_{i+1,1}$ depends on the values ϕ_{i0} and $\phi_{i+1,0}$. The exact solution of the transport equation is such that the initial value at point $i\,\delta x$ should influence only the values on the characteristic line represented as a dotted line in figure 5.10. The slope of the characteristic line is $1/c$; the slope of the line joining the points $(i\,\delta x, 0)$ and $((i+1)\delta x, \delta t)$ is clearly $\delta t/\delta x$. In the figure this second line has a larger slope than the first one and the stability condition (5.18) is violated, since

$$\frac{\delta t}{\delta x} > \frac{1}{c}.$$

From a physical point of view this makes no sense, since in this case the numerical scheme is such that the initial value at point $i\,\delta x$ is influencing the value at a point *above* the characteristic line. In other words, the "speed" of the numerical scheme, $\delta x/\delta t$, should not be smaller than the transport speed c to ensure stability.

All of these considerations are nothing more than intuitive arguments. The instability problem may be analyzed rigorously in different ways. One approach is related to Fourier analysis and is illustrated in the next example. Another approach, based on matrix theoretic arguments, will be illustrated in section 5.3, where we consider the heat equation in more detail. It should also be noted that in some cases a financial interpretation of instability may be given (see section 8.2.1).

Example 5.2 Consider again the transport equation, but with different initial values:

$$\phi(x,0) = f(x) = \epsilon \, \cos\!\left(\frac{\pi x}{\delta x}\right).$$

Since we know that the exact solution is $\phi(x,t) = f(x - ct)$, we see that the solution will be bounded everywhere, just like the initial values. Note also that after discretization we have a peculiar set of initial values on the grid:

$$\phi_{i,0} = \epsilon \, \cos\!\left(\frac{\pi i \, \delta x}{\delta x}\right) = \epsilon \, (-1)^i.$$

Going forward one layer of nodes in time, applying the scheme (5.17) yields

$$
\begin{aligned}
\phi_{i,1} &= \left(1 - \frac{c}{\rho}\right)\epsilon\,(-1)^i + \frac{c}{\rho}\epsilon\,(-1)^{i-1} = \left(1 - \frac{c}{\rho}\right)\epsilon\,(-1)^i - \frac{c}{\rho}\epsilon\,(-1)^i \\
&= \epsilon\,(-1)^i\left(1 - 2\frac{c}{\rho}\right).
\end{aligned}
$$

By the same token,

$$
\begin{aligned}
\phi_{i,2} &= \left(1 - \frac{c}{\rho}\right)\epsilon\,(-1)^i\left(1 - 2\frac{c}{\rho}\right) + \frac{c}{\rho}\epsilon\,(-1)^{i-1}\left(1 - 2\frac{c}{\rho}\right) \\
&= \epsilon\,(-1)^i\left(1 - 2\frac{c}{\rho}\right)^2,
\end{aligned}
$$

and in general we get

$$\phi_{ij} = \epsilon\,(-1)^i\left(1 - 2\frac{c}{\rho}\right)^j.$$

We see that the if the stability condition (5.18) is violated, i.e., if $c/\rho > 1$, we have

$$|1 - 2c/\rho| > 1$$

and the initial data are amplified by a factor that goes to infinity for increasing j. □

5.3   EXPLICIT AND IMPLICIT METHODS FOR SECOND-ORDER PDEs

Let us consider the heat equation in dimensionless form:

$$\frac{\partial \phi}{\partial t} = \frac{\partial^2 \phi}{\partial x^2}.$$

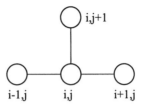

Fig. 5.11   Computational diagram of the explicit method for the heat equation.

With some work, the Black-Scholes equation can be transformed into this form, so it is worthwhile to investigate this equation in some detail. We also assume that the domain of interest is $x \in (0,1)$ and $t \in (0,\infty)$; actually, in a practical scheme, we will also limit the domain with respect to time, $t \in (0,T)$. We have initial conditions for $t = 0$ and boundary conditions at $x = 0$ and $x = 1$ for any $t > 0$. We discretize with respect to x with a step δx, such that $N \, \delta x = 1$, and with respect to t with a step δt, such that $M \, \delta t = T$. Note that this results in a grid with $(N+1) \times (M+1)$ points.

5.3.1   Solving the heat equation by an explicit method

A first possibility for coping with this equation is to approximate the derivative with respect to time by a forward approximation, and the second derivative by the approximation (5.10). This yields

$$\frac{\phi_{i,j+1} - \phi_{ij}}{\delta t} = \frac{\phi_{i+1,j} - 2\phi_{ij} + \phi_{i-1,j}}{(\delta x)^2}.$$

It is easy to see that we may use this equation by solving for the unknown value $\phi_{i,j+1}$:

$$\phi_{i,j+1} = \rho\phi_{i-1,j} + (1 - 2\rho)\phi_{ij} + \rho\phi_{i+1,j}, \qquad (5.19)$$

where $\rho = \delta t/(\delta x)^2$.

Starting from the initial conditions $(j = 0)$, we may solve the equation for increasing values of $j = 1, \ldots, M$. Note that for each j, i.e., for each layer in time, we must use equation (5.19) to find out $N-1$ values for $i = 1, \ldots, N-1$, as the remaining two are given by the boundary conditions. Since the unknown values are given by an explicit expression, this approach is called *explicit*. It can be represented by the computational diagram in figure 5.11.

Example 5.3 Consider the following initial data:

$$\phi(x,0) = f(x) = \begin{cases} 2x & 0 \le x \le 0.5 \\ 2(1-x) & 0.5 \le x \le 1, \end{cases}$$

and boundary conditions

$$\phi(0,t) = \phi(1,t) = 0 \qquad \forall t.$$

```
% HeatExpl.m
function sol = HeatExpl(deltax, deltat, tmax)
N = round(1/deltax);
M = round(tmax/deltat);
sol = zeros(N+1,M+1);
rho = deltat / (deltax)^2;
rho2 = 1-2*rho;
vetx = 0:deltax:1;
for i=2:ceil((N+1)/2)
   sol(i,1) = 2*vetx(i);
   sol(N+2-i,1) = sol(i,1);
end
for j=1:M
   for i=2:N
      sol(i,j+1) = rho*sol(i-1,j) + ...
         rho2*sol(i,j) + rho*sol(i+1,j);
   end
end
```

Fig. 5.12 MATLAB code for solving the heat equation by the explicit method.

The MATLAB code for solving the heat equation for this initial condition is shown in figure 5.12. Let us solve the equation with $\delta x = 0.1$ and $\delta t = 0.001$.

```
>> dx = 0.1;
>> dt = 0.001;
>> tmax = dt*100;
>> sol=HeatExpl(dx, dt, tmax);
>> subplot(2,2,1);
>> plot(0:dx:1,sol(:,1))
>> axis([0 1 0 1])
>> subplot(2,2,2);
>> plot(0:dx:1,sol(:,11))
>> axis([0 1 0 1])
>> subplot(2,2,3);
>> plot(0:dx:1,sol(:,51))
>> axis([0 1 0 1])
>> subplot(2,2,4);
>> plot(0:dx:1,sol(:,101))
>> axis([0 1 0 1])
```

The result, plotted in figure 5.13, looks reasonable, as the heat is progressively diffused and lost through the endpoints. At this point the reader may wish to refer back to figure 1.16, which depicts the value of a call option when the

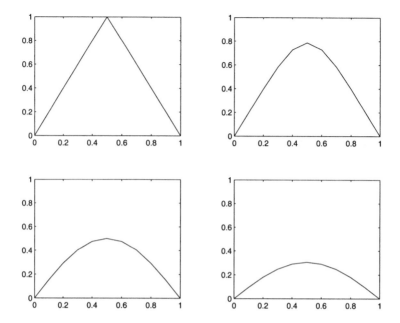

Fig. 5.13 Numerical solution of the heat equation with $\delta x = 0.1$ and $\delta t = 0.001$, by the explicit method, for $t = 0$, $t = 0.01$, $t = 0.05$, $t = 0.1$.

expiration date is approached. The only difference between figures 5.13 and 1.16 is that time goes forward for the heat equation, and it goes backward for the Black-Scholes equation; in fact, for an option we have a final condition rather than an initial one. Apart from this difference, the two solutions are qualitatively similar, as the boundary condition is a kinky function which is smoothed going forward or backward in time. This is a characteristic of parabolic equations, which smooth the irregularities of the boundary conditions. On the contrary, these are propagated by hyperbolic equations and, as we have seen, by the transport equation.

We have used a discretization step $\delta t = 0.001$ with respect to time. Maybe this is more than enough and we would like to increase the precision with respect to space, rather than time. Let us try solving the heat equation with a coarser time-discretization, $\delta t = 0.01$.

```
>> dt = 0.01;
>> sol=HeatExpl(dx, dt, tmax);
>> sol(:,5)
ans =
         0
    0.2000
   -0.0000
    1.4000
```

```
      -1.2000
       2.6000
      -1.2000
       1.4000
      -0.0000
       0.2000
            0
>> sol(:,10)
ans =

            0
     -70.2000
     150.0000
    -233.4000
     301.2000
    -326.6000
     301.2000
    -233.4000
     150.0000
     -70.2000
            0
```

We see that the solution does not make any sense; first, it assumes negative values, which should not be the case for intuitive physical reasons; then
it shows an evident instability. The point is that here we have chosen discretization steps such that $\rho = 1$. In the following we show that for stability,
the condition $0 < \rho \le 0.5$ is required. □

How can we figure out a way to understand what condition should be
required on the discretization steps to ensure numerical stability? In the
case of the transport equation we have used one approach, based on Fourier
analysis. Here we illustrate a matrix theoretic approach. The explicit method
of equation (5.19), together with the boundary conditions

$$\phi_{0,j} = f_0(j\,\delta t) = f_{0j} \qquad \phi_{1,j} = f_N(j\,\delta t) = f_{Nj}$$

can be represented in matrix terms as

$$\Phi_{j+1} = \mathbf{A}\Phi_j + \rho\mathbf{g}_j \qquad j = 0, 1, 2, \ldots,$$

where

$$\mathbf{A} = \begin{bmatrix} 1-2\rho & \rho & 0 & \cdots & 0 & 0 \\ \rho & 1-2\rho & \rho & \cdots & 0 & 0 \\ 0 & \rho & 1-2\rho & \cdots & 0 & 0 \\ \vdots & \vdots & \vdots & \ddots & \vdots & \vdots \\ 0 & 0 & 0 & \cdots & \rho & 1-2\rho \end{bmatrix}$$

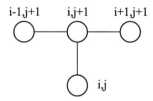

Fig. 5.14   Computational diagram of the implicit method for the heat equation.

$$\Phi_j = \begin{bmatrix} \phi_{1,j} \\ \phi_{2,j} \\ \vdots \\ \phi_{N-1,j} \end{bmatrix}, \qquad g_j = \begin{bmatrix} f_{0,j} \\ 0 \\ \vdots \\ 0 \\ f_{N,j} \end{bmatrix}.$$

Note that $\mathbf{A} \in \mathbb{R}^{N-1,N-1}$ is a tridiagonal matrix. Recalling the convergence analysis that we carried out in section 2.2.4 for iterative algorithms, it is easy to see that the scheme will be stable when

$$\|\mathbf{A}\|_\infty \leq 1.$$

Now when $0 < \rho \leq 1/2$, then $1 - 2\rho \geq 0$, and

$$\|\mathbf{A}\|_\infty = \rho + (1 - 2\rho) + \rho = 1.$$

But if $\rho > 1/2$, then $\mid 1 - 2\rho \mid = 2\rho - 1$, and

$$\|\mathbf{A}\|_\infty = \rho + 2\rho - 1 + \rho = 4\rho - 1 > 1.$$

To satisfy the stability condition, we may be forced to keep δt small and to adopt a fine grid, which may require too much computational effort. An alternative approach is based on implicit methods.

5.3.2   Solving the heat equation by an implicit method

This time, let us approximate the derivative with respect to time by a backward approximation. This yields

$$\frac{\phi_{ij} - \phi_{i,j-1}}{\delta t} = \frac{\phi_{i+1,j} - 2\phi_{ij} + \phi_{i-1,j}}{(\delta x)^2},$$

and we end up with the following equation:

$$-\rho \phi_{i-1,j} + (1 + 2\rho)\phi_{ij} - \rho \phi_{i+1,j} = \phi_{i,j-1}, \qquad (5.20)$$

where again $\rho = \delta t/(\delta x)^2$. Now, if we proceed forward in time, we have an

```
% HeatImpl.m
function sol = HeatImpl(deltax, deltat, tmax)
N = round(1/deltax);
M = round(tmax/deltat);
sol = zeros(N+1,M+1);
rho = deltat / (deltax)^2;
B = diag((1+2*rho) * ones(N-1,1)) - ...
   diag(rho*ones(N-2,1),1) - diag(rho*ones(N-2,1),-1);
vetx = 0:deltax:1;
for i=2:ceil((N+1)/2)
   sol(i,1) = 2*vetx(i);
   sol(N+2-i,1) = sol(i,1);
end
for j=1:M
   sol(2:N,j+1) = B \ sol(2:N,j);
end
```

Fig. 5.15 MATLAB code for the implicit method.

equation linking three unknown values to one known value (see the computational diagram of figure 5.14). Thus, the unknown values are given implicitly, which is where the "implicit method" name comes from; a scheme like this is often referred to as *fully implicit*. We have to solve a system of linear equations for each time layer. Since boundary conditions are given, we have $N-1$ equations in $N-1$ unknowns. In matrix terms, we have to solve a set of systems like

$$\mathbf{B}\Phi_{j+1} = \Phi_j + \rho\mathbf{g}_j \qquad j = 0, 1, 2, \ldots, \tag{5.21}$$

where $\mathbf{B} \in \mathbb{R}^{N-1,N-1}$ is a tridiagonal matrix,

$$\mathbf{B} = \begin{bmatrix} 1+2\rho & -\rho & 0 & \cdots & 0 & 0 \\ -\rho & 1+2\rho & -\rho & \cdots & 0 & 0 \\ 0 & -\rho & 1+2\rho & \cdots & 0 & 0 \\ \vdots & \vdots & \vdots & \ddots & \vdots & \vdots \\ 0 & 0 & 0 & \cdots & -\rho & 1+2\rho \end{bmatrix}.$$

Example 5.4 The MATLAB code for the implicit method to solve the heat equation is illustrated in figure 5.15 (here $\mathbf{g}_j = \mathbf{0}$). Note that we are not exploiting the fact that the matrix $\mathbf{B}$ is tridiagonal, as we simply leave to MATLAB the solution of the system of linear equations; the techniques described in section 2.2.3 could and should be used here. We may verify that the case $\rho = 1$ does not cause any trouble (see the plots in figure 5.16):

```
>> dx=0.1;
```

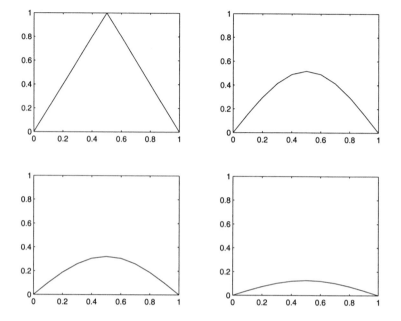

Fig. 5.16 Numerical solution of the heat equation with $\delta x = 0.1$ and $\delta t = 0.01$, by the implicit method, for $t = 0$, $t = 0.05$, $t = 0.1$, $t = 0.2$.

```
>> dt=0.01;
>> tmax=dt*100;
>> sol=HeatImpl(dx,dt,tmax);
>> subplot(2,2,1)
>> plot(0:dx:1,sol(:,1))
>> axis([0 1 0 1])
>> subplot(2,2,2)
>> plot(0:dx:1,sol(:,6))
>> axis([0 1 0 1])
>> subplot(2,2,3)
>> plot(0:dx:1,sol(:,11))
>> axis([0 1 0 1])
>> subplot(2,2,4)
>> plot(0:dx:1,sol(:,21))
>> axis([0 1 0 1])
```

In fact, we may prove that the implicit method is unconditionally stable.   □

To prove that the implicit method of equation (5.21) is stable, we may rewrite the scheme as

$$\Phi_{j+1} = \mathbf{B}^{-1}(\Phi_j + \rho \mathbf{g}_j),$$

from which it is easy to see that stability depends on the spectral radius $\rho(\mathbf{B}^{-1})$. In this case, we may work directly on the spectral radius, rather than on a matrix norm. The scheme will be stable if the eigenvalues of $\mathbf{B}^{-1}$ are less than 1 in absolute value; to see that this is indeed the case, we may rewrite the matrix as follows:

$$\mathbf{B} = \mathbf{I} + \rho\mathbf{T},$$

where

$$\mathbf{T} = \begin{bmatrix} 2 & -1 & 0 & \cdots & 0 & 0 \\ -1 & 2 & -1 & \cdots & 0 & 0 \\ 0 & -1 & 2 & \cdots & 0 & 0 \\ \vdots & \vdots & \vdots & \ddots & \vdots & \vdots \\ 0 & 0 & 0 & \cdots & -1 & 2 \end{bmatrix}. \tag{5.22}$$

It can be shown that the eigenvalues of $\mathbf{T} \in \mathbb{R}^{N-1,N-1}$ are

$$\lambda_k = 4\sin^2\left(\frac{k\pi}{2N}\right) \qquad k = 1, 2, \ldots, N-1.$$

If you don't believe it, check it with MATLAB:

```
>> N=6;
>> T = diag(2*ones(N-1,1)) - diag(ones(N-2,1),1) - ...
   diag(ones(N-2,1),-1);
>> sort(eig(T))
ans =
    0.2679
    1.0000
    2.0000
    3.0000
    3.7321
>> sort(4*sin((1:N-1)*pi/(2*N)).^2)
ans =
    0.2679    1.0000    2.0000    3.0000    3.7321
```

Now we recall a couple of facts from matrix algebra, which are easily proved:

- If λ is an eigenvalue of the matrix $\mathbf{T}$, $1 + \rho\lambda$ is an eigenvalue of the matrix $\mathbf{I} + \rho\mathbf{T}$.

- If β is an eigenvalue of the matrix $\mathbf{B}$, β^{-1} is an eigenvalue of the matrix $\mathbf{B}^{-1}$.

Putting all together, we may conclude that the eigenvalues of $\mathbf{B}^{-1}$ are

$$\alpha_k = \frac{1}{1 + 4\rho\sin^2\left(\frac{k\pi}{2N}\right)} < 1 \qquad k = 1, 2, \ldots, N-1,$$

and the fully implicit scheme is unconditionally stable.

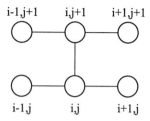

Fig. 5.17 Computational diagram of the Crank-Nicolson method for the heat equation.

5.3.3 Solving the heat equation by the Crank-Nicolson method

So far, we have seen methods involving three points on one time layer and one on a neighboring layer. It is natural to wonder if a better scheme may be obtained by considering three points on both layers. One way to do this is to consider the point $(x_i, t_{j+1/2}) = (x_i, t_j + \delta t/2)$ and to approximate the derivatives at that point using values in the six neighboring points in the grid. By using Taylor expansions as we did in section 5.2, we may conclude that

$$\frac{\partial^2 \phi}{\partial x^2}(x_i, t_{j+1/2}) = \frac{1}{2}\left[\frac{\partial^2 \phi}{\partial x^2}(x_i, t_{j+1}) + \frac{\partial^2 \phi}{\partial x^2}(x_i, t_j)\right] + O(\delta x^2)$$

and

$$\frac{\partial \phi}{\partial t}(x_i, t_{j+1/2}) = \frac{\phi(x_i, t_{j+1}) - \phi(x_i, t_j)}{\delta t} + O(\delta t^2).$$

Using these two approximations together with the usual ones, we get the Crank-Nicolson scheme:

$$-\rho\phi_{i-1,j+1} + 2(1+\rho)\phi_{i,j+1} - \rho\phi_{i+1,j+1} = \rho\phi_{i-1,j} + 2(1-\rho)\phi_{ij} + \rho\phi_{i+1,j}, \quad (5.23)$$

which is represented in figure 5.17. The fundamental feature of this scheme is that the error is both $O(\delta x^2)$ and $O(\delta t^2)$; this implies that less computational effort is required to obtain a satisfactory degree of accuracy in the numerical solution.

The Crank-Nicolson scheme may be analyzed in a more general framework. We may think of using a convex combination of two approximations of the second-order derivative in the finite difference scheme:

$$\frac{\phi_{i,j+1} - \phi_{ij}}{\delta t} = \frac{1}{(\delta x)^2}\left[\lambda(\phi_{i-1,j+1} - 2\phi_{i,j+1} + \phi_{i+1,j+1})\right.$$
$$\left. + (1-\lambda)(\phi_{i-1,j} - 2\phi_{ij} + \phi_{i+1,j})\right] \quad (5.24)$$

for $0 \le \lambda \le 1$. Note that we get the explicit scheme by choosing $\lambda = 0$, the implicit scheme for $\lambda = 1$, and the Crank-Nicolson scheme for $\lambda = 1/2$.

To see that the Crank-Nicolson scheme is unconditionally stable, we may proceed just as with the first implicit scheme. We may rewrite equation (5.23) in matrix form:

$$\mathbf{C}\Phi_{j+1} = \mathbf{D}\Phi_j + \rho(\mathbf{g}_{j+1} + \mathbf{g}_j),$$

where

$$\mathbf{C} = \begin{bmatrix} 2(1+\rho) & -\rho & 0 & \cdots & 0 & 0 \\ -\rho & 2(1+\rho) & -\rho & \cdots & 0 & 0 \\ 0 & -\rho & 2(1+\rho) & \cdots & 0 & 0 \\ \vdots & \vdots & \vdots & \ddots & \vdots & \vdots \\ 0 & 0 & 0 & \cdots & -\rho & 2(1+\rho) \end{bmatrix}$$

$$\mathbf{D} = \begin{bmatrix} 2(1-\rho) & \rho & 0 & \cdots & 0 & 0 \\ \rho & 2(1-\rho) & \rho & \cdots & 0 & 0 \\ 0 & \rho & 2(1-\rho) & \cdots & 0 & 0 \\ \vdots & \vdots & \vdots & \ddots & \vdots & \vdots \\ 0 & 0 & 0 & \cdots & \rho & 2(1-\rho) \end{bmatrix}.$$

Then, using matrix the same matrix $\mathbf{T}$ of equation (5.22) again, we may see that the eigenvalues of $\mathbf{C}^{-1}\mathbf{D}$ are

$$\alpha_k = \frac{2 - 4\sin^2\left(\frac{k\pi}{2N}\right)}{2 + 4\sin^2\left(\frac{k\pi}{2N}\right)} < 1 \qquad k = 1, 2, \ldots, N-1,$$

which shows that the scheme is unconditionally stable.

5.4 CONVERGENCE, CONSISTENCY, AND STABILITY

We have developed finite difference schemes and we have informally noted that there is some truncation error that tends to zero as the discretization steps tend to zero. We would expect that this ensures the convergence of the solution to the difference equations to the solution of the differential equation. However, the counterexample of section 5.2.1 shows that the matter is not so trivial, since we should consider carefully the interplay of three concepts: convergence, stability, and consistency. The point is that the solution of the finite difference equations for discretization steps $\delta x, \delta t \to 0$ could converge to a function which is not the solution of the PDE. A rigorous analysis of these concepts and their relationships is beyond the scope of the book, but we would like to give at least a glimpse into these topics.

An initial value problem such as the familiar heat equation is defined over a space/time domain

$$\mathcal{D} \times (0 < t < \infty).$$

The problem can be cast in a more abstract way as

$$L\phi = f,$$

where L is a differential operator, f is a known function, and ϕ is the unknown function we seek to determine. When we set up a discrete grid $\mathcal{G}_\Delta$, we also discretize the operator L by an operator L_Δ. Given a function ψ and a point $(\mathbf{P}_i, t_j) \in \mathcal{G}_\Delta$, we may consider the truncation error

$$t_\psi(\mathbf{P}_i, t_j) = L\psi(\mathbf{P}_i, t_j) - L_\Delta \psi(\mathbf{P}_i, t_j).$$

If, when the grid is refined and the discretization steps tend to zero, this truncation error tends to zero,[5] the numerical scheme is said to be *consistent*. This essentially says that the finite difference representation we use tends to the PDEs we are interested in.

The stability issue is concerned basically with whether or not the difference between the numerical solution and the exact solution remains bounded as time progresses. To be more specific, consider the heat equation of section 5.3. Let ϕ_{ij} be the solution of the finite difference scheme and $\phi(x, t)$ the correct solution of the PDE. We may investigate

- the behavior of $|\phi_{ij} - \phi(i\,\delta x, j\,\delta t)|$ as $j \to \infty$ for fixed discretization steps δx and δt,

- or the behavior of $|\phi_{ij} - \phi(i\,\delta x, j\,\delta t)|$ as $\delta x, \delta t \to 0$ for a fixed value of $j\,\delta t$.

The first issue is related to stability; the second issue is related to convergence. To ensure the convergence of the numerical solution to the exact solution, the consistency condition is not enough. However, it can be shown (Lax's equivalence theorem; see [5]) that for a well-posed linear initial value problem, stability is a necessary and sufficient condition for convergence of a consistent numerical scheme. As the following example shows, the numerical scheme of section 5.2.1 is not stable, and this is why it fails to converge.

Example 5.5 For the sake of convenience, let us recall the numerical scheme of section 5.2.1 for the transport equation with constant velocity c:

$$\phi_{i,j+1} = \left(1 + \frac{c}{\rho}\right)\phi_{ij} - \frac{c}{\rho}\phi_{i+1,j},$$

where $\rho = \delta x / \delta t$. We may apply the same Fourier analysis of stability that we applied in example 5.2. Leaving the details as an exercise, we may see that in this case

$$\phi_{ij} = \epsilon\,(-1)^i \left(1 + 2\frac{c}{\rho}\right)^j.$$

Since c and ρ are both positive, we see that ϕ_{ij} goes to infinity as $j \to \infty$. Hence, the scheme is unconditionally unstable and convergence is not ensured even if the discretization steps tend to zero. ▯

---

[5]This should be made more precise, as the space and time discretization steps could tend to zero in an arbitrary way, or with some relationship between them.

S5.1   CLASSIFICATION OF SECOND-ORDER PDEs AND CHARACTERISTIC CURVES

The classification of second-order quasilinear equations into hyperbolic, parabolic, and elliptic equations derives from a solution method depending on characteristics. This method reduces the solution of a PDE to the solution of a set of ordinary differential equations along certain curves. It is best to understand the idea behind the method in the case of a first-order equation such as the transport equation

$$\frac{\partial u}{\partial t} + c(x)\frac{\partial u}{\partial x} = 0 \qquad (5.25)$$

along with the boundary condition $u(x,0) = f(x)$.

Note that expression (5.25) is simply the directional derivative of the function $u(x,t)$ along the direction $[c(x),1]^T$. To see this, we recall that this directional derivative may be expressed as

$$(\nabla u)^T \begin{bmatrix} c(x) \\ 1 \end{bmatrix} = \begin{bmatrix} \frac{\partial u}{\partial x} & \frac{\partial u}{\partial t} \end{bmatrix}^T \begin{bmatrix} c(x) \\ 1 \end{bmatrix}.$$

Since this directional derivative is zero, we may argue that there is a set of curves along which the function $u(x,t)$ is constant. We may find these curves by deriving an ordinary differential equation. Let us consider the restriction of $u(x,t)$ on such a curve; parameterizing with respect to the time variable t, we may consider the function $v(t) = u[x(t),t]$. But $v(t)$ is constant on the curve, hence we may write

$$\frac{dv(t)}{dt} = \frac{\partial u}{\partial x}\frac{dx}{dt} + \frac{\partial u}{\partial t} = 0.$$

This condition and equation (5.25) lead to

$$\frac{dx}{dt} = -\frac{\partial u/\partial t}{\partial u/\partial x} = c(x), \qquad (5.26)$$

which is an ordinary differential equation defining a *characteristic curve* along which the solution is constant. To see how we can take advantage of this, let us assume $c(x) = x$. In this case the differential equation (5.26) is easy to solve with given initial conditions. The characteristic curve passing through point $(x_0, 0)$ is

$$x(t) = x_0 e^t. \qquad (5.27)$$

To obtain a generic value $u(x,t)$, we must spot the characteristic curve passing through point (x,t); by inverting equation (5.27) we see that the same characteristic passes through point $(x_0, 0)$, where $x_0 = xe^{-t}$. But since u is constant on the characteristic, we have

$$u(x,t) = u(x_0, 0) = u(xe^{-t}, 0) = f(xe^{-t}),$$

which is the solution we were seeking.

We have just regarded a characteristic curve as a curve along which the solution of a PDE is constant. Actually, characteristic curves for a PDE are more than this. They may be considered as curves on which the solution of a PDE may be transformed into the solution of a system of ordinary differential equations. They are also linked to changes of coordinates that allow us to rewrite quasilinear second-order PDEs into a canonical form; in fact, pursuing this idea we may solve the Black-Scholes equation by reducing it to the more familiar heat equation (see, e.g., [9, pp. 100-101]).

Second-order equations are classified based on the existence of characteristic curves: parabolic equations have one characteristic curve passing through each point; elliptic equations have none; hyperbolic equations have two characteristic curves passing through each point, which may be exploited for computational purposes. Further details may be found in the references, but for our purposes the most important point is that in hyperbolic equations the irregularities are propagated along the characteristics, whereas in parabolic equations the irregularities are smoothed. This is relevant for our applications, as option payoffs may be nondifferentiable or even discontinuous at expiration; one such case is a digital option, which pays $1 if the stock price is above a certain threshold, and $0 otherwise.

For further reading

- Partial differential equations are a large and complicated topic. For an introduction including both classical and advanced concepts, see, e.g., [3].[6]

- Another book covering PDEs in a relatively general setting is [1], which also includes many pieces of MATLAB code.

- A classical reference on finite difference methods for PDEs is [6]. See also [4] and [7].

- Advanced issues, including the important Lax theorem, are covered in [5].

- To see extensive examples of PDEs in action to tackle financial engineering problems, see [8] or [9].

---

[6]An errata sheet for this book is available at `www.math.neu.edu/~mcowen/mathindex.html`.

REFERENCES

1. J. Cooper. *Introduction to Partial Differential Equations with MATLAB.* Birkhäuser, Berlin, 1998.

2. D.G. Luenberger. *Investment Science.* Oxford University Press, New York, 1998.

3. R. McOwen. *Partial Differential Equations: Methods and Applications.* Prentice Hall, Upper Saddle River, NJ, 1996.

4. K.W. Morton and D.F. Mayers. *Numerical Solution of Partial Differential Equations: An Introduction.* Cambridge University Press, Cambridge, 1996.

5. R.D. Richtmyer and K.W. Morton. *Difference Methods for Initial Value Problems (2nd ed.).* Wiley, New York, 1967. Reprinted in 1994 by Krieger, New York.

6. G.D. Smith. *Numerical Solution of Partial Differential Equations: Finite Difference Methods (3rd ed.).* Oxford University Press, Oxford, 1985.

7. J.W. Thomas. *Numerical Partial Differential Equations: Finite Difference Methods.* Springer-Verlag, New York, 1995.

8. P. Wilmott. *Derivatives: The Theory and Practice of Financial Engineering.* Wiley, Chichester, West Sussex, England, 1999.

9. P. Wilmott. *Quantitative Finance (vols. I and II).* Wiley, Chichester, West Sussex, England, 2000.

Part III

Applications to Finance

6

Optimization models for portfolio management

The content of this chapter is a little different from the rest of the book, as MATLAB will play practically no role here. We will present some very simple models for portfolio optimization, based on mixed-integer programming, stochastic programming, and global optimization. The aim is to illustrate the use of the more advanced optimization methods discussed in chapter 3. As before, the treatment is limited to finite-dimensional models, and we do not present continuous-time formulations which lead to optimal control and stochastic dynamic programming models. These do play a prominent role in the financial literature but are beyond the scope of this introductory book. Yet the reader will gain at least an intuitive grasp of what (relatively) unconventional optimization tools may offer.

When dealing with an optimization model in practice, one should use three basic tools:

1. The solver, i.e., the implementation of the numerical method we select to cope with the model (the concepts behind commercial solvers have been discussed in chapter 3).

2. An algebraic modeling language, i.e., a formalism which acts as a bridge between the pencil and paper formulation and the solver (this is one of the topics of this chapter).

3. A surrounding environment, including a database from which the data may be loaded, and a graphical interface and visualization tool, which may help in interpreting the results and carrying out some sensitivity analysis.

Only the first building block is strictly necessary. However, using only a relatively low level optimization library, where you have to load directly the data structures required by the solver, is an awkward and error-prone task, making both model development and maintenance very hard. This is why algebraic modeling languages such as AMPL and GAMS have been developed. Such tools are essentially solver independent, since the same model may be passed to different solvers, and enable the user to express the model constraints and objectives in much the same way as one would do on paper. To the best of our knowledge, there is no commercial interface between modeling tools and MATLAB at present (one could be available in the future). Furthermore, while the recent versions of MATLAB provide the user with excellent user-interface and visualization tools, including the possibility of accessing databases to gather financial data, the solution procedures are not able to cope with mixed-integer problems. Despite notable improvements with respect to previous implementations, the Optimization toolbox is probably not competitive with specialized optimization solvers such as CPLEX and OSL. So it is difficult to cope with large-scale problems arising from stochastic formulations. While it is possible that the gap will be at least partially filled in the future, at present we cannot formulate and solve large-scale portfolio optimization models in MATLAB, with the exception of the classical mean-variance formulations. Nevertheless, we think it is important to understand why mixed-integer and stochastic programming may be important for portfolio optimization, and how the model may easily be expressed in an algebraic language.

The motivations behind the models outlined in this chapter are the well known limitations of classical mean-variance portfolio models:

- They are single-period models and do not consider the possibility of portfolio rebalancing within the planning horizon, so a dynamic multi-period model could be devised as an alternative.

- They are based on possibly unrealistic assumptions on the return distributions and/or the utility function (the returns should be normal, or the utility function should be quadratic).

- They assume that the covariance matrix is constant in time, whereas the covariance may change (usually right when there is a crash and you would like to exploit diversification).

Even if you do not consider these issues and you want to settle for a classical mean-variance modeling framework, it may well be the case that in a diversified portfolio you have assets with very small weights; while these assets may be beneficial from the diversification point of view, they may complicate portfolio management, and may be detrimental when you consider transaction costs. All these considerations apply to stock portfolios. When you consider fixed-income assets, you have also to recognize the limitations of simple immunization strategies based on duration and convexity. Apart from their inherent assumptions on the term structure of interest rates, they are not able to address the issue of what to do when bonds reach their maturity within the planning horizon. Since not all bonds have the same maturity, a dynamic model may be helpful.

There is no easy and simple answer to all of these issues. In this chapter we show how they may be (at least partially) addressed by operations research models. In section 6.1 we see how binary decision variables may be used to restrict the asset allocation in order to obtain a more "desirable" portfolio. In section 6.2 we introduce multistage stochastic programming models through a toy example; we will also show how these models may be formulated in both AMPL and GAMS. Then we outline how complicating features such as transaction costs and stochastic liabilities may be modeled; we also comment on the thorny issue of scenario generation for stochastic programming models. Finally, section 6.3 is devoted to optimization of a fixed-mix portfolio. As this leads to a nonconvex (continuous or discrete) model, useful solution approaches in this case may be branch and bound or metaheuristics such as tabu search and genetic algorithms.

6.1 MIXED-INTEGER PROGRAMMING MODELS

We discussed methods to solve mixed-integer linear programming (MILP) models in section 3.5.1. Considering that MILP models may be quite difficult to deal with, there is no wonder that their use in finance is not quite widespread. Nevertheless, recent progress in both computing hardware and solution methods have made their use in management practice much more common. Indeed, there are a variety of issues in portfolio management, which are neglected in classical mean-variance models, that could be tackled fruitfully within an integer programming framework:

- Limited diversification portfolio

- Minimum portfolio weights for assets

- Minimum transaction lots

- Fixed and piecewise linear transaction costs

An efficient mean-variance portfolio may include a large set of assets, and some of them may account for a tiny part of the overall asset allocation. While this is, at least in principle, beneficial for diversification, there are a few downsides in a too diversified portfolio. One issue is the amount of transaction costs we have to pay, making small transactions unattractive. Another issue is the effort that is required in analyzing the historical data for too many assets, in order to control the portfolio risk. We could extend the mean-variance model by constraining the portfolio cardinality, i.e., the number of assets included.

Writing a constraint stating that at most k assets out of the I available may be included in the portfolio is easily accomplished by introducing, for each asset $i = 1, \ldots, I$, the following binary variable:

$$\delta_i = \begin{cases} 1 & \text{if asset } i \text{ is included in the portfolio,} \\ 0 & \text{otherwise.} \end{cases}$$

Then all we have to do is to add the following constraints to the model:

$$x_i \leq M_i \delta_i \qquad \forall i \tag{6.1}$$

$$\sum_{i=1}^{I} \delta_i \leq k,$$

where M_i is an upper bound on the weight of asset i. This is actually the same trick we have described in section 3.1.5 to model fixed costs in the fixed-charge problem. Another possibility would be to enforce a minimal limit to the asset weight if positive. Note that this is not equivalent to requiring that $m_i \leq x_i \leq M_i$. Rather, we want something like

$$x_i \in \{0\} \cup [m_i, M_i],$$

which is a nonconvex set (recall that the union of convex sets need not be convex). This requirement cannot be enforced within a continuous linear or quadratic programming model. However, it is easy to extend constraint (6.1):

$$m_i \delta_i \leq x_i \leq M_i \delta_i \qquad \forall i.$$

This is what we called a *semicontinuous variable* in chapter 3. By the way, x_i need not be the weight in a portfolio; it could be the amount of stock traded, in which case m_i would be the minimal tradeable lot. We could even go further and require, in such a case, that x_i is a general integer variable, in order to avoid the additional costs involved in trading odd lots. Putting all of this together, we can trace the efficient frontier by solving a set of mixed-integer quadratic programs like the following:

$$\min \quad \sum_{i=1}^{I} \sum_{j=1}^{I} \sigma_{ij} x_i x_j$$

$$\text{s.t.} \quad \sum_{i=1}^{I} \overline{r}_i x_i \geq \overline{r}_T$$

$$\sum_{i=1}^{I} x_i = 1$$

$$m_i \delta_i \leq x_i \leq M_i \delta_i \qquad \forall i$$

$$\sum_{i=1}^{I} \delta_i \leq k$$

$$w_i \geq 0, \quad \delta_i \in \{0, 1\} \qquad \forall i,$$

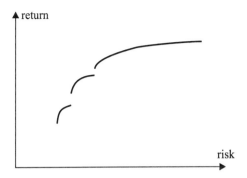

Fig. 6.1 Qualitative sketch of a cardinality-constrained efficient frontier.

where $\bar{r}_i$ is the expected return of asset i, σ_{ij} is the covariance between the returns of assets i and j, and $\bar{r}_T$ is a target return. By varying the target return we would trace the efficient frontier. It is also important to realize that the efficient frontier will be qualitatively different from the usual one, which was illustrated in figure 1.7. A qualitative sketch of the cardinality-constrained efficient frontier is illustrated in figure 6.1. This plot may be understood by imagining of tracing the efficient sets for each portfolio consisting of a subset of cardinality k, and then patching all of them together. One difficulty with the formulation above is that it is a mixed-integer *quadratic*, rather than linear, problem. In principle, and in practice as well, it can be solved by the same branch and bound algorithm illustrated in section 3.5.1; the only difference is that the lower bounds are computed by solving a quadratic programming problem. Unfortunately, not many commercial codes are available to tackle such problems; furthermore, the computational requirements could turn out to be prohibitive for a large-scale application. Still, different alternatives may be tried.

- We may trace only the relevant part of the efficient set, given our risk aversion.

- In [2] ad hoc methods are discussed for mixed-integer quadratic programming; taking a route like this may be advantageous, but it requires writing our own code.

- Another possibility is to simplify the model by reducing the data requirements, e.g., by assuming that all the correlations are equal. See [17] for an approach like this, and for additional references as well.

- Metaheuristics such as genetic algorithms and simulated annealing (section 3.6) may also be used [4].

- If one wants to use MILP codes, it is also possible to devise a different representation of risk. In [11] the use of the mean absolute deviation has been advocated:

$$E\left[\left|\sum_{i=1}^{I} R_i x_i - E\left[\sum_{j=1}^{I} R_j x_j\right]\right|\right],$$

where R_i is the random return of asset i. This definition is quite similar to variance; an absolute deviation is used rather than a squared deviation. This objective may be translated in linear terms, and MILP methods, exact or heuristic, may be applied. Suppose in fact that we have a set of historical returns r_{it} for each asset in time periods $t = 1, \ldots, T$. Then we may estimate $E[R_i] = \bar{r}_i = (1/T) \sum_{t=1}^{T} r_{it}$ and set

$$E\left[\sum_{j=1}^{N} R_j x_j\right] = \sum_{j=1}^{N} \bar{r}_j x_j.$$

By the same token, we may approximate the objective function as

$$E\left\{\left|\sum_{i=1}^{N} R_i x_i - \sum_{j=1}^{N} \bar{r}_j x_j\right|\right\} = \frac{1}{T} \sum_{t=1}^{T} \sum_{i=1}^{N} |(r_{it} - \bar{r}_i) x_i|.$$

This objective function may be expressed in linear form by introducing a set of auxiliary variables y_t. The model will include, among other things, the following objective function and constraints:

$$\min \quad \frac{1}{T} \sum_{t=1}^{T} y_t$$

$$\text{s.t.} \quad y_t + \sum_{i=1}^{N} (r_{it} - \bar{r}_i) x_i \geq 0 \qquad \forall t$$

$$y_t - \sum_{i=1}^{N} (r_{it} - \bar{r}_i) x_i \geq 0 \qquad \forall t.$$

For instance, this approach is taken in [14], where minimum transaction lots are dealt with.

- Finally, the MILP model may not really be aimed at building a portfolio from scratch. Rather, one could devise a target portfolio by whatever technique, subject to variety of constraints related to critical market exposure and liquidity. Then the target is approximated by enforcing some practical requirements, such as minimizing the number of assets included in the real portfolio. This is the approach taken in [1] to cope with a real-life case.

A final important remark is that the difficulty of solving a mixed-integer problem depends on the strength of its relaxation (see section 3.5.1). The least one should do is to reduce the M_i bounds in constraints like (6.1). Thanks to careful modeling, computational times on the order of a few minutes are reported in [1] for problems involving something like 1500 assets (using what is now an old version of CPLEX).

A last point is that classical mean-variance models neglect transaction costs. This is debatable in a single-period model, and is even more questionable in a multiple-period model, since excessive trading may disrupt any advantage gained by optimizing the portfolio. The simplest idea is to use a linear model of the transaction cost; i.e., if we trade an amount x_i of an asset, we pay a proportional cost $\alpha_i x_i$, where the proportionality constant may depend on the asset liquidity. This results in a linear programming model, and one such formulation is given in section 6.2.3. However, a linear model fails to account for the dependence of transaction costs on the volume traded. Different assumptions can be made, depending on the nature of the traded asset, leading to different model formulations. In the case of fixed transaction costs, we may simply adopt the binary variable trick used earlier and treat it as a fixed cost. If transaction costs are nonlinear, they may be approximated by piecewise linear functions along the lines we illustrated in section 3.1.5. If we assume that transaction costs increase marginally with the traded volume (maybe because the asset is highly illiquid and it is difficult to deal with the sale/purchase order), the function is convex and can be dealt with by ordinary LP methods. However, in the case of concave costs, this is no longer the case, and mixed-integer models must be used. See also [13] for an example of how a model involving fixed transaction costs may be tackled.

6.2 MULTISTAGE STOCHASTIC PROGRAMMING MODELS

The best way to introduce multistage stochastic models is by using a simple asset and liability model. We use the same basic problem and data as [3, pp. 20-28]. We have an initial wealth W_0 now, and in the future we will have to pay an amount L, which is our only liability. We should devise an investment strategy to meet the liability; if possible, we would like to end up with a final wealth larger than L; however, we should account properly for risk aversion, since there could be some chance to end up with a terminal wealth which is not sufficient to pay for the liability, in which case we will have to borrow some money. A nonlinear, strictly concave utility function of the difference between the terminal wealth and the liability would do the job, but this would lead to a nonlinear programming model. So we may build a piecewise linear utility function like that illustrated in figure 6.2. The utility is zero when the terminal wealth W matches the liability exactly. If the slope r penalizing the shortfall is larger than q, this function is concave, but not strictly.

The portfolio consists of a set of I assets. For simplicity, we assume that we may rebalance it only at a discrete set of time instants $t = 1, \ldots, T$, with no transaction cost; the initial portfolio is chosen at time $t = 0$, and the

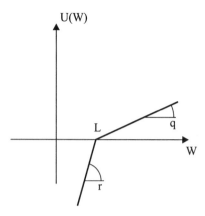

Fig. 6.2 Piecewise linear concave utility function.

liability must be paid at time $T+1$. Time period t is the period between time instants $t-1$ and t. In order to represent uncertainty, we may build a tree like that in figure 6.3, which is a generalization of the two-stage tree of figure 3.5. Each node n_k corresponds to an event, where we should take some decision. We have an initial node n_0 corresponding to time $t = 0$. Then for each event node, we have two branches; each branch is labeled by a conditional probability of occurrence, $P\{n_k \mid n_i\}$, where $n_i = a(n_k)$ is the immediate ancestor of node n_k. Here, we have two nodes at time $t = 1$ and four at time $t = 2$, where we may rebalance our portfolio on the basis of the previous asset returns. Finally, in the eight nodes corresponding to $t = 3$, we just compare our final wealth to the liability and we evaluate our utility function. To each node of the tree we must associate the set of asset returns during the corresponding time period. A scenario consists of an event sequence, i.e., a sequence of asset returns. We have eight scenarios in figure 6.3. For instance, scenario 2 consists of the node sequence (n_0, n_1, n_3, n_8). The probability of each scenario depends on the conditional probability of each node on its path. If each branch at each node is equiprobable, i.e., the conditional probability is $1/2$, each scenario in the figure has probability $1/8$. The branching factor may be arbitrary in principle; the more branches we use, the better our ability to model uncertainty; unfortunately, the number of nodes grows exponentially with the number of stages, and the computational effort as well.

At each node in the tree, we must take a set of decisions. In practice, the decisions we are interested in and must be implemented here and now are those corresponding to the first node of the tree; the other (recourse) decision variables are instrumental to the aim of devising a robust plan, but they are not implemented in practice, as the multistage model is solved on a rolling horizon basis. This suggests that in order to model the uncertainty as accurately as possible with a limited computational effort, a possible idea is to branch many paths from the initial node, and less from the subsequent nodes.

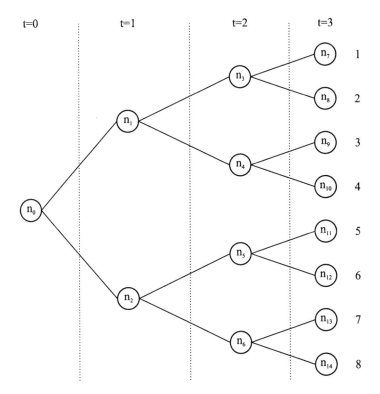

Fig. 6.3 Event tree.

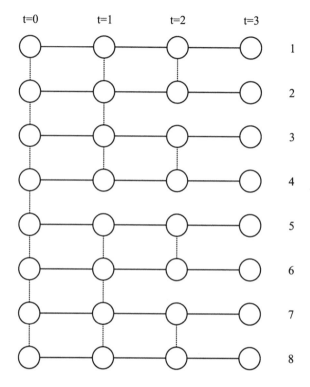

Fig. 6.4 Split-variable view of an event tree.

Each decision at each stage may depend on the information gathered so far, but not on the future; this requirement is called *nonanticipativity condition*. There are two basic ways to build a multistage stochastic programming model: the split-variable and the compact formulations, which are described in the next sections. They depend on how the nonanticipativity requirement is modeled. The suitability of each modeling approach also depends on the solution algorithm.

6.2.1 Split-variable formulation

In the split-variable approach, the decision variables are defined as follows:

- x_{it}^s is the amount invested in asset i at the beginning of time period t in scenario s.

By the same token, R_{it}^s is the (total) return of asset i in scenario $s = 1, \ldots, S$ during time period t. It is important to understand that if we define the decision variables in this way, we must enforce the nonanticipativity constraint explicitly. The issue may be understood by looking at figure 6.4. We have a set of decision variables for each node; however, the decision variables corresponding to different scenarios at the same time t must be equal if the two scenarios are indistinguishable at time t. This is represented by the dotted lines in figure 6.4. To begin with, the initial portfolio must be the same for all scenarios. Hence:

$$x_{i0}^s = x_{i0}^{s'} \qquad i = 1, \ldots, I; \; s, s' = 1, \ldots, S.$$

Now consider time $t = 1$ and node n_1 of the original event tree as depicted in figure 6.3; the scenarios $s = 1, 2, 3, 4$ pass through this node and are indistinguishable at time $t = 1$. Hence, we must have

$$x_{i1}^1 = x_{i1}^2 = x_{i1}^3 = x_{i1}^4 \qquad i = 1, \ldots, I.$$

In fact, node n_1 corresponds to four nodes in the split view of the tree. By the same token, at time $t = 2$ we have constraints like

$$x_{i2}^5 = x_{i2}^6 \qquad i = 1, \ldots, I.$$

More generally, it is customary to denote by $\{s\}_t$ the set of scenarios which are not distinguishable from s up to time t. For instance:

$$
\begin{aligned}
\{1\}_0 &= \{1, 2, 3, 4, 5, 6, 7, 8\} \\
\{2\}_1 &= \{1, 2, 3, 4\} \\
\{5\}_2 &= \{5, 6\}.
\end{aligned}
$$

Then the nonanticipativity constraints may be written in general as

$$x_{it}^s = x_{it}^{s'} \qquad \forall i, t, s, s' \in \{s\}_t.$$

This is not the only way of expressing the nonanticipativity requirement, and selection of the best approach depends on the solution algorithm one wants to select. Now we may write the following model for the basic asset-liability management problem:

$$\max \quad \sum_s p^s (q w_+^s - r w_-^s) \tag{6.2}$$

$$\text{s.t.} \quad \sum_{i=1}^I x_{i0}^s = W_0 \qquad \forall s \in S \tag{6.3}$$

$$\sum_{i=1}^I R_{it}^s x_{i,t-1}^s = \sum_{i=1}^I x_{it}^s \qquad \forall s \in S; \; t = 1, \ldots, T \tag{6.4}$$

$$\sum_{i=1}^I R_{i,T+1}^s x_{iT}^s = L + w_+^s - w_-^s \qquad \forall s \in S \tag{6.5}$$

$$x_{it}^s = x_{it}^{s'} \qquad \forall i, t, s, s' \in \{s\}_t$$

$$x_{it}^s, w_+^s, w_-^s \geq 0.$$

Here w_+^s is the surplus at the end of the planning horizon, with reward q, and w_-^s is the shortfall, with penalty r. The objective function (6.2) is the expected value of the utility function; p^s is the probability of each scenario; the utility function is concave if $r > q$. Equation (6.3) states that our initial wealth W_0 is allocated among the different assets. The portfolio rebalancing constraints (6.4) say that the wealth at time t is reallocated. In equation (6.5) we evaluate how we did, by comparing the final wealth with the liability L, and setting the proper surplus and shortfall values. Then we add nonanticipativity and nonnegativity constraints. Note that since the variables w_+^s and w_-^s are restricted by nonnegativity constraints, we will have $w_+^s \cdot w_-^s = 0$ in the optimal solution (i.e., only one variable may be different than 0 in each scenario). The nonnegativity requirements on x_{it}^s may be relaxed if we allow short selling.

In this modeling approach we introduce a large set of variables, which are then linked by nonanticipativity constraints. Hence, one could wonder if this really makes sense. The answer depends on the solution algorithm. If one wants to adopt an algorithm like the L-shaped decomposition, the compact formulation explained in the following section must be used. The split-variable approach may be exploited with interior point methods aimed at stochastic programming. Furthermore, relaxing the nonanticipativity constraints by a set of Lagrangian multipliers, we obtain a set of independent subproblems, one per scenario (much in the same vein as example 3.9). Pursuing this idea leads to scenario aggregation algorithms.

Representing the split-variable formulation in AMPL   The split-variable formulation is easily expressed in an algebraic language like AMPL. It is customary to set up two files: the first contains the model structure, which is illustrated in figure 6.5, and the other contains the data for a particular model instance, as illustrated in figure 6.6.

The way we express a model in AMPL is almost self-explanatory. All the characters after the # character are treated as a comment; note also that in an algebraic language, one prefers longer names than in the usual mathematical notation. We have to define the following entities.

- The sets involved in our formulation, introduced by the keyword **set**. Here there is one set, **assets**, declared in the model file (figure 6.5), consisting of **stocks** and **bonds**, as declared in the data file (figure 6.6). Stocks here play the role of the riskier but potentially more rewarding asset, whereas bonds are safer.

- The numerical parameters, introduced by the keyword **param**. They are declared in the model file of figure 6.5, but their numerical value is given in the data file of figure 6.6. In our model most parameters are scalars: the initial wealth (55), the number of scenarios (8), the number of periods (3), the target wealth (80), the reward for exceeding the target wealth (1), and the penalty for falling short of the target (4). The asset returns are a tabular parameter, indexed by asset, scenario,

```
set assets;           # different assets
param initwealth;     # initial wealth
param scenarios;      # number of scenarios
param T;              # number of time periods
param target;    # target value (liability) at time T
param reward;    # reward for wealth beyond target value
param penalty;   # penalty for not meeting the target
param return{assets, 1..scenarios, 1..T};
# return of each asset during each period in each scenario
param prob{1..scenarios}; # probability of each scenario
# this indexed set points out which scenarios
# are linked at each period t in 0..T-1
set links{0..T-1} within {1..scenarios, 1..scenarios};

# VARIABLES
var invest{assets,1..scenarios,0..T-1} >= 0; # amount invested
var above_target{1..scenarios}>=0; # amt above final target
var below_target{1..scenarios}>=0; # amt below final target

# OBJECTIVE
maximize exp_value :
    sum{i in 1..scenarios}
        prob[i]*(reward*above_target[i]
        - penalty*below_target[i]);

# CONSTRAINTS
# initial wealth is allocated
subject to budget{i in 1..scenarios} :
sum{k in assets} (invest[k,i,0]) = initwealth;
# portfolio rebalancing
subject to balance{j in 1..scenarios, t in 1..T-1} :
    (sum{k in assets} return[k,j,t]*invest[k,j,t-1]) =
    sum{k in assets} invest[k,j,t];
# check final wealth
subject to scenario_value{j in 1..scenarios} :
(sum{k in assets} return[k,j,T]*invest[k,j,T-1])
- above_target[j] + below_target[j] = target;
# this makes all investments nonanticipative
subject to linkscenarios
    {k in assets, t in 0..T-1, (s1,s2) in links[t]} :
    invest[k,s1,t] = invest[k,s2,t];
```

Fig. 6.5 AMPL model.

```
set assets := stocks bonds;
param initwealth := 55;
param scenarios := 8;
param T := 3;

set links[0] := (1,2) (2,3) (3,4) (4,5) (5,6) (6,7) (7,8);
set links[1] := (1,2) (2,3) (3,4) (5,6) (6,7) (7,8);
set links[2] := (1,2) (3,4) (5,6) (7,8);

param target := 80;
param reward := 1;
param penalty := 4;

param return :=
[stocks, 1, *] 1 1.25 2 1.25 3 1.25
[stocks, 2, *] 1 1.25 2 1.25 3 1.06
[stocks, 3, *] 1 1.25 2 1.06 3 1.25
[stocks, 4, *] 1 1.25 2 1.06 3 1.06
[stocks, 5, *] 1 1.06 2 1.25 3 1.25
[stocks, 6, *] 1 1.06 2 1.25 3 1.06
[stocks, 7, *] 1 1.06 2 1.06 3 1.25
[stocks, 8, *] 1 1.06 2 1.06 3 1.06
[bonds, 1, *] 1 1.14 2 1.14 3 1.14
[bonds, 2, *] 1 1.14 2 1.14 3 1.12
[bonds, 3, *] 1 1.14 2 1.12 3 1.14
[bonds, 4, *] 1 1.14 2 1.12 3 1.12
[bonds, 5, *] 1 1.12 2 1.14 3 1.14
[bonds, 6, *] 1 1.12 2 1.14 3 1.12
[bonds, 7, *] 1 1.12 2 1.12 3 1.14
[bonds, 8, *] 1 1.12 2 1.12 3 1.12;

param prob default 0.125;
```

Fig. 6.6 Data for the AMPL model.

and time period. Note how a tridimensional table is specified in the data file. A notation like [stocks, 1, *] means that values of the third index, to which the wildcard corresponds, will be listed together with the corresponding entries. The scenarios are quite simple for the example:

- For each up branch, the returns are 1.25 for stocks and 1.14 for bonds.

- For each down branch, the returns are 1.06 for stocks and 1.12 for bonds.

The last parameter, prob, defines the probabilities for each scenario; here the probability is set to 1/8 for each scenario, using the default keyword to streamline the notation.

- The decision variables are introduced by the var keyword and correspond clearly to the variables of the model (see figure 6.5).

- Then the objective function is expressed and the solver is instructed to maximize its value. Note how the sum notation is used to express sums over an index in a very natural way.

- The constraints are introduced by the subject to keywords. For each constraint we list a label (which may be used to get the dual variables for each constraint after solving the model), then the index values for which the constraint should be replicated (which corresponds to the $\forall s, \ldots$ in the mathematical notation), and finally, we express the constraints themselves.

Solving the model yields the solution, illustrated in figure 6.7, which is clearly nonanticipative, as it should be. The solution might be obtained through CPLEX, OSL, or any other solver for which an interface is available.

6.2.2 Compact formulation

The split-variable formulation is based on a large number of variables, which are then linked together by the nonanticipativity constraints. This may be useful for algorithms based on decomposition with respect to the scenarios; but if we want to apply a generalization of the L-shaped method (section 3.7) to multistage stochastic programs, the model must be written in a different way. A more compact formulation may be obtained directly by associating decision variables to the nodes in the tree. Let us introduce the following notation:

- N is the set of event nodes; in our case

$$N = \{n_0, n_1, n_2, \ldots, n_{14}\}.$$

```
invest [bonds,*,*]
:       0        1        2        :=
1    13.5207   2.16814     0
2    13.5207   2.16814     0
3    13.5207   2.16814   71.4286
4    13.5207   2.16814   71.4286
5    13.5207   22.368    71.4286
6    13.5207   22.368    71.4286
7    13.5207   22.368      0
8    13.5207   22.368      0

   [stocks,*,*]
:       0        1        2        :=
1    41.4793   65.0946   83.8399
2    41.4793   65.0946   83.8399
3    41.4793   65.0946     0
4    41.4793   65.0946     0
5    41.4793   36.7432     0
6    41.4793   36.7432     0
7    41.4793   36.7432    64
8    41.4793   36.7432    64
;
```

Fig. 6.7 Solution of the AMPL model.

- Each node $n \in N$, apart from the root node n_0, has a unique direct ancestor, denoted by $a(n)$: for instance, $a(n_3) = n_1$.

- There is a set $S \subset N$ of leaf (terminal) nodes; in our case

$$S = \{n_7, \ldots, n_{14}\};$$

for each node $s \in S$ we have surplus and shortfall variables w_+^s and w_-^s.

- There is a set $T \subset N$ of intermediate nodes, where portfolio rebalancing may occur after the initial allocation in node n_0; in our case

$$T = \{n_1, \ldots, n_6\};$$

for each node $n \in \{n_0\} \cup T$ there is an investment variable x_{in} corresponding to the amount invested in asset i at node n.

With this notation the model may be written as follows:

$$\max \quad \sum_{s \in S} p^s (q w_+^s - r w_-^s)$$

$$\text{s.t.} \quad \sum_{i=1}^{I} x_{i,n_0} = W_0$$

$$\sum_{i=1}^{I} R_{i,n} x_{i,a(n)} = \sum_{i=1}^{I} x_{in} \qquad \forall n \in T$$

$$\sum_{i=1}^{I} R_{is} x_{i,a(s)} = L + w_+^s - w_-^s \qquad \forall s \in S$$

$$x_{in}, w_+^s, w_-^s \geq 0,$$

where $R_{i,n}$ is the total return for asset i during the period that *leads to* node n, and p^s is the probability of reaching the terminal node $s \in S$; this probability is the product of all the conditional probabilities on the path that leads from node n_0 to s.

Representing the compact formulation in GAMS The compact formulation could be expressed easily in AMPL. We give here an example using GAMS (see figure 6.8). For simplicity, we have used only one file, merging the problem structure and the instance data. You should bear in mind that this is bad practice.

While the notation is different from AMPL, the basic concepts are similar, and only a few points are worth mentioning.

- Each set may be labeled by a comment string like `'nodes'` for the set N. The comment string is free, but using `'nodes'` is important if you want to solve the model through the IBM OSL stochastic solver, which is based on a variant of L-shaped decomposition and is interfaced to GAMS. In this way you instruct the GAMS solver interface to generate the model taking the tree structure into proper account. This is also important when defining variables and expressing constraints: those involving the nodes are treated in a special way, which is transparent to the user.

- The notation `ROOT(N)` simply says that the set `ROOT` is a subset of the set N.

- The `PARAMETER` section features a few streamlined and self-explanatory notations which are used in assigning the probability for each terminal node in the tree and the returns for each node.

- The `anc(N,N)` parameter is set to 1 for each ordered pair of nodes which are linked by an ancestor relationship. This is exploited to write the constraints. For instance, the notation

```
balance(N,T)$anc(N,T)..
```

```
SETS
  I          'assets' / bonds , stocks /,
  N          'nodes' / n0 , n1*n14 / ,
  ROOT(N)    'root node' / n0 / ,
  T(N)       'rebalancing nodes' / n1*n6 / ,
  S(N)       'leaf nodes' / n7*n14 / ;
PARAMETER
  prob(S) / n7*n14 0.125 / , initwealth / 55 / ,
  target / 80 / , reward / 1 / , penalty / 4 / ,
  ret(I,N)
    / bonds . (n1, n3, n5, n7, n9, n11, n13) 1.14
      bonds . (n2, n4, n6, n8, n10, n12, n14) 1.12
      stocks . (n1, n3, n5, n7, n9, n11, n13) 1.25
      stocks . (n2, n4, n6, n8, n10, n12, n14) 1.06 / ,
  anc(N,N)
    / n0.n1 1 , n0.n2 1 , n1.n3 1 , n1.n4 1 ,
      n2.n5 1 , n2.n6 1 , n3.n7 1 , n3.n8 1 ,
      n4.n9 1 , n4.n10 1 , n5.n11 1 , n5.n12 1 ,
      n6.n13 1 , n6.n14 1 / ;

FREE VARIABLE z;
POSITIVE VARIABLES
    invest(I,N), above_target(N), below_target(N);

EQUATIONS obj, initial(N), balance(N,N), final(N,N);

obj..
  z =E= sum{S , prob(S)*
    (reward*above_target(S) - penalty*below_target(S)) };
initial(ROOT)..
  sum(I, invest(I,ROOT)) =E= initwealth;
balance(N,T)$anc(N,T)..
  sum{I, invest(I,T)} =E= sum{I, ret(I,T) * invest(I,N)};
final(N,S)$anc(N,S)..
  sum{I, ret(I,S)*invest(I,N)} -
  above_target(S) + below_target(S) =E= target;

model stoch / all / ;
option lp=oslse;
solve stoch using lp max z;
```

Fig. 6.8   GAMS model.

is used to specify that the constraint should be written only for nodes N and T, such that N is the ancestor of T.

- Note that there is no direct way of expressing the objective. We have to introduce an auxiliary variable z, which is related to the objective through an equality constraint. Equality constraints use the =E= keyword rather than the = sign. We specify that z should be maximized by the statement

$$\text{solve stoch using lp max z;}$$

- The instruction option lp=oslse; tells the systyem that the OSL stochastic solver should be used; without that instruction, the standard simplex algorithm would be used; it is also possible to select an interior point method. As with AMPL, a wide array of solvers may be selected.

6.2.3 Sample asset and liability management model formulation

To give the reader an idea of how to build nontrivial financial planning models, we generalize a bit the compact formulation of the preceding section. The assumptions and the limitations behind the model are the following:

- We are given a set of initial holdings for each asset; this is a more realistic assumption, since we should use the model to rebalance the portfolio periodically according to a rolling horizon strategy.

- We take linear transaction costs into account; the transaction cost is a percentage c of the traded value, both for buying and selling.

- We want to maximize the expected utility of the terminal wealth.

- There is a stream of uncertain liabilities that we have to meet.

- We do not consider the possibility of lending and borrowing money; we assume all of the available wealth at each rebalancing period is invested in the available assets; actually, the possibility of investing in a risk-free asset is implicit in the model.

- The strategy is self-financing, since we do not consider the possibility of investing new wealth at each rebalancing date (as would be the case, e.g., for a pension fund).

Some of the limitations of the model may easily be relaxed. The important point we make is that when transaction costs are involved, we have to introduce new decision variables to express the amount of assets held, sold, and bought at each rebalancing date. We use a notation which is similar to that used in the compact formulation:

- N is the set of nodes in the tree; n_0 is the initial node.

- The (unique) predecessor of node $n \in N \backslash \{n_0\}$ is denoted by $a(n)$; the the set of terminal nodes is denoted by S; as in the previous formulation, each of these nodes corresponds to a scenario, which is the unique path leading from n_0 to $s \in S$, with probability p^s.

- $T = N \backslash (\{n_0\} \cup S)$ is the set of intermediate trading nodes.

- L^n is the liability we have to meet in node $n \in N$.

- c is the percentage transaction cost.

- $\overline{h}_i^{n_0}$ is the initial holding for asset $i = 1, \ldots, I$ at the initial node.

- P_i^n is the price for asset i at node n.

- z_i^n is the amount of asset i purchased at node n.

- y_i^n is the amount of asset i sold at node n.

- x_i^n is the amount of asset i we hold at node n, after rebalancing.

- W^s is the wealth at node $s \in S$.

- $U(W)$ is the utility for wealth W.

Based on this notation, we may write the following model:

$$\max \quad \sum_{s \in S} p_s U(W^s) \tag{6.6}$$

$$\text{s.t.} \quad x_i^{n_0} = \overline{h}_i^{n_0} + z_i^{n_0} - y_i^{n_0} \quad \forall i \tag{6.7}$$

$$x_i^n = x_i^{a(n)} + z_i^n - y_i^n \quad \forall i, \forall n \in T \tag{6.8}$$

$$(1 - c) \sum_{i=1}^{I} P_i^n y_i^n - (1 + c) \sum_{i=1}^{I} P_i^n z_i^n = L^n \quad \forall n \in T \cup \{n_0\} \tag{6.9}$$

$$W^s = \sum_{i=1}^{I} P_i^s x_i^{a(s)} - L^s \quad \forall s \in S \tag{6.10}$$

$$x_i^n, z_i^n, y_i^n, W^s \geq 0. \tag{6.11}$$

The objective (6.6) is the expected utility of the terminal wealth; if we approximate this nonlinear concave function by a piecewise linear concave function, we get an LP problem (see section 3.1.5). Equation (6.7) expresses the initial asset balance, taking the current holdings into account; the asset balance at intermediate trading dates is taken into account by equation (6.8). Equation (6.9) makes sure that enough cash is generated by selling assets in order to meet the liabilities; we may also reinvest the proceeds of what we sell in new asset holdings; note how the transaction costs are expressed for selling and purchasing. Equation (6.10) is used to estimate the final wealth; note that here we have not taken into account the need to sell assets to generate cash to meet the last liability. If we assume that the entire portfolio is liquidated at the end of the planning horizon, we could rewrite equation (6.10) as

$$W^s = (1 - c) \sum_{i=1}^{I} P_i^s x_i^{a(s)} - L^s.$$

In practice, we would repeatedly solve the model on a rolling horizon basis, so the exact expression of the objective function is a bit debatable.

This model can be generalized in a number of ways, which are left as an exercise to the reader. The most important point is that we have assumed that the liabilities must be met. This may be a very hard constraint; if extreme scenarios are included in the formulation, as they should be, it may well be the case that the model above is infeasible. So the formulation should be relaxed in a sensible way; we could consider the possibility of borrowing cash; we could also introduce suitable penalties for not meeting the liabilities. In principle, we could also require that the probability of not meeting the liabilities is small enough; this leads to chance-constrained formulations, for which we refer the reader to the literature.

6.2.4   Scenario generation for multistage stochastic programming

The quality of the solution obtained by solving a multistage stochastic program depends on how well the scenario tree models the inherent uncertainty influencing the decision problem. Financial scenarios may be generated by building a time series model and simulating some sample paths. A very simple model is the vector autoregressive model (VAR, which should not be confused with value-at-risk). Let $\mathbf{h}_t$ be a vector of economic and financial variables at time t. An example of a VAR model is

$$\mathbf{h}_t = \mathbf{c} + \mathbf{\Omega} \mathbf{h}_{t-1} + \boldsymbol{\epsilon}_t \qquad t = 1, \ldots, T,$$

where $\mathbf{c}$ and $\mathbf{\Omega}$ are model parameters, and $\boldsymbol{\epsilon} \sim N(\mathbf{0}, \boldsymbol{\Sigma})$ is a vector of jointly normal random variables with zero mean and covariance matrix $\boldsymbol{\Sigma}$. Each h_{it}, $i = 1, \ldots, N$ may be linked to any variable of interest, such as inflation, asset returns, etc. Some transformation may be exploited to build a good model; for instance, in [12] the VAR approach is used where $h_{it} = \ln(1 + r_{it})$, and r_{it} is the discrete-time rate of change of the quantity of interest.

Quite sophisticated models may be built, e.g., to take mean reversion and stochastic volatility into account. Then, scenarios may be generated by random sampling, much in the vein of Monte Carlo simulation. The problem here is that the number of scenarios we may afford is quite limited, since the computational effort of solving a stochastic program grows exponentially with the number of time stages and the branching factor at each node of the tree.

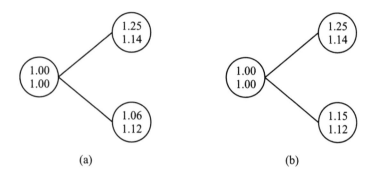

Fig. 6.9   Two simple scenario trees for asset price paths.

But a few random samples are unlikely to capture the uncertainty correctly, making the model prone to sampling uncertainty. To ease the difficulty, variance reduction techniques, such as antithetic sampling, may be tried; the use of importance sampling is advocated in [9].

An alternative to random sampling is fitting some properties of the joint distribution of the data. For instance, if we know the covariance of asset returns, we may generate the scenarios in such a way that the samples reflect the covariance. Other moments may be fitted, and it is also possible to require the inclusion of extreme scenarios. This approach is proposed in [8] and [12], and it requires the solution of a nonlinear programming model to generate data paths that fit the required statistics as much as possible.

The considerations above apply to any stochastic program. When we consider an application in finance, there is still another issue: arbitrage. Consider the data of the toy problem we have considered in section 6.2. Are they sensible data? To understand the issue, consider the two simple trees depicted in figure 6.9. The first one corresponds to the scenarios we have used in the example. The only difference is that here we are considering the asset prices for the two assets; since the initial prices are 1 for both assets, the total returns are just the prices in the two scenarios. Sensible scenarios should not only reflect the information we have but should also rule out arbitrage opportunities. One way to define an arbitrage opportunity is the following. We have an arbitrage opportunity if there exists a portfolio which is guaranteed to have a nonnegative value at the end of the holding period in any scenario, but which has a negative value at the beginning. Formally, let $\mathbf{p} \in \mathbb{R}_+^n$ be the vector of the initial prices for n assets, $\mathbf{x} \in \mathbb{R}^n$ the portfolio holdings for each asset, and $\mathbf{R} \in \mathbb{R}^{m,n}$ the return of each asset in each of the m scenarios (i.e., R_{ij} is the return of asset j in scenario i). Then an arbitrage opportunity is a portfolio $\mathbf{x}$ such that

$$\mathbf{R}\mathbf{x} \geq \mathbf{0} \quad \text{and} \quad \mathbf{p}^T\mathbf{x} < 0. \tag{6.12}$$

Another form of arbitrage opportunity is the following:[1]

$$\mathbf{Rx} \geq \mathbf{0} \quad \text{and} \quad \mathbf{p}^T \mathbf{x} = 0, \tag{6.13}$$

where at least one inequality is strict. In other words, we are sure that we will not lose any money in any scenario and there is at least one scenario in which we gain something.

In order to exploit an arbitrage opportunity to gain an infinite profit, we should be able to do some short selling; if the optimization model forbids short selling, we will not see a blatant error such as an unbounded solution, but what we get could be not very sensible anyway.

It is easy to see that the scenario tree in figure 6.9b leads to an arbitrage opportunity like (6.13). With those asset prices, an initial portfolio has zero value if

$$x_1 + x_2 = 0.$$

We may use this condition to express the final portfolio value in the two scenarios:

$$1.25x_1 + 1.14x_2 = (1.25 - 1.14)x_1$$
$$1.15x_1 + 1.12x_2 = (1.15 - 1.12)x_1.$$

It is easy to see that we should sell the second asset short, so that $x_1 > 0$, to get an arbitrage opportunity. The same does not hold in the case of figure 6.9a.

But how can we be sure that a set of scenarios is arbitrage-free? An answer is given by the following theorem.

THEOREM 6.1 *There is no arbitrage opportunity of the form (6.12) if and only if there is a vector* $\mathbf{y}$ *such that*

$$\mathbf{R}^T \mathbf{y} = \mathbf{p} \quad \text{and} \quad \mathbf{y} \geq \mathbf{0}.$$

Proof. Consider the following linear programming problem:

$$\begin{aligned} \max \quad & \mathbf{0}^T \mathbf{y} \\ \text{s.t.} \quad & \mathbf{R}^T \mathbf{y} = \mathbf{p} \\ & \mathbf{y} \geq \mathbf{0}. \end{aligned}$$

If this problem is solvable, so is its dual:

$$\begin{aligned} \min \quad & \mathbf{p}^T \mathbf{x} \\ \text{s.t.} \quad & \mathbf{Rx} \geq \mathbf{0}. \end{aligned}$$

---

[1]See [10, chapter 2] for a discussion about the relationships between the two forms of arbitrage.

But in this case, the optimal objective values are both equal to zero. Then we see that if there exists a feasible vector $\mathbf{y}$ for the primal problem, we cannot have $\mathbf{p}^T\mathbf{x} < 0$.   ☐

On the one hand, the theorem suggests a way to make scenarios arbitrage free. We could simply add a node in such a way that the conditions of the theorem are met. The full details of this idea are given in [5]. It should be noted that finding the best way to generate scenarios is still an open issue, as we may well generate arbitrage-free scenarios which do not fit the assumed distributions at all. On the other hand, reasoning along the lines of the theorem we may get a grasp on the relationships between the absence of arbitrage opportunities and the existence of risk-neutral probability measures.

To begin with, we should note that if a vector $\mathbf{p}$ of initial prices satisfies theorem 6.1, then any vector $\lambda\mathbf{p}$, $\lambda > 0$, does, too. So there is a degree of freedom in pricing; in fact, we have only considered risky assets. What if we consider a risk-free asset with a risk-free rate r? To characterize arbitrage when a risk-free asset is available, let us consider a two-stage scenario tree: the initial node is 0 and there are N nodes at the second stage. Let P_{i0} the current price of asset i, $i = 1, \ldots, I$, and P_{in} the price if scenario n, $n = 1, \ldots, N$, occurs. For each asset, we may define the discounted gain for asset i in scenario n, with respect to the risk-free asset (see [16]):

$$R_{in}^* = \frac{P_{in}}{1+r} - P_{i0} \qquad \forall i, n.$$

Note that if a discounted gain is positive, it means that the risky asset has performed better than the risk-free asset. Given a set of portfolio holdings x_i, we may define the overall discounted gain in node n:

$$g_n^* = \sum_{i=1}^{I} R_{in}^* x_i,$$

which is the realization of the random variable G^* in scenario n. Now it is intuitive that an arbitrage opportunity may be characterized by the conditions[2]

$$g_n^* \geq 0 \qquad \forall n$$
$$E[G^*] > 0.$$

This means that the portfolio is expected to gain more than the risk-free asset on the average, but it cannot gain less in any possible scenario. To find a condition ruling out arbitrage, we may try to reason as in theorem 6.1. We may rewrite the arbitrage conditions as:

$$\sum_{n=1}^{N} \sum_{i=1}^{I} R_{in}^* x_i = 1$$

---

[2]See [16] for more details.

$$\sum_{i=1}^{I} R_{in}^* x_i \geq 0 \qquad \forall n.$$

The first condition may look a bit arbitrary, but its purpose is to make sure that at least one of the g_n^* is strictly positive; since an arbitrage opportunity may be scaled arbitrarily, setting the double sum value to 1 serves the purpose. Now, to apply linear programming duality, we should rewrite these conditions in the standard form:

$$\mathbf{Ax} = \mathbf{b}$$
$$\mathbf{x} \geq \mathbf{0}.$$

(6.14)

We may simply express each portfolio holding, which may be negative if short-selling is allowed, as

$$x_i = x_i^+ - x_i^- \qquad x_i^+, x_i^+ \geq 0,$$

and introduce a set of nonnegative auxiliary variables x_{I+n}, $n = 1, \ldots, N$:

$$x_{I+n} = \sum_{i=1}^{I} R_{in}^* x_i = \sum_{i=1}^{I} \left(R_{in}^* x_i^+ - R_{in}^* x_i^- \right) \qquad \forall n.$$

So we have a vector of nonnegative decision variables:

$$\mathbf{x} = \begin{bmatrix} x_1^+ & x_1^- & x_2^+ & \cdots & x_I^- & x_{I+1} & \cdots & x_{I+N} \end{bmatrix}^T.$$

Now the existence of an arbitrage is linked to the existence of a solution to the system (6.14), where

$$\mathbf{A} = \begin{bmatrix} 0 & 0 & 0 & \cdots & 0 & 1 & 1 & \cdots & 1 \\ R_{11}^* & -R_{11}^* & R_{21}^* & \cdots & -R_{I1}^* & -1 & 0 & \cdots & 0 \\ R_{12}^* & -R_{12}^* & R_{22}^* & \cdots & -R_{I2}^* & 0 & -1 & \cdots & 0 \\ \vdots & \vdots & \vdots & \ddots & \vdots & \vdots & \vdots & \ddots & \vdots \\ R_{1N}^* & -R_{1N}^* & R_{2N}^* & \cdots & -R_{IN}^* & 0 & 0 & \cdots & -1 \end{bmatrix}$$

and

$$\mathbf{b} = [1, 0, \ldots, 0]^T.$$

If there is a feasible solution of (6.14), there cannot be a solution of the following system:

$$\mathbf{A}^T \mathbf{y} \leq \mathbf{0}$$
$$\mathbf{b}^T \mathbf{y} > 0.$$

(6.15)

This is a direct consequence of linear programming duality. In fact, the existence of a solution of system (6.15) would imply that there is direction $\hat{\mathbf{y}}$

along which we may arbitrarily increase the objective function $\mathbf{b}^T\mathbf{y}$ without violating the constraints $\mathbf{A}^T\mathbf{y} \leq \mathbf{c}$, for an arbitrary vector $\mathbf{c}$. Hence, the dual linear program would be unbounded, and the primal could not be feasible.

Seeing it the other way around, if there is a solution to system (6.15), there is no arbitrage opportunity. It is possible to find an important interpretation of system (6.15), taking the forms of $\mathbf{A}$ and $\mathbf{b}$ into account. Let us denote the dual variable corresponding to the first primal constraint by y_0; we also have a dual variable y_n for each primal constraint corresponding to scenario n. Now let us write the dual constraints $\mathbf{A}^T\mathbf{y} \leq \mathbf{0}$ explicitly. For each asset i, we have a pair of inequalities:

$$\sum_{n=1}^{N} R_{in}^* y_n \leq 0$$

$$-\sum_{n=1}^{N} R_{in}^* y_n \leq 0.$$

Together, they imply that for all assets i we have

$$\sum_{n=1}^{N} R_{in}^* y_n = 0. \tag{6.16}$$

Furthermore, considering the last n columns of matrix $\mathbf{A}$, we also have

$$y_0 - y_n \leq 0 \qquad \forall n.$$

This, together with second condition in system (6.15), has the following implications:

$$\mathbf{b}^T\mathbf{y} > 0 \;\Rightarrow\; y_0 > 0 \;\Rightarrow\; y_n > 0 \qquad \forall n.$$

Let us rescale the dual solution as follows:

$$\pi_n = \frac{y_n}{\sum_{k=1}^{N} y_k} \qquad \forall n. \tag{6.17}$$

We see that the vector $\boldsymbol{\pi}$ may be interpreted as a probability measure, since the components are nonnegative and their sum is 1. Moreover, it is a *risk-neutral* probability measure, according to which any scenario is possible (it has strictly positive probability) and any asset gains the risk-free return on the average. To see this, we may plug equation (6.17) into equation (6.16) to obtain

$$\sum_{n=1}^{N} R_{in}^* \pi_n = 0.$$

This means that according to this probability measure the expected discounted gain for any asset is zero, which in turn implies

$$\mathrm{E}_\pi[P_i] = (1+r)P_{i0}.$$

Now we may see a little better why risk-neutral probability measures play a role in option pricing under the no-arbitrage assumption, at least in a two-period economy with discrete states of the world. Rigorous treatment with continuous time and continuous asset prices requires the tools of stochastic calculus.

6.3 FIXED-MIX MODEL BASED ON GLOBAL OPTIMIZATION

In the multistage stochastic programming models we have illustrated in section 6.2, we have assumed that the portfolio could be rebalanced at specified time instants. A different type of model is obtained if we assume that the asset mix is held constant over the whole period. This means that the proportion of wealth that we allocate to each asset is kept constant; thus, we trade according a sell-high/buy-low strategy. Using the same notation as in section 6.2.1, we have a discrete set of scenarios, each with a probability p_s, $s = 1 \ldots, S$, where the returns are represented by R_{it}^s. Now, the decision variables are simply the proportion of wealth allocated to each asset, denoted by x_i; note that since there is no recourse action, the scenarios need not be structured according to a tree, as the nonanticipativity condition is satisfied by the definition of the decision variables. The model we describe here is due to [15], to which we refer the reader for further information and for computational experiments, and is basically an extension of the mean-variance framework: no liability is considered, and we base our objective function on the terminal wealth.

Let W_0 be the initial wealth. Then the wealth at the end of time period 1 in scenario s will be

$$W_1^s = W_0 \sum_{i=1}^{I} R_{i1}^s x_i.$$

Note that the wealth is scenario dependent, but the asset allocation is not. In general, when we consider two consecutive time periods, we have

$$W_t^s = W_{t-1}^s \sum_{i=1}^{I} R_{it}^s x_i \qquad \forall t, s.$$

The wealth at the end of the planning horizon is

$$W_T^s = W_0 \prod_{t=1}^{T} \left(\sum_{i=1}^{I} R_{it}^s x_i \right) \qquad \forall s.$$

Within a mean-variance framework, we may build a quadratic utility function depending on the terminal wealth. Given a parameter λ linked to our risk aversion, the objective function will be something like

$$\max \quad \lambda \, \mathrm{E}[W_T] - (1 - \lambda) \, \mathrm{Var}(W_T).$$

To express the objective function, we must recall that $\text{Var}(X) = \text{E}[X^2] - \text{E}^2[X]$, and we may write the model as

$$\max \quad \lambda W_0 \sum_{s=1}^{S} p^s \left[\prod_{t=1}^{T} \left(\sum_{i=1}^{I} R_{it}^s x_i \right) \right]$$

$$+ (1 - \lambda) W_0^2 \left\{ \left[\sum_{s=1}^{S} p^s \left[\prod_{t=1}^{T} \left(\sum_{i=1}^{I} R_{it}^s x_i \right) \right] \right]^2 \right.$$

$$\left. - \sum_{s=1}^{S} p^s \left[\prod_{t=1}^{T} \left(\sum_{i=1}^{I} R_{it}^s x_i \right) \right]^2 \right\}$$

$$\text{s.t.} \quad \sum_{i=1}^{I} x_i = 1$$

$$0 \le x_i \le 1.$$

This looks like a very complex problem; however, while the objective function is a bit messy, the constraints are quite simple. The real difficulty is that this is a nonconvex problem. To see why, just note that the objective turns out to be a polynomial in the decision variables; since polynomials may have many minima and maxima, we have a nonlinear nonconvex problem.

The problem may be tackled by the branch and bound methods described in section 3.5. In particular, the idea of bounding a nonconvex function by a convex underestimator is used in [15]. If complicating features are added to the model, this may turn out a quite difficult mixed-integer nonlinear problem; in this case, the use of metaheuristics such as tabu search may be the best option [7].

It is useful to interpret this approach within an integration framework of simulation and optimization. Actually, simulation is separated from optimization, since scenarios are generated beforehand; we evaluate the solutions on the same set of scenarios, which is consistent with variance reduction by common random numbers. After the optimization, simulation could be used to evaluate the solution we obtain on a larger set of scenarios, possibly including stress test scenarios; in other words, we may carry out an out-of-sample analysis to check the robustness of the solution. This is easily accomplished for a fixed-mix policy, but not for a dynamic policy, as this would require the repeated solution of difficult multistage stochastic programs. In fact, even if a fixed-mix policy is in principle an inferior policy with respect to a dynamic one, it may be more robust in practice; what's more important, it is easier to prove its robustness with respect to an arbitrary set of scenarios, and to persuade a manager to adopt it.

Selection of the best portfolio management policy is actually an open issue, but it is worth noting that the fixed-mix policy is only the simplest policy structure that we may consider for the integration of simulation and optimization. More complex policies could be devised, depending on a set of numerical parameters, whose value may be set by the integration of simulation and optimization methods.

For further reading

In the literature

- The use of mixed-integer programming models in portfolio management is the subject of an increasing number of papers including [1], [2], [4], [13], [14], and [17].

- An excellent source for several papers on the application of stochastic programming in finance is [18].

- Scenario generation is one of the topic covered in [5], [8], and [12]. A few papers may also be downloaded from

    `http://www.few.eur.nl/few/people/kouwenberg`.

- For a thorough discussion on arbitrage and risk-neutral probability measures, see [10] and [16].

- The AMPL language is described in the book [6].

- GAMS is the subject of an online book available at

    `http://ageco.tamu.edu/faculty/mccarl`.

- Global optimization techniques for optimization of a fixed-mix portfolio are discussed in [15]; the model is extended and tackled by metaheuristics in [7].

On the Web

- The GAMS modeling environment is described at `http://www.gams.de`.

- The corresponding AMPL site is `http://www.ampl.com`.

- Many pointers to stochastic programming, including financial applications, can be found by browsing `http://mat.gsia.cmu.edu`.

- Some researchers maintain Web pages on stochastic programming, providing all sorts of link. One is

    `http://www.math.ku.dk/~caroe/stocprog/index.html`.

REFERENCES

1. D. Bertsimas, C. Darnell, and R. Stoucy. Portfolio Construction through Mixed-Integer Programming at Grantham, Mayo, Van Otterloo and Company. *Interfaces*, 29:49–66, 1999.

2. D. Bienstock. Computational Study of a Family of Mixed-Integer Quadratic Programming Problems. *Mathematical Programming*, 74:121–140, 1996.

3. J.R. Birge and F. Louveaux. *Introduction to Stochastic Programming.* Springer-Verlag, New York, 1997.

4. T.-J. Chang, N. Meade, J.E. Beasley, and Y.M. Sharaiha. Heuristics for Cardinality Constrained Portfolio Optimization. *Computers and Operations Research*, 27:1271–1302, 2000.

5. C. Dert. *Asset Liability Management for Pension Funds: A Multistage Chance Constrained Programming Approach.* Ph.D. thesis, Erasmus University, Rotterdam, The Netherlands, 1995.

6. R. Fourer, D.M. Gay, and B.W. Kernighan. *AMPL: A Modeling Language for Mathematical Programming.* Boyd and Fraser, Danvers, MA, 1993.

7. F. Glover, J.M. Mulvey, and K. Hoyland. Solving Dynamic Stochastic Control Problems in Finance Using Tabu Search with Variable Scaling. In I.H. Osman and J.P. Kelly, editors, *Meta-Heuristics: Theory and Applications*, pages 429–448. Kluwer Academic, Dordrecht, The Netherlands, 1996.

8. K. Hoyland and S.W. Wallace. Generating Scenario Trees for Multistage Problems. *Management Science*, 47:295–307, 2001.

9. G. Infanger. *Planning under Uncertainty: Solving Large-Scale Stochastic Linear Programs.* Boyd and Fraser, Danvers, MA, 1994.

10. Jr. J.E. Ingersoll. *Theory of Financial Decision Making.* Rowman & Littlefield, Totowa, NJ, 1987.

11. H. Konno and H. Yamazaki. Mean-Absolute Deviation Portfolio Optimization Model and Its Application to Tokyo Stock Market. *Management Science*, 37:519–531, 1991.

12. R.P. Kouwenberg. *Dynamic Asset Liability Management.* Ph.D. thesis, Erasmus University, Rotterdam, The Netherlands, 2000.

13. M.S. Lobo, M. Fazel, and S. Boyd. Portfolio Optimization with Linear and Fixed Transaction Costs and Bounds on Risk. Unpublished manuscript (available at http://www.stanford.edu/~boyd), 1999.

14. R. Mansini and M.G. Speranza. Heuristic Algorithms for the Portfolio Selection Problem with Minimum Transaction Lots. *European Journal of Operational Research*, 114:219–233, 1999.

15. C.D. Maranas, I.P. Androulakis, C.A. Floudas, A.J. Berger, and J.M. Mulvey. Solving Long-Term Financial Planning Problems via Global Optimization. *Journal of Economic Dynamics and Control*, 21:1405–1425, 1997.

16. S.R. Pliska. *Introduction to Mathematical Finance: Discrete Time Models*. Blackwell Publishers, Malden, MA, 1997.

17. J.K. Sankaran and A.A. Patil. On the Optimal Selection of Portfolios under Limited Diversification. *Journal of Banking and Finance*, 23:1655–1666, 1999.

18. W.T. Ziemba and J.M. Mulvey, editors. *Worldwide Asset and Liability Modeling*. Cambridge University Press, Cambridge, 1998.

7

Option valuation by Monte Carlo simulation

Monte Carlo simulation is an important tool in computational finance. It may be used to evaluate portfolio management rules, to price options, to simulate hedging strategies, and to estimate value-at-risk. The main advantages of this tool are its generality, relative ease of use, and flexibility: for instance, it may take stochastic volatility and many complicating features of exotic options into account, and it lends itself to treating high-dimensional problems, where the partial differential equations framework may be inefficient. It is difficult to apply simulation to American options, as simulation goes forward in time, and establishing an optimal exercise policy requires nontrivial considerations, which in some sense require going backward in time; however, the application of simulation to such options is the subject of ongoing research.[1]

Apart from the last consideration, the real issue with Monte Carlo simulation is its computational burden. An increasing number of replications is needed to refine the confidence interval of the estimates we are interested in. The problem may be partially solved by variance reduction techniques or by resorting to low-discrepancy sequences. The aim of this chapter is to illustrate the application of these techniques to a few examples, including some path-dependent options. When possible, we will compare the results of simulation with analytical formulas. Clearly, our aim in doing so is a purely didactic one. If you have to compute the area of a rectangular room, you just multiply the room length times the room width; you would never count how many times a

---

[1]Pricing American options by simulation is beyond the scope of an introductory book; see, e.g., [3] or [9].

standard tile fits the surface. However, you should learn to do so in an easy case, as this may come handy when no analytical formula is available.

We first illustrate in section 7.1 how asset price paths may be simulated based on the geometric Brownian motion model. In section 7.2 we illustrate how crude Monte Carlo may be used to price a vanilla European option; then we illustrate the use of variance reduction techniques and low-discrepancy sequences. More complex and exotic options are introduced in section 7.3. Exotic options are the source of interesting problems that may be tackled by numerical methods: section 7.4 deals with barrier options, whereas in section 7.5 an Asian option is treated.

7.1 SIMULATING ASSET PRICE DYNAMICS

In section 1.4.1 we discussed possible models for asset price dynamics. For convenience, we recall here the geometric Brownian motion model for the asset price $S(t)$, with drift μ and volatility σ:

$$dS = \mu S\,dt + \sigma S\,dz,$$

where dz is a standard Wiener process. An equivalent expression is

$$d\ln S = \left(\mu - \frac{1}{2}\sigma^2\right) dt + \sigma\,dz. \tag{7.1}$$

We also recall that, setting $\nu = \mu - \sigma^2/2$, we have

$$
\begin{aligned}
\mathrm{E}[\ln S(t)/\ln S(0)] &= \nu t & (7.2)\\
\mathrm{Var}[\ln S(t)/\ln S(0)] &= \sigma^2 t & (7.3)\\
\mathrm{E}[S(t)/S(0)] &= e^{\mu t} & (7.4)\\
\mathrm{Var}[S(t)/S(0)] &= e^{2\mu t}(e^{\sigma^2 t}-1). & (7.5)
\end{aligned}
$$

Equation (7.1) is particularly useful as it can be integrated exactly, yielding

$$S(t) = S(0)\,\exp\!\left(\nu t + \sigma \int_0^t dz\right).$$

To simulate the path of the asset price over an interval $(0,T)$, we must discretize time with a time step δt. From the last equation, and recalling the properties of the standard Wiener process (see section 1.4.1), we get

$$S(t+\delta t) = S(t)\,\exp\!\left(\nu\,\delta t + \sigma\sqrt{\delta t}\,\epsilon\right), \tag{7.6}$$

where $\epsilon \sim N(0,1)$ is a standard normal random variable. Based on equation (7.6), it is easy to generate sample paths for the asset price.

```
% AssetPaths.m
function SPaths=AssetPaths(S0,mu,sigma,T,NSteps,NRepl)
SPaths = zeros(NRepl, 1+NSteps);
SPaths(:,1) = S0;
dt = T/NSteps;
nudt = (mu-0.5*sigma^2)*dt;
sidt = sigma*sqrt(dt);
for i=1:NRepl
   for j=1:NSteps
      SPaths(i,j+1)=SPaths(i,j)*exp(nudt + sidt*randn);
   end
end
```

Fig. 7.1   Naive code to generate asset price paths by Monte Carlo simulation.

Example 7.1 A straightforward code to generate sample paths of asset prices is given in figure 7.1. The function `AssetPaths` yields a matrix of asset paths, where the replications are stored row by row and columns correspond to time instants. We have to provide the function with the initial price S0, the drift mu, the volatility `sigma`, the time horizon T, the number of time steps `NSteps`, and the number of replications `NRepl`. Note that the function takes the drift parameter μ as input and then it computes the parameter ν.

For instance, let us generate and plot three one-year sample paths for an asset with an initial price \$50, drift 0.1, and volatility 0.3 (on a yearly basis), assuming that the time step is one day:[2]

```
>> randn('seed',0);
>> paths=AssetPaths(50,0.1,0.3,1,365,3);
>> plot(1:length(paths),paths(1,:))
>> hold on
>> plot(1:length(paths),paths(2,:))
>> hold on
>> plot(1:length(paths),paths(3,:))
```

The result is plotted in figure 7.2. If you start the random number generator for standard normals `randn` with another seed, you will get different results. The code in figure 7.1 is in some sense naive, as it written in a nonvectorized form. In general, MATLAB code is more efficient if we avoid `for` loops. In

---

[2]We assume here that a year consists of 365 trading days. How to treat nontrading days is a bit controversial (see, e.g., [7, pp. 255-257]).

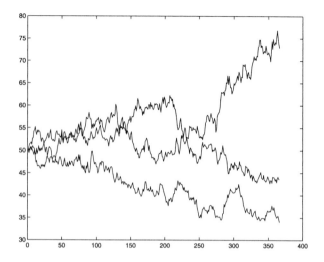

Fig. 7.2 Sample paths generated by Monte Carlo simulation.

order to vectorize the code, it is convenient to rewrite equation (7.6) as

$$\ln S(t + \delta t) - \ln S(t) = \nu \, \delta t + \sigma \sqrt{\delta t} \, \epsilon.$$

So we may generate the *differences* in the logarithm of the asset prices and then use the cumsum function with an optional parameter set to 2 in order to compute the cumulative sums over the rows (the default is summing over columns). The resulting function AssetPaths1 is illustrated in figure 7.3. We may see the improvement in speed:

```
>> tic, paths=AssetPaths(50,0.1,0.3,1,100,1000);, toc
elapsed_time =
    17.5200
>> tic, paths=AssetPaths1(50,0.1,0.3,1,100,1000);, toc
elapsed_time =
     0.7700
```

The speedup is evident. Actually, there is a trade-off between vectorization and memory requirements. It may happen that a fully vectorized code requires huge matrices, which do not fit the main memory of the computer. In such a case, using disk space as virtual memory may slow the execution. As usual, a suitable compromise must be sought.   ▯

```
% AssetPaths1.m
function SPaths=AssetPaths1(S0,mu,sigma,T,NSteps,NRepl)
dt = T/NSteps;
nudt = (mu-0.5*sigma^2)*dt;
sidt = sigma*sqrt(dt);
Increments = nudt + sidt*randn(NRepl, NSteps);
LogPaths = cumsum([log(S0)*ones(NRepl,1) , Increments] , 2);
SPaths = exp(LogPaths);
```

Fig. 7.3   Improved code to generate asset price paths.

7.2   PRICING A VANILLA EUROPEAN OPTION BY MONTE CARLO SIMULATION

In option pricing, we need to estimate the expected value of the discounted payoff of the option:

$$f = e^{-rT}\hat{E}[f_T],$$

where f_T is the payoff at the maturity date T, a constant risk-free rate r is assumed, and the expectation $\hat{E}[\cdot]$ is taken with respect to a risk-neutral measure. For our simple examples, this means that the drift μ for the asset price must be replaced by the risk-free rate r (see section 1.4.5). Depending on the nature of the option at hand, we may need to generate the full sample paths, or simply the terminal asset price. We give an introduction to path-dependent options in the next section.

As a starting point, we may price a vanilla European call option. This may seem a little pointless, as the Black-Scholes formula is a much easier way to price such an option. On the one hand, this is a good introductory exercise; on the other one, the value of a vanilla European option may be used as a control variate to reduce variance, in which case it is necessary to compare its correct value with the estimated value.

In this case, the path we need consists of just two points: the initial price, and the price at expiration, so we do not need to use the AssetPath1 function. We must generate the option payoffs according to the expression

$$\max\{0, S(0)e^{(r-\sigma^2/2)T+\sigma\sqrt{T}\epsilon} - X\},$$

where X is the strike price. The code to price the call option is displayed in figure 7.4, and it also gives a confidence interval obtained by the normfit function. It should be emphasized that from a rigorous point of view, using normfit is *not correct*, as this function assumes normally distributed data and uses Student's t distribution in computing the confidence interval for the

```
% BlsMC.m
function [Price, CI] = BlsMC(S0,X,r,T,sigma,NRepl)
nuT = (r - 0.5*sigma^2)*T;
siT = sigma * sqrt(T);
DiscPayoff = exp(-r*T)*max(0, S0*exp(nuT+siT*randn(NRepl,1))-X);
[Price, VarPrice, CI] = normfit(DiscPayoff);
```

Fig. 7.4   Code to price a vanilla European call by Monte Carlo simulation.

mean.[3] The option payoffs are not normally distributed, but their sum will be in the limit, due to the central limit theorem. So using normfit is just an easy way to get an *approximate* confidence interval.

Let us price a European call with initial asset price $S(0) = \$50$, strike price $X = \$52$, expiring in five months, when the annual risk-free rate is 10% and the volatility is 40%.

```
>> blsprice(50,52,0.1,5/12,0.4)
ans =
    5.1911
>> randn('seed',0);
>> [price,CI] = BlsMC(50,52,0.1,5/12,0.4,1000)
price =
    5.4445
CI =
    4.8776
    6.0115
>> randn('seed',0);
>> [price,CI] = BlsMC(50,52,0.1,5/12,0.4,200000)
price =
    5.1780
CI =
    5.1393
    5.2167
```

The comparison with the correct price obtained by the Black-Scholes formula, and the width of the 95% confidence intervals, show that a large number of replications is needed to get an acceptable degree of accuracy. So we could try to improve our estimates with variance reduction techniques or low-discrepancy sequences as an alternative.

---

[3]Student's distribution yields larger confidence intervals than the normal distribution for a small number of samples; for a large number of samples the two confidence intervals are approximately the same.

```
% BlsMCAV.m
function [Price, CI] = BlsMCAV(S0,X,r,T,sigma,NRepl)
nuT = (r - 0.5*sigma^2)*T;
siT = sigma * sqrt(T);
Veps = randn(NRepl,1);
Payoff1 = max( 0 , S0*exp(nuT+siT*Veps) - X);
Payoff2 = max( 0 , S0*exp(nuT+siT*(-Veps)) - X);
DiscPayoff = exp(-r*T) * 0.5 * (Payoff1+Payoff2);
[Price, VarPrice, CI] = normfit(DiscPayoff);
```

Fig. 7.5   Using antithetic variates to price a vanilla European call by Monte Carlo simulation.

7.2.1   Using antithetic variates to price a vanilla European option

A code using antithetic variates is shown in figure 7.5. We simply generate a stream of standard normal variates and use the same sequence, with a change in sign, in the antithetic run. Each pair of antithetic samples is averaged and used as an estimator. To check the variance reduction obtained by using antithetic sampling, we should use half the replications we used with crude Monte Carlo:

```
>> randn('seed',0);
>> [price,CI] = BlsMCAV(50,52,0.1,5/12,0.4,100000)
price =
    5.1971
CI =
    5.1656
    5.2286
>> randn('seed',0);
>> [price,CI] = BlsMCAV(50,52,0.1,5/12,0.2,100000)
price =
    2.6337
CI =
    2.6198
    2.6476
>> blsprice(50,52,0.1,5/12,0.2)
ans =
    2.6318
```

From the first run we see that some improvement is indeed obtained with respect to crude Monte Carlo. Furthermore, we also see from the other runs that the difficulty in getting an accurate estimate may also depend on the

```
% BlsTrMC.m
function [Price, CI] = BlsTrMC(S0,X,r,T,sigma,Sb,NRepl)
nuT = (r - 0.5*sigma^2)*T;
siT = sigma * sqrt(T);
StockPrice = S0*exp(nuT+siT*randn(NRepl,1));
Clip = find(StockPrice > Sb);
StockPrice(Clip) = 0;
DiscPayoff = exp(-r*T) * max( 0 , StockPrice - X);
[Price, VarPrice, CI] = normfit(DiscPayoff);
```

Fig. 7.6 Pricing a call option with truncated payoff by crude Monte Carlo.

underlying asset volatility. An important point to bear in mind is that anti-thetic sampling may not yield a variance reduction when some monotonicity condition is not satisfied (see section 4.4.1). In the case of a vanilla call option, the higher the random variates, the higher the payoff. In the next section, we describe a type of option in which antithetic sampling may not work.

7.2.2 Using antithetic variates to price a European option with truncated payoff

Consider a *truncated* payoff call option. The idea is that the payoff is obtained by truncating the payoff of a vanilla call as follows:

$$f[S(T)] = \begin{cases} S(T) - X & \text{if } S(T) \in [X, S_b] \\ 0 & \text{otherwise.} \end{cases}$$

The stock price S_b acts as a barrier, canceling the option if $S(T) > S_b$. Readers familiar with barrier options (described in the following) should note that here the option is canceled only if the barrier is crossed at maturity; nothing happens if the barrier is crossed before the maturity. We see that the payoff is not a monotonic function of the stock price; hence, we may expect that antithetic sampling is not guaranteed to work.

The functions BlsTrMC and BlsTrMCAV, whose code is given in figures 7.6 and 7.7, respectively, may be used to see the improvement obtained by anti-thetic sampling.

```
>> randn('seed',0)
>> [Price, CI] = BlsTrMC(50,50,0.1,5/12,0.4,70,200000)
Price =
    3.2185
CI =
    3.1961
```

```
% BlsTrMCAV.m
function [Price, CI] = BlsTrMCAV(S0,X,r,T,sigma,Sb,NRepl)
nuT = (r - 0.5*sigma^2)*T;
siT = sigma * sqrt(T);
Veps = randn(NRepl,1);
StockPrice1 = S0*exp(nuT+siT*Veps);
StockPrice2 = S0*exp(nuT-siT*Veps);
Clip1 = find(StockPrice1 > Sb);
Clip2 = find(StockPrice2 > Sb);
StockPrice1(Clip1) = 0;
StockPrice2(Clip2) = 0;
Payoff1 = max( 0 , StockPrice1 - X);
Payoff2 = max( 0 , StockPrice2 - X);
DiscPayoff = exp(-r*T) * 0.5 * (Payoff1+Payoff2);
[Price, VarPrice, CI] = normfit(DiscPayoff);
```

Fig. 7.7 Pricing a call option with truncated payoff by antithetic sampling.

```
      3.2410
>> CI(2) - CI(1)
ans =
    0.0449
>> randn('seed',0)
>> [Price, CI] = BlsTrMCAV(50,50,0.1,5/12,0.4,70,100000)
Price =
    3.2261
CI =
    3.2087
    3.2436
>> CI(2) - CI(1)
ans =
    0.0350
```

Indeed, this run seems to suggest that although some variance reduction may be obtained, the improvement is not very impressive. In some extreme situations, when the monotonicity condition is not satisfied, it may well be the case that antithetic sampling actually increases variance.

7.2.3 Using control variates to price a vanilla European option

In this case the stock price is a natural control variate, as both its expected value and the variance at the expiration of the option are known. To apply the

```
% BlsMCCV.m
function [Price, CI] = BlsMCCV(S0,X,r,T,sigma,NRepl,NPilot)
nuT = (r - 0.5*sigma^2)*T;
siT = sigma * sqrt(T);
% compute parameters first
StockVals = S0*exp(nuT+siT*randn(NPilot,1));
OptionVals = exp(-r*T) * max( 0 , StockVals - X);
MatCov = cov(StockVals, OptionVals);
VarY = S0^2 * exp(2*r*T) * (exp(T * sigma^2) - 1);
c = - MatCov(1,2) / VarY;
ExpY = S0 * exp(r*T);
%
NewStockVals = S0*exp(nuT+siT*randn(NRepl,1));
NewOptionVals = exp(-r*T) * max( 0 , NewStockVals - X);
ControlVars = NewOptionVals + c * (NewStockVals - ExpY);
[Price, VarPrice, CI] = normfit(ControlVars);
```

Fig. 7.8 Using control variates to price a vanilla European call by Monte Carlo simulation.

method, we must compute an estimation of the covariance between the option value and the underlying asset price. The MATLAB code is illustrated in figure 7.8. The BlsMCCV function requires as an additional input parameter the number NPilot of pilot replications we want to run to estimate the covariance. Note that the first set of pilot replications is discarded to avoid biasing the estimator.

```
>> randn('seed',0)
>> [P,CI] = blsMC(50,52,0.1,5/12,0.4,200000)
P =
    5.1780
CI =
    5.1393
    5.2167
>> blsprice(50,52,0.1,5/12,0.4)
ans =
    5.1911
>> randn('seed',0)
>> [P,CI] = blsMCCV(50,52,0.1,5/12,0.4,195000,5000)
P =
    5.1881
CI =
    5.1710
```

```
% BlsHalton.m
function Price = BlsHalton(S0,X,r,T,sigma,NPoints,Base1,Base2)
nuT = (r - 0.5*sigma^2)*T;
siT = sigma * sqrt(T);
% Use Box Muller to generate standard normals
H1 = GetHalton(ceil(NPoints/2),Base1);
H2 = GetHalton(ceil(NPoints/2),Base2);
VLog = sqrt(-2*log(H1));
Norm1 = VLog .* cos(2*pi*H2);
Norm2 = VLog .* sin(2*pi*H2);
Norm = [Norm1 ; Norm2];
%
DiscPayoff = exp(-r*T) * max( 0 , S0*exp(nuT+siT*Norm) - X);
Price = mean(DiscPayoff);
```

Fig. 7.9   Using Halton sequences to price a vanilla European call.

5.2053

From these runs it would seem that there is some reduction in variance by using control variates. No conclusion may be drawn from a single run, but it is worth noting that control variates and antithetic sampling may easily be combined. This is left as an exercise for the reader.

7.2.4   Using Halton low-discrepancy sequences to price a vanilla European option

A last exercise we may try with a vanilla European call is using a low-discrepancy sequence. We use here the simplest sequence, the Halton sequence. To generate normal variates, we may use the Box-Muller algorithm, which we described in section 4.2.4. We recall the Box-Muller algorithm here for convenience. To generate two independent standard normal variates, we should first generate two independent random numbers U_1 and U_2, and then set

$$\begin{aligned} X &= \sqrt{-2\ln U_1}\cos(2\pi U_2) \\ Y &= \sqrt{-2\ln U_1}\sin(2\pi U_2). \end{aligned}$$

Rather than generating pseudorandom numbers, we may use two Halton sequences with two prime numbers as bases. This is accomplished by the code displayed in figure 7.9.

```
>> blsprice(50,52,0.1,5/12,0.4)
```

```
ans =
    5.1911
>> BlsHalton(50,52,0.1,5/12,0.4,5000,2,7)
ans =
    5.1970
>> BlsHalton(50,52,0.1,5/12,0.4,5000,11,7)
ans =
    5.2173
>> BlsHalton(50,52,0.1,5/12,0.4,5000,2,4)
ans =
    6.2485
```

The first run shows the potential of low-discrepancy sequences; note that we have used the simplest sequence, and better results might be obtained with more refined methods such as Sobol or Faure sequences (see [8]). From the second run we also see that the quality of the estimate may depend on the choice of the bases; the third run shows that using a nonprime number as a basis yields a poor result.

7.3 INTRODUCTION TO EXOTIC AND PATH-DEPENDENT OPTIONS

The variety of options that have been conceived in the past years seems to have no limit. You have options on stocks, commodities, and even options on options. Interest rate derivatives play a fundamental role in interest rate risk management. Some options are rather peculiar and are traded over-the-counter for specific needs.[4]

Exotic options on stocks may be designed by introducing a certain degree of path dependency. The idea is that unlike a vanilla European option, the payoff depends not only on the underlying asset price at expiration, but also on its whole path. In the following we briefly describe barrier, Asian, and lookback options. They are of particular interest in learning and testing numerical methods.

7.3.1 Barrier options

In barrier options a specific asset price S_b is selected as a barrier value. During the life of the option, this barrier may be crossed or not. In knock-out options, the contract is canceled if the barrier value is crossed at any time during the whole life; on the contrary, knock-in options are activated only if the barrier is crossed. The barrier S_b may be above or below the current asset price S_0:

---

[4]This means that they are not traded on an organized exchange.

if $S_b > S_0$, we have an up option; if $S_b < S_0$, we have a down option. These features may be combined with the payoffs of call and put options to define a set of barrier options.

For instance, a down-and-out put option is a put option that becomes void if the asset price falls below the barrier S_b; in this case $S_b < S_0$, and $S_b < X$. The rationale behind such an option is that the risk for the option writer is reduced. So it is reasonable to expect that a down-and-out put option is cheaper than a vanilla one. From the point of view of the option holder this means that the potential payoff is reduced; however, if you are interested in options to manage risk, and not as a speculator, this also means that you may get cheaper insurance. By the same token, an up-and-out call option may be defined.

Now consider a down-and-in put option. This option is activated only if the barrier level $S_b < S_0$ is crossed. Holding both a down-and-out and a down-and-in put option is equivalent to holding a vanilla put option. So we have the following parity relationship:

$$P = P_{\text{di}} + P_{\text{do}},$$

where P is the price of the vanilla put, and P_{di}, P_{do} are the prices for the down-and-in and the down-and-out options, respectively. Sometimes a rebate is paid to the option holder if the barrier is crossed and option is canceled; in such a case the parity relationship above is not correct.

In principle, the barrier might be monitored continuously; in practice periodic monitoring may be applied (e.g., the price could be checked each day at the close of the trading). This may affect the price, as a lower monitoring frequency makes crossing the barrier less likely.

Analytical pricing formulas are available for certain barrier options. As an example, consider a down-and-out put with strike price X, expiring in T time units, with a barrier set to S_b. The following formulas are known (see, e.g., [15, pp. 250-251]), where S_0, r, σ have the usual meaning.

$$P = Xe^{-rT}\{N(d_4) - N(d_2) - a[N(d_7) - N(d_5)]\} - S_0\{N(d_3) - N(d_1) - b[N(d_8) - N(d_6)]\},$$

where

$$a = \left(\frac{S_b}{S_0}\right)^{-1+2r/\sigma^2}$$

$$b = \left(\frac{S_b}{S_0}\right)^{1+2r/\sigma^2}$$

and

$$d_1 = \frac{\ln(S_0/X) + (r + \sigma^2/2)T}{\sigma\sqrt{T}}$$

```
% DownOutPut.m
function P = DownOutPut(S0,X,r,T,sigma,Sb)
a = (Sb/S0)^(-1 + (2*r / sigma^2));
b = (Sb/S0)^(1 + (2*r / sigma^2));
d1 = (log(S0/X) + (r+sigma^2 / 2)* T) / (sigma*sqrt(T));
d2 = (log(S0/X) + (r-sigma^2 / 2)* T) / (sigma*sqrt(T));
d3 = (log(S0/Sb) + (r+sigma^2 / 2)* T) / (sigma*sqrt(T));
d4 = (log(S0/Sb) + (r-sigma^2 / 2)* T) / (sigma*sqrt(T));
d5 = (log(S0/Sb) - (r-sigma^2 / 2)* T) / (sigma*sqrt(T));
d6 = (log(S0/Sb) - (r+sigma^2 / 2)* T) / (sigma*sqrt(T));
d7 = (log(S0*X/Sb^2) - (r-sigma^2 / 2)* T) / (sigma*sqrt(T));
d8 = (log(S0*X/Sb^2) - (r+sigma^2 / 2)* T) / (sigma*sqrt(T));
P = X*exp(-r*T)*(normcdf(d4)-normcdf(d2) - ...
    a*(normcdf(d7)-normcdf(d5))) ...
    - S0*(normcdf(d3)-normcdf(d1) - ...
    b*(normcdf(d8)-normcdf(d6)));
```

Fig. 7.10 Implementing the analytical pricing formula for a down-and-out put option.

$$d_2 = \frac{\ln(S_0/X) + (r - \sigma^2/2)T}{\sigma\sqrt{T}}$$

$$d_3 = \frac{\ln(S_0/S_b) + (r + \sigma^2/2)T}{\sigma\sqrt{T}}$$

$$d_4 = \frac{\ln(S_0/S_b) + (r - \sigma^2/2)T}{\sigma\sqrt{T}}$$

$$d_5 = \frac{\ln(S_0/S_b) - (r - \sigma^2/2)T}{\sigma\sqrt{T}}$$

$$d_6 = \frac{\ln(S_0/S_b) - (r + \sigma^2/2)T}{\sigma\sqrt{T}}$$

$$d_7 = \frac{\ln(SX/S_b^2) - (r - \sigma^2/2)T}{\sigma\sqrt{T}}$$

$$d_8 = \frac{\ln(SX/S_b^2) - (r + \sigma^2/2)T}{\sigma\sqrt{T}}.$$

A MATLAB code implementing these formulas is given in figure 7.10.

```
>> [Call, Put] = blsprice(50,50,0.1,5/12,0.4);
>> Put
Put =
    4.0760
>> DOPut(50,50,0.1,5/12,0.4,40)
```

```
ans =
    0.5424
>> DOPut(50,50,0.1,5/12,0.4,35)
ans =
    1.8481
>> DOPut(50,50,0.1,5/12,0.4,30)
ans =
    3.2284
>> DOPut(50,50,0.1,5/12,0.4,1)
ans =
    4.0760
```

We see that the down-and-out put is cheaper than the vanilla put; the price of the barrier option tends to that of the vanilla put as S_b tends to zero. It is also interesting to see what happens with respect to volatility:

```
>> [Call, Put] = blsprice(50,50,0.1,5/12,0.4);
>> Put
Put =
    4.0760
>> [Call, Put] = blsprice(50,50,0.1,5/12,0.3);
>> Put
Put =
    2.8446
>> DOPut(50,50,0.1,5/12,0.4,40)
ans =
    0.5424
>> DOPut(50,50,0.1,5/12,0.3,40)
ans =
    0.8792
>> DOPut(50,50,0.1,5/12,0.4,30)
ans =
    3.2284
>> DOPut(50,50,0.1,5/12,0.3,30)
ans =
    2.7294
```

For a vanilla put, less volatility implies a lower price, as there is less uncertainty; for the barrier option, less volatility *may* imply a higher price since breaching the barrier may be less likely. We see that the dominating effect depends on the barrier level.

In the formula above, it is assumed that barrier monitoring is continuous. A possibly more realistic assumption is that the asset price is checked against the barrier only periodically, e.g., at the end of each trading day. In this case we should expect that the price for a down-and-out option is increased, since breaching the barrier is less likely. An approximate correction has been suggested (see [4] or [9, p. 266]). The idea is using the analytical formula above, correcting the barrier as follows:

$$S_b \Rightarrow S_b e^{\pm 0.5826 \cdot \sigma \sqrt{\delta t}},$$

where the term 0.5826 derives from the Riemann zeta function, δt is time elapsing between two consecutive monitoring, and the sign $\pm$ depends on the option type. For a down-and-out put we should take the minus sign, as the barrier level should be lowered to reflect the reduced likelihood of crossing the barrier. For instance, if we monitor the barrier each day, the prices above change approximately as follows:

```
>> DOPut(50,50,0.1,5/12,0.4,40)
ans =
    0.5424
>> DOPut(50,50,0.1,5/12,0.4,40*exp(-0.5826*0.4*sqrt(1/12/30)))
ans =
    0.6380
>> DOPut(50,50,0.1,5/12,0.4,30)
ans =
    3.2284
>> DOPut(50,50,0.1,5/12,0.4,30*exp(-0.5826*0.4*sqrt(1/12/30)))
ans =
    3.3056
```

We have assumed here that each month consists of 30 days.

7.3.2   Asian options

Barrier options exhibit a weak degree of path dependency. A stronger degree of path dependency is typical of Asian options, as the payoff depends on the average asset price over the option life.

Different Asian options may be devised, depending on how the average is computed. Sampling may be discrete or (in principle) continuous. Furthermore, the average may be arithmetic or geometric. The discrete arithmetic average is

$$A_{\text{da}} = \frac{1}{n} \sum_{i=1}^{n} S(t_i),$$

where t_i, $i = 1, \ldots, n$, are the discrete sampling times. The geometric average is

$$A_{\text{dg}} = \left[\prod_{i=1}^{n} S(t_i) \right]^{1/n}.$$

If continuous sampling is assumed, we get

$$A_{\text{ca}} = \frac{1}{T} \int_{0}^{T} S(t) \, dt$$

$$A_{\text{cg}} = \exp\left[\frac{1}{T}\int_0^T \ln S(t)\,dt\right].$$

Given some way to measure the average A, you may use it to define a rate or a strike. An average rate call has a payoff given by

$$\max\{A - X, 0\},$$

whereas for an average strike call we have

$$\max\{S(T) - A, 0\}.$$

By the same token, we may define an average rate put:

$$\max\{X - A, 0\},$$

or an average strike put:

$$\max\{A - S(T), 0\}.$$

Early exercise features may also be defined in the contract; in this case, applying Monte Carlo simulation is not easy.

7.3.3 Lookback options

Lookback options come in many forms, just like Asian options. The basic difference is that a maximum (or a minimum) value is monitored during the option life. Assuming continuous monitoring, we may measure the maximum and the minimum asset price:

$$S_{\max} = \max_{t \in [0,T]} S(t)$$
$$S_{\min} = \min_{t \in [0,T]} S(t).$$

A European style lookback call has a payoff given by

$$S(T) - S_{\min},$$

whereas in the case of a lookback put we have

$$S_{\max} - S(T).$$

Just as in the Asian option case, you may use the maximum and minimum to define rates or strikes, and you may also add early exercise features. Assuming continuous monitoring, some analytical pricing formulas are known for lookback options.

```
% DOPutMC.m
function [P,CI,NCrossed] = DOPutMC(S0,X,r,T,sigma,Sb,NSteps,NRepl)
% Generate asset paths
[Call,Put] = blsprice(S0,X,r,T,sigma);
Payoff = zeros(NRepl,1);
NCrossed = 0;
for i=1:NRepl
   Path=AssetPaths1(S0,r,sigma,T,NSteps,1);
   crossed = any(Path <= Sb);
   if crossed == 0
      Payoff(i) = max(0, X - Path(NSteps+1));
   else
      Payoff(i) = 0;
      NCrossed = NCrossed + 1;
   end
end
[P,aux,CI] = normfit( exp(-r*T) * Payoff);
```

Fig. 7.11 Crude Monte Carlo simulation for a discrete barrier option.

7.4 PRICING A DOWN-AND-OUT PUT

In this section we consider a weakly path-dependent option, a down-and-out put option, under the assumption that the barrier is checked at the end of each trading day. We have seen in section 7.3.1 how the analytical formula for continuous monitoring can be adjusted to reflect discrete monitoring; we will use the function DOPut to check the result of Monte Carlo simulation. An important point is that in practice barrier options may be very sensitive to stochastic volatility; Monte Carlo simulation could be used together with a model of stochastic volatility to price a barrier option. A code implementing a crude Monte Carlo simulation is given in figure 7.11. The parameter NSteps is used to determine how many times the stock price should be checked against the barrier level S_b. The payoff is set to 0 whenever the barrier is crossed. Note that we always simulate the complete path even if the barrier is crossed during the option life: some part of the path is actually useless, but the point is that in this way we can generate the paths by a vectorized code without for loops. The DOPutMC function also returns the number NCrossed of paths in which the barrier has been crossed.

Let us price an option with two months to maturity, assuming that each month consists of 30 days and that the barrier is checked each day. The barrier S_b is $40.

```
>> DOPut(50,50,0.1,2/12,0.4,40*exp(-0.5826*0.4*sqrt(1/12/30)))
ans =
    1.3629
>> randn('seed',0)
>> [P,CI,NCrossed]=DOPutMC(50,50,0.1,2/12,0.4,40,60,50000)
P =
    1.3600
CI =
    1.3393
    1.3808
NCrossed =
       7392
```

From section 7.2.2 we know that antithetic sampling may be not very effective, because the payoff is nonmonotonic with respect to the asset price at expiration. Things are more complicated here, as the complete asset price path matters. Control variates may also be used; a natural candidate as a control variate is the price of a vanilla put, which may be computed by the Black-Scholes formula. Here we try a different approach, i.e., variance reduction by conditioning, which was explained in section 4.4.4. To this end, it is convenient to consider the price P_{di} of the down-and-in put, since we know that

$$P_{\text{do}} = P - P_{\text{di}}.$$

Assume that we discretize the option life in time intervals of width δt (in our case, 1 day), so that $T = M\delta t$, and consider the asset price path for days i, $i = 1, \ldots, M$:

$$\mathbf{S} = \{S_1, S_2, \ldots, S_M\}.$$

Based on this path, we estimate the option price as

$$P_{\text{di}} = e^{-rT}\mathrm{E}[I(\mathbf{S})(X - S_M)^+],$$

where the indicator function I is

$$I(\mathbf{S}) = \begin{cases} 1 & \text{if } S_j < S_b \text{ for some } j \\ 0 & \text{otherwise.} \end{cases}$$

Now let j^* be the index for the time instant at which the barrier is first crossed; by convention, let $j^* = M + 1$ if the barrier is not crossed during the option life. At time $j^*\delta t$ the option is activated, and from now on it behaves just like a vanilla put. So, conditional on the crossing time $t^* = j^*\delta t$ and the price S_{j^*}, we may use the Black-Scholes formula to estimate the payoff:

$$\mathrm{E}\left[I(\mathbf{S})(X - S_M)^+ \mid j^*, S_{j^*}\right] = e^{r(T-t^*)}B_p(S_{j^*}, X, T - t^*),$$

where $B_p(S_{j^*}, X, T - t^*)$ is the Black-Scholes price for a vanilla put with strike price X, initial underlying price S_{j^*}, and time to maturity $T - t^*$; the exponential term takes discounting into account. Given a simulated path **S**, this suggests using the following estimator:

$$I(\mathbf{S})e^{-rt^*}B_p(S_{j^*}, X, T - t^*).$$

Unlike antithetic sampling, conditional Monte Carlo exploits specific knowledge about the problem; the more we know, the less we leave to numerical integration. The function DOPutMCCond in figure 7.12 implements this variance reduction method. The only point worth noting is that for efficiency reasons, it is advisable to call the blsprice function only once with a vector argument rather than once per replication. So, when the barrier is crossed, we record the time Times at which the down-and-in put has been activated, and the stock price StockVals. When the barrier is not crossed, the estimator is simply 0. Also note that the vectors passed to blsprice have NCrossed elements, whereas the size of the vector Payoff containing the estimator values is NRepl.

```
>> DOPut(50,52,0.1,2/12,0.4,30*exp(-0.5826*0.4*sqrt(1/12/30)))
ans =
    3.8645
>> randn('seed',0)
>> [P,CI,NCrossed] = DOPutMC(50,52,0.1,2/12,0.4,30,60,200000)
P =
    3.8751
CI =
    3.8545
    3.8957
NCrossed =
   249
>> randn('seed',0)
>> [P,CI,NCrossed] = DOPutMCCond(50,52,0.1,2/12,0.4,30,60,200000)
P =
    3.8651
CI =
    3.8617
    3.8684
NCrossed =
   249
```

This run shows that variance reduction by conditioning may indeed be helpful, but we should not get too excited. To begin with, one lucky run does not prove anything. Even worse, we have run a huge number of replications (200,000), but the barrier has been crossed only in 249 replications. This means that most of the replications are a wasted effort. In other words, with the data for this option, crossing the barrier is a rare event. This is a typical case in which importance sampling may help (see section 4.4.6).

```
% DOPutMCCond.m
function [Pdo,CI,NCrossed] = ...
   DOPutMCCond(S0,X,r,T,sigma,Sb,NSteps,NRepl)
dt = T/NSteps;
[Call,Put] = blsprice(S0,X,r,T,sigma);
% Generate asset paths and payoffs for the down and in option
NCrossed = 0;
Payoff = zeros(NRepl,1);
Times = zeros(NRepl,1);
StockVals = zeros(NRepl,1);
for i=1:NRepl
   Path=AssetPaths1(S0,r,sigma,T,NSteps,1);
   tcrossed = min(find( Path <= Sb ));
   if not(isempty(tcrossed))
      NCrossed = NCrossed + 1;
Times(NCrossed) = (tcrossed-1) * dt;
StockVals(NCrossed) = Path(tcrossed);
   end
end
if (NCrossed > 0)
   [Caux, Paux] = blsprice(StockVals(1:NCrossed),X,r,...
      T-Times(1:NCrossed),sigma);
   Payoff(1:NCrossed) = exp(-r*Times(1:NCrossed)) .* Paux;
end
[Pdo, aux, CI] = normfit(Put - Payoff);
```

Fig. 7.12 Conditional Monte Carlo simulation for a discrete barrier option.

One possible idea is changing the drift of the asset price in such a way that crossing the barrier is more likely.[5] We should go a step back and consider what we do in order to generate an asset price path **S**. For each time step, we generate a normal variate Z_j with expected value

$$\nu = \left(r - \frac{\sigma^2}{2}\right)\delta t$$

and variance $\sigma^2\,\delta t$. All these variates are mutually independent, and the asset price is generated by setting

$$\ln S_j - \ln S_{j-1} = Z_j.$$

Let **Z** be the vector of the random normals, and let $f(\mathbf{Z})$ be its joint density. If we use the modified expected value

$$\nu - b,$$

we may expect that the barrier will be crossed more often. Let $g(\mathbf{Z})$ be the joint density for the normal variates generated with this modified expected value. Then we must find out a correction term to come up with the correct importance sampling estimator. Combining importance sampling with the conditional expectation we have just described, we get

$$
\begin{aligned}
\mathrm{E}_f\!\left[I(\mathbf{S})(X-S_M)^+\big|\,j^*,S_{j^*}\right] &= \mathrm{E}_g\!\left[\frac{f(\mathbf{Z})I(\mathbf{S})(X-S_M)^+}{g(\mathbf{Z})}\bigg|\,j^*,S_{j^*}\right]\\
&= \frac{f(z_1,\dots,z_{j^*})}{g(z_1,\dots,z_{j^*})}\,\mathrm{E}_g\!\left[\frac{f(Z_{j^*+1},\dots,Z_M)}{g(Z_{j^*+1},\dots,Z_M)}I(\mathbf{S})(X-S_M)^+\bigg|\,j^*,S_{j^*}\right]\\
&= \frac{f(z_1,\dots,z_{j^*})}{g(z_1,\dots,z_{j^*})}\mathrm{E}_f\!\left[I(\mathbf{S})(X-S_M)^+\big|\,j^*,S_{j^*}\right]\\
&= \frac{f(z_1,\dots,z_{j^*})}{g(z_1,\dots,z_{j^*})}\,e^{r(T-t^*)}B_p(S_{j^*},X,T-t^*).
\end{aligned}
$$

In practice, we should generate the normal variates with expected value $(\nu-b)$, and multiply the conditional estimator by a likelihood ratio. The only open problem is how to compute the likelihood ratio. In appendix B we consider the joint distribution of a multivariate normal with expected value $\boldsymbol{\mu}$ and covariance matrix $\boldsymbol{\Sigma}$:

$$f(\mathbf{z}) = \frac{1}{(2\pi)^{n/2}\,|\,\boldsymbol{\Sigma}\,|^{1/2}}\,e^{-\frac{1}{2}(\mathbf{Z}-\boldsymbol{\mu})^T\boldsymbol{\Sigma}^{-1}(\mathbf{Z}-\boldsymbol{\mu})}.$$

In our case, due to the mutual independence of the random variates Z_j, the covariance matrix is a diagonal matrix with elements $\sigma^2\,\delta t$, and the vector of

---

[5]The treatment here follows the approach of [13].

the expected values has components

$$\mu = \left(r - \frac{\sigma^2}{2} \right) \delta t$$

for the density f and $\mu - b$ for the density g. So we have

$$\frac{f(z_1, \ldots, z_{j^*})}{g(z_1, \ldots, z_{j^*})}$$

$$= \exp \left\{ -\frac{1}{2} \sum_{k=1}^{j^*} \left(\frac{z_k - \mu}{\sigma \sqrt{\delta t}} \right)^2 \right\} \exp \left\{ \frac{1}{2} \sum_{k=1}^{j^*} \left(\frac{z_k - \mu + b}{\sigma \sqrt{\delta t}} \right)^2 \right\}$$

$$= \exp \left\{ -\frac{1}{2\sigma^2 \delta t} \sum_{k=1}^{j^*} \left[(z_k - \mu)^2 - (z_k - \mu + b)^2 \right] \right\}$$

$$= \exp \left\{ -\frac{1}{2\sigma^2 \delta t} \sum_{k=1}^{j^*} \left[-2(z_k - \mu)b - b^2 \right] \right\}$$

$$= \exp \left\{ -\frac{1}{2\sigma^2 \delta t} \left[-2b \sum_{k=1}^{j^*} z_k + 2j^* \mu b - j^* b^2 \right] \right\}$$

$$= \exp \left\{ \frac{b}{\sigma^2 \delta t} \sum_{k=1}^{j^*} z_k - \frac{j^* b}{\sigma^2} \left(r - \frac{\sigma^2}{2} \right) + \frac{j^* b^2}{2\sigma^2 \delta t} \right\}.$$

The resulting code is illustrated in figure 7.13. The function DOPutMCCondIS is similar to DOPutMCCond; the difference is that we must generate the asset price path and record the normal variates in vector vetZ, so that we may compute the likelihood ratio which is stored in the vector ISRatio. We compute the Black-Scholes price only at the end of the main loop. Finding the parameter b is a matter of trial and error. In the function DOPutMCCondIS we assume that the user provides a percentage bp, and the modified expected value is computed as

```
(1 - bp)(r-0.5*sigma^2)*dt
```

So the parameter b is given as a percentage of the correct expected value. Note that we may use a value for bp which is larger than 1, to lower the drift rate at will. Now we may experiment a bit with importance sampling.

```
>> randn('seed',0)
>> [P,CI,NCrossed] = DOPutMC(50,52,0.1,2/12,0.4,30,60,10000)
P =
    3.8698
CI =
```

```
% DOPutMCCondIS.m
function [Pdo,CI,NCrossed] = ...
    DOPutMCCondIS(S0,X,r,T,sigma,Sb,NSteps,NRepl,bp)
dt = T/NSteps;
nudt = (r-0.5*sigma^2)*dt;
b = bp*nudt;
sidt = sigma*sqrt(dt);
[Call,Put] = blsprice(S0,X,r,T,sigma);
% Generate asset paths and payoffs for the down and in option
NCrossed = 0;
Payoff = zeros(NRepl,1);
Times = zeros(NRepl,1);
StockVals = zeros(NRepl,1);
ISRatio = zeros(NRepl,1);
for i=1:NRepl
   % generate normals
vetZ = nudt - b + sidt*randn(1,NSteps);
LogPath = cumsum([log(S0), vetZ]);
Path = exp(LogPath);
   jcrossed = min(find( Path <= Sb ));
   if not(isempty(jcrossed))
      NCrossed = NCrossed + 1;
      TBreach = jcrossed - 1;
Times(NCrossed) = TBreach * dt;
StockVals(NCrossed) = Path(jcrossed);
      ISRatio(NCrossed) = exp( TBreach*b^2/2/sigma^2/dt +...
         b/sigma^2/dt*sum(vetZ(1:TBreach)) - ...
         TBreach*b/sigma^2*(r - sigma^2/2));
   end
end
if (NCrossed > 0)
   [Caux, Paux] = blsprice(StockVals(1:NCrossed),X,r,...
      T-Times(1:NCrossed),sigma);
   Payoff(1:NCrossed) = exp(-r*Times(1:NCrossed)) .* Paux ...
      .* ISRatio(1:NCrossed);
end
[Pdo, aux, CI] = normfit(Put - Payoff);
```

Fig. 7.13 Using conditional Monte Carlo and importance sampling for a discrete barrier option.

```
      3.7778
      3.9618
NCrossed =
      12
>> randn('seed',0)
>> [P,CI,NCrossed] = DOPutMCCondIS(50,52,0.1,2/12,0.4,30,60,10000,0)
P =
      3.8661
CI =
      3.8513
      3.8810
NCrossed =
      12
>> randn('seed',0)
>> [P,CI,NCrossed] = DOPutMCCondIS(50,52,0.1,2/12,0.4,30,60,10000,20)
P =
      3.8651
CI =
      3.8570
      3.8733
NCrossed =
      43
>> randn('seed',0)
>> [P,CI,NCrossed] = DOPutMCCondIS(50,52,0.1,2/12,0.4,30,60,10000,50)
P =
      3.8634
CI =
      3.8596
      3.8671
NCrossed =
     225
>> randn('seed',0)
>> [P,CI,NCrossed] = DOPutMCCondIS(50,52,0.1,2/12,0.4,30,60,10000,200)
P =
      3.8637
CI =
      3.8629
      3.8645
NCrossed =
        8469
```

Calling DOPutMCCondIS with the parameter bp set to zero is just like calling DOPutMCCond; by increasing bp we see that the barrier is crossed in more and more replications, and the quality of the estimate is improved. Note that this

```
% AsianMC.m
function [P,CI] = AsianMC(S0,X,r,T,sigma,NSamples,NRepl)
Payoff = zeros(NRepl,1);
for i=1:NRepl
   Path=AssetPaths1(S0,r,sigma,T,NSamples,1);
   Payoff(i) = max(0, mean(Path(2:(NSamples+1))) - X);
end
[P,aux,CI] = normfit( exp(-r*T) * Payoff);
```

Fig. 7.14   Monte Carlo simulation for an Asian option.

does not necessarily imply that the larger b, the better; suggestions for setting this parameter are given in [13].

7.5   PRICING AN ASIAN OPTION

We consider here pricing an Asian average rate call option with discrete arithmetic averaging. So the option payoff is

$$\max\left\{\frac{1}{N}\sum_{i=1}^{N} S(t_i) - X, 0\right\},$$

where the option maturity is T years, $t_i = i\,\delta t$, and $\delta t = T/N$. In a crude Monte Carlo approach, we must simply generate asset price paths and estimate the discounted payoff as usual. The code is illustrated in figure 7.14; the only thing worth noting is that NSamples is the number N of sampled points in computing the average, whereas the number of replications is NRepl.

This crude Monte Carlo sampling may be improved by using control variates (see [12, chapter 9]). As a control variate, we may take the following sum of the asset prices:

$$Y = \sum_{i=0}^{N} S(t_i).$$

This is a suitable control variate since it is clearly correlated with the option payoff, and we may compute its expected value (under a risk-neutral measure):

$$
\begin{aligned}
\mathrm{E}[Y] &= \mathrm{E}\left[\sum_{i=0}^{N} S(t_i)\right] = \sum_{i=0}^{N} \mathrm{E}[S(i\,\delta t)] \\
&= \sum_{i=0}^{N} S(0)e^{ri\,\delta t} = S(0)\sum_{i=0}^{N}[e^{r\,\delta t}]^i = S(0)\frac{1 - e^{r(N+1)\delta t}}{1 - e^{r\,\delta t}},
\end{aligned}
$$

```
% AsianMCCV.m
function [P,CI] = AsianMCCV(S0,X,r,T,sigma,NSamples,NRepl,NPilot)
% pilot replications to set control parameter
TryPath=AssetPaths1(S0,r,sigma,T,NSamples,NPilot);
StockSum = sum(TryPath,2);
PP = mean(TryPath(:,2:(NSamples+1)) , 2);
TryPayoff = exp(-r*T) * max(0, PP - X);
MatCov = cov(StockSum, TryPayoff);
c = - MatCov(1,2) / var(StockSum);
dt = T / NSamples;
ExpSum = S0 * (1 - exp((NSamples + 1)*r*dt)) / (1 - exp(r*dt));
% MC run
ControlVars = zeros(NRepl,1);
for i=1:NRepl
    StockPath = AssetPaths1(S0,r,sigma,T,NSamples,1);
    Payoff = exp(-r*T)*max(0, mean(StockPath(2:(NSamples+1)))-X);
    ControlVars(i) = Payoff + c * (sum(StockPath) - ExpSum);
end
[P,aux,CI] = normfit(ControlVars);
```

Fig. 7.15 Monte Carlo simulation with control variates for an Asian option.

where we have used the formula

$$\sum_{i=0}^{N} \alpha^i = \frac{1 - \alpha^{N+1}}{1 - \alpha}.$$

The MATLAB code in figure 7.15 implements this variance reduction strategy. The user must fix the number of pilot replications NPilot needed to set the control parameter in the control variates procedure. The following runs give an idea of the improvement we may obtain:

```
>> rand('seed',0)
>> [P,CI] = AsianMC(50,50,0.1,5/12,0.4,5,50000)
P =
    3.9802
CI =
    3.9283
    4.0321
>> CI(2) - CI(1)
ans =
    0.1038
>> [P,CI] = AsianMCCV(50,50,0.1,5/12,0.4,5,45000,5000)
```

```
% HaltonPaths.m
function SPaths=HaltonPaths(S0,mu,sigma,T,NSteps,NRepl)
dt = T/NSteps;
nudt = (mu-0.5*sigma^2)*dt;
sidt = sigma*sqrt(dt);
NRepl = 2*ceil(NRepl/2); % make sure it's even
% Use Box Muller to generate standard normals
RandMat = zeros(NRepl, NSteps);
seeds = myprimes(2*NSteps);
Base1 = seeds(1:NSteps);
Base2 = seeds((NSteps+1):(2*NSteps));
for i=1:NSteps
   H1 = GetHalton(NRepl/2,Base1(i));
   H2 = GetHalton(NRepl/2,Base2(i));
   VLog = sqrt(-2*log(H1));
   Norm1 = VLog .* cos(2*pi*H2);
   Norm2 = VLog .* sin(2*pi*H2);
   RandMat(:,i) = [Norm1 ; Norm2];
end
Increments = nudt + sidt*RandMat;
LogPaths = cumsum([log(S0)*ones(NRepl,1) , Increments] , 2);
SPaths = exp(LogPaths);
```

Fig. 7.16 Generating asset price paths by Halton sequences.

```
P =
    3.9738
CI =
    3.9511
    3.9965
>> CI(2) - CI(1)
ans =
    0.0454
```

Another tool we might use to improve the estimate is a low-discrepancy sequence. As in section 7.2.4, we use simple Halton sequences here. This case is a little more complicated. For a path-independent option, we need just two Halton sequences, on which we may apply the Box-Muller transformation. For the discretely sampled Asian option, we should use more sequences; we need two sequences for each time point in which we sample the asset price to calculate the arithmetic average. A possible MATLAB function to generate the asset paths is given in figure 7.16. The function HaltonPaths is similar

```
% AsianHalton.m
function P = AsianHalton(S0,X,r,T,sigma,NSamples,NRepl)
Payoff = zeros(NRepl,1);
Path=HaltonPaths(S0,r,sigma,T,NSamples,NRepl);
Payoff = max(0, mean(Path(:,2:(NSamples+1)),2) - X);
P = mean( exp(-r*T) * Payoff);
```

Fig. 7.17 Pricing an Asian option by Halton sequences.

to the function `AssetPaths1` of figure 7.3. To come up with the necessary bases, we use the `myprimes` function to generate the first N prime numbers; this function is illustrated in figure A.2 in appendix A. If N is the number of sampled time points, we need $2 \times N$ prime numbers, which are stored in the `Base1` and `Base2` vectors used in the Box-Muller transformation.

Given the asset paths, the function `AsianHalton` of figure 7.17 computes the discounted average payoff in a straightforward way. Here are a few examples of its use, which should be compared with the estimates we have obtained by crude Monte Carlo and Monte Carlo with variance reduction by control variates.

```
>> AsianHalton(50,50,0.1,5/12,0.4,5,1000)
ans =
    4.0289
>> AsianHalton(50,50,0.1,5/12,0.4,5,2000)
ans =
    3.9979
>> AsianHalton(50,50,0.1,5/12,0.4,5,3000)
ans =
    3.9619
>> AsianHalton(50,50,0.1,5/12,0.4,5,10000)
ans =
    3.9441
>> AsianHalton(50,50,0.1,5/12,0.4,5,30000)
ans =
    3.9742
```

It is worth noting that:

1. We have used simple Halton sequences, which are arguably not the best low-discrepancy sequences (that would be too simple).

2. We have taken a rather crude approach to generate normal variates and asset paths (see, e.g., [1] to appreciate the issues of path generation within quasi-Monte Carlo).

3. We have done nothing to figure out the best way of selecting the bases.

Despite these limitations, the results show that even a very crude quasi-Monte Carlo simulation may achieve a remarkable degree of accuracy with a limited number of samples. Unfortunately, there is no simple way to define a confidence interval for low-discrepancy sequences (see, e.g., [14]).

Actually, a possible sign of trouble is that the estimates we get wander a little with the increasing number of replications. However, this little experiment helps in understanding why research is done on low-discrepancy sequences. A caveat is that if the number of sampled time points increases, tending to continuous sampling, it may be difficult to generate good asset paths (we would need a large number of bases). Simulating stochastic volatilities or interest rates would compound with this issue, and in such a case Monte Carlo could be a better option (not considering the finite difference approach illustrated in chapter 8).

For further reading

In the literature

- An early paper on using Monte Carlo simulation in option pricing is [2]. An updated survey is given in [3].

- Readers interested in applications to interest rate derivatives may see [5].

- Interesting sources on the use of low-discrepancy sequences for derivatives pricing are [10] and [11]. See also [1] and [14] for specific issues such as path generation in high-dimensional problems and quantifying the estimation error.

- Another interesting paper on quasi-Monte Carlo simulation in finance is [8], where Faure low-discrepancy sequences, which we did not consider, are discussed.

- In this chapter we have only considered applications to option pricing. However, another important application field for Monte Carlo simulation is estimating the value-at-risk. In [6], and related references, you may find some information on the use of variance reduction methods to speed up VaR computations.

On the Web

- A Web page related to Monte Carlo and quasi-Monte Carlo methods is http://www.mcqmc.org.

- Some information on using low-discrepancy sequences in finance can be also obtained by browsing the following pages:

http://www.cs.columbia.edu/~traub

http://www.cs.columbia.edu/~ap/html/information.html

REFERENCES

1. F. Åkesson and J.P. Lehoczky. Path Generation for Quasi-Monte Carlo Simulation of Mortgage-Backed Securities. *Management Science*, 46:1171–1187, 2000.

2. P. Boyle. Options: A Monte Carlo Approach. *Journal of Financial Economics*, 4:323–338, 1977.

3. P. Boyle, M. Broadie, and P. Glasserman. Monte Carlo Methods for Security Pricing. *Journal of Economics Dynamics and Control*, 21:1267–1321, 1997.

4. M. Broadie, P. Glasserman, and S.G. Kou. A Continuity Correction for Discrete Barrier Options. *Mathematical Finance*, 7:325–349, 1997.

5. L. Clewlow and C. Strickland. *Implementing Derivatives Models*. Wiley, Chichester, West Sussex, England, 1998.

6. P. Glasserman, P. Heidelberger, and P. Shahabuddin. Variance Reduction Techniques for Estimating Value-at-Risk. *Management Science*, 46:1349–1364, 2000.

7. J.C. Hull. *Options, Futures, and Other Derivatives (4th ed.)*. Prentice Hall, Upper Saddle River, NJ, 2000.

8. C. Joy, P.P. Boyle, and K.S. Tan. Quasi-Monte Carlo Methods in Numerical Finance. *Management Science*, 42:926–938, 1996.

9. Y.K. Kwok. *Mathematical Models of Financial Derivatives*. Springer-Verlag, Berlin, 1998.

10. S.H. Paskov. New Methodologies for Valuing Derivatives. In S.R. Pliska and M.A.H. Dempster, editors, *Mathematics of Derivative Securities*, pages 545–582. Cambridge University Press, Cambridge, 1997.

11. S.H. Paskov and J.F. Traub. Faster Valuation of Financial Derivatives. *Journal of Portfolio Management*, 22:113–120, Fall 1995.

12. S. Ross. *An Introduction to Mathematical Finance: Options and Other Topics*. Cambridge University Press, Cambridge, 1999.

13. S.M. Ross and J.G. Shanthikumar. Monotonicity in Volatility and Efficient Simulation. *Probability in the Engineering and Informational Sciences*, 14:317–326, 2000.

14. K.S. Tan and P.P. Boyle. Applications of Randomized Low Discrepancy Sequences to the Valuation of Complex Securities. *Journal of Economic Dynamics and Control*, 24:1747–1782, 2000.

15. P. Wilmott. *Quantitative Finance (vols. I and II)*. Wiley, Chichester, West Sussex, England, 2000.

8

Option valuation by finite difference methods

In this chapter we give a few simple examples of how the PDE framework may be exploited in option pricing. The idea is applying the finite difference methods illustrated in chapter 5 to solution of the Black-Scholes PDE. We start in section 8.1 by recalling derivatives approximation schemes and by pointing out how suitable boundary conditions may be set up in order to model a specific option. In section 8.2 we apply a straightforward explicit scheme to the pricing of a vanilla European option; as we already know, this scheme is prone to numerical instabilities, which we may also interpret from a financial point of view. In section 8.3 we see how a fully implicit method may overcome the instability issue. The Crank-Nicolson method, which may be regarded as a hybrid between the explicit and the fully implicit approach, is applied in section 8.4 to a barrier option. Finally, in section 8.5 we see how iterative overrelaxation methods may be exploited to tackle an American option with a fully implicit method, which is not trivial due to the presence of a free boundary due to the possibility of early exercise.

8.1 APPLYING FINITE DIFFERENCE METHODS TO THE BLACK-SCHOLES EQUATION

We have seen in section 1.4.2 that the value at time t of an option written on an underlying asset whose price is $S(t)$ is a function $f(t, S)$ satisfying the partial differential equation

$$\frac{\partial f}{\partial t} + rS\frac{\partial f}{\partial S} + \frac{1}{2}\sigma^2 S^2 \frac{\partial^2 f}{\partial S^2} = rf \qquad (8.1)$$

with suitable boundary conditions that characterize the type of option. Different equations may be written if the hypotheses are changed and if path dependency is introduced, but this equation is the starting point to learn how to apply numerical methods based on finite differences for option pricing.

As we have seen in chapter 5, to solve a PDE by finite difference methods we must set up a discrete grid, in this case with respect to time and asset prices. Let T be the option expiration date and $S_{\max}$ a suitably large asset price that it cannot be reached by $S(t)$ within the time horizon we consider. We need $S_{\max}$, since the domain for the PDE is unbounded with respect to asset prices, but we must bound it in some way for computational purposes; so $S_{\max}$ plays the role of $+\infty$. The grid consists of points (t, S) such that

$$\begin{aligned} t &= 0, \delta t, 2\,\delta t, \ldots, N\,\delta t = T, \\ S &= 0, \delta S, 2\,\delta S, \ldots, M\,\delta S = S_{\max}. \end{aligned}$$

We will use the grid notation $f_{i,j} = f(i\,\delta t, j\,\delta S)$.

Let us recall the different ways we have to approximate the partial derivatives in equation (8.1):

- Forward difference:

$$\frac{\partial f}{\partial S} = \frac{f_{i,j+1} - f_{i,j}}{\delta S} \qquad \frac{\partial f}{\partial t} = \frac{f_{i+1,j} - f_{i,j}}{\delta t}$$

- Backward difference:

$$\frac{\partial f}{\partial S} = \frac{f_{i,j} - f_{i,j-1}}{\delta S} \qquad \frac{\partial f}{\partial t} = \frac{f_{i,j} - f_{i-1,j}}{\delta t}$$

- Symmetric difference:

$$\frac{\partial f}{\partial S} = \frac{f_{i,j+1} - f_{i,j-1}}{2\,\delta S} \qquad \frac{\partial f}{\partial t} = \frac{f_{i+1,j} - f_{i-1,j}}{2\,\delta t}$$

- As to the second derivative, we have

$$\begin{aligned} \frac{\partial^2 f}{\partial S^2} &= \left(\frac{f_{i,j+1} - f_{i,j}}{\delta S} - \frac{f_{i,j} - f_{i,j-1}}{\delta S} \right) \Big/ \delta S \\ &= \frac{f_{i,j+1} + f_{i,j-1} - 2f_{i,j}}{\delta S^2}. \end{aligned}$$

Depending on which combination of schemes we use in discretizing the equation, we end up with different approaches, explicit or implicit, which we experiment with in the following sections.

Another issue which we must take care of is setting the boundary conditions. The terminal condition at expiration is

$$f(T, S) = \max\{S - X, 0\} \qquad \forall S$$

for a call with strike price X, and

$$f(T, S) = \max\{X - S, 0\} \qquad \forall S$$

for a put. When we consider boundary conditions with respect to asset prices, the problem is not so trivial, since we have to solve the equation numerically on a bounded region, whereas the domain is unbounded with respect to asset prices. We may use a few examples to clarify this issue.

Example 8.1 Let us consider first a vanilla European put option. When the asset price $S(t)$ is very large, the option is worthless, since we may be sure that it will stay out-of-the-money:

$$f(t, S_{\max}) = 0.$$

When the asset price is $S(t) = 0$, we may say that given our model for asset dynamics, the asset price will remain zero. So the payoff at expiration will be X; discounting back to time t, we have

$$f(t, 0) = Xe^{-r(T-t)}.$$

In grid notation:

$$\begin{aligned}
f_{N,j} &= \max[X - j\,\delta S, 0] & j &= 0, 1, \ldots, M \\
f_{i,0} &= Xe^{-r(N-i)\delta t} & i &= 0, 1, \ldots, N \\
f_{i,M} &= 0 & i &= 0, 1, \ldots, N.
\end{aligned}$$ ▯

Example 8.2 We may deal with a vanilla European call by reasoning as in example 8.1. When the asset price is $S(t) = 0$, the option will expire worthless:

$$f(t, 0) = 0.$$

For a large asset price $S(t)$, we may be sure that it will be in-the-money at expiration and we will get a payoff $S(T) - X$. The value at time t requires discounting back the term X and considering that the arbitrage-free price at time t for the underlying asset is simply $S(t)$. Then a suitable boundary condition is

$$f(S_{\max}, t) = S_{\max} - Xe^{-r(T-t)}.$$

In grid notation:

$$\begin{aligned}
f_{N,j} &= \max[j\,\delta S - X, 0] & j &= 0, 1, \ldots, M \\
f_{i,0} &= 0 & i &= 0, 1, \ldots, N \\
f_{i,M} &= M\,\delta S - Xe^{-r(N-i)\delta t} & i &= 0, 1, \ldots, N.
\end{aligned}$$ ▯

When dealing with barrier options, things may be easier. In the case of a knock-out option, such as a down-and-out put, the option value is 0 on the barrier. Similarly, in the case of an up-and-out call, with the additional advantage that the domain we must consider is bounded. American options are more complex to deal with because of the early exercise boundary; we should take into account for which asset prices and at which times (if any) it is optimal to exercise the option. Thus we have a free boundary that must be discovered in the solution process. A variety of boundary conditions must be required for exotic options; figuring out the correct boundary conditions and approximating them within the numerical scheme is an option-dependent issue.

8.2   PRICING A VANILLA EUROPEAN OPTION BY AN EXPLICIT METHOD

As a first attempt to solve equation (8.1), let us consider a vanilla European put option. We approximate the derivative with respect to S by a central difference and the derivative with respect to time by a backward difference. This is not the only possible choice; we observe that any choice must be somehow compatible with the boundary conditions. The result is the following set of equations:

$$\frac{f_{i,j} - f_{i-1,j}}{\delta t} + rj\,\delta S\,\frac{f_{i,j+1} - f_{i,j-1}}{2\,\delta S}$$
$$+ \frac{1}{2}\sigma^2 j^2\,\delta S^2\,\frac{f_{i,j+1} + f_{i,j-1} - 2f_{i,j}}{\delta S^2} = rf_{i,j}, \qquad (8.2)$$

to be solved with the boundary conditions of example 8.1. It should be noted that since we have a set of terminal conditions, the equations should be solved backward in time. Let $i = N$ in equation (8.2); given the terminal condition, we have one unknown quantity, $f_{N-1,j}$, expressed as a function of three known quantities. If we imagine going backward in time the same consideration holds for each time layer. Rewriting the equations, we get an explicit scheme:

$$f_{i-1,j} = a_j^* f_{i,j-1} + b_j^* f_{i,j} + c_j^* f_{i,j+1} \qquad i = 0, 1, \ldots, N-1,\ j = 1, 2, \ldots, M-1,$$
$$(8.3)$$

where

$$a_j^* = \frac{1}{2}\delta t(\sigma^2 j^2 - rj)$$
$$b_j^* = 1 - \delta t(\sigma^2 j^2 + r)$$
$$c_j^* = \frac{1}{2}\delta t(\sigma^2 j^2 + rj).$$

This scheme is rather straightforward to implement in MATLAB. The code is illustrated in figure 8.1, and it requires the value $S_{\max}$ as well as the dis-

```
% EuPutExpl1.m
function price = EuPutExpl1(S0,X,r,T,sigma,Smax,dS,dt)
% set up grid and adjust increments if necessary
M = round(Smax/dS);
dS = Smax/M;
N = round(T/dt);
dt = T/N;
matval = zeros(M+1,N+1);
vetS = linspace(0,Smax,M+1)';
veti = 0:N;
vetj = 0:M;
% set up boundary conditions
matval(:,N+1) = max(X-vetS,0);
matval(1,:) = X*exp(-r*dt*(N-veti));
matval(M+1,:) = 0;
% set up coefficients
a = 0.5*dt*(sigma^2*vetj - r).*vetj;
b = 1- dt*(sigma^2*vetj.^2 + r);
c = 0.5*dt*(sigma^2*vetj + r).*vetj;
% solve backward in time
for i=N:-1:1
   for j=2:M
      matval(j,i) = a(j)*matval(j-1,i+1) + b(j)*matval(j,i+1)+ ...
         c(j)*matval(j+1,i+1);
   end
end
% find closest point to S0 on the grid and return price
% possibly with a linear interpolation
jdown = floor(S0/dS);
jup = ceil(S0/dS);
if jdown == jup
   price = matval(jdown+1,1);
else
   price = matval(jdown+1,1) + ...
      (S0 - jdown*dS)*(matval(jup+1,1) - matval(jup+1,1))/dS;
end
```

Fig. 8.1 MATLAB code to price a European vanilla put by a straightforward explicit scheme.

cretization steps. The only point requiring some care is that in the mathematical notation it is convenient to uses indexes starting from 0, whereas matrix indexes start from 1 in MATLAB. Moreover, if the initial asset price does not lie on the grid, we must interpolate between the two neighboring points. We have used here a crude linear interpolation; more sophisticated splines could be a better alternative, especially if we are interested in approximating option price sensitivities (as it is always the case in practice).

```
>> [c,p] = blsprice(50,50,0.1,5/12,0.4);
>> p
p =
     4.0760
>> EuPutExpl1(50,50,0.1,5/12,0.4,100,2,5/1200)
ans =
     4.0669
>> [c,p] = blsprice(50,50,0.1,5/12,0.3);
>> p
p =
     2.8446
>> EuPutExpl1(50,50,0.1,5/12,0.3,100,2,5/1200)
ans =
     2.8288
```

We see that the numerical method gives fairly accurate results. We might try to improve them by using a finer grid.

```
>> EuPutExpl1(50,50,0.1,5/12,0.3,100,1.5,5/1200)
ans =
     3.1414
>> EuPutExpl1(50,50,0.1,5/12,0.3,100,1,5/1200)
ans =
 -2.8271e+022
```

What we see here is another example of the numerical instability that we have analyzed in chapter 5. One possibility to avoid the trouble is to resort to implicit methods. Another one is to carry out a stability analysis and to derive bounds on the discretization steps. We will not pursue the second way here, which would be quite similar to what we have done in chapter 5 for the simpler transport and heat equations. Rather, in the next section we describe a financial interpretation of instability, which suggests still another possibility: rewriting the equation with a change of variables.

8.2.1 Financial interpretation of the instability of the explicit method

In the explicit scheme we obtain an option value $f(t, S)$ as a combination of the values $f(t + \delta t, S + \delta S)$, $f(t + \delta t, S)$, and $f(t + \delta t, S - \delta S)$. This

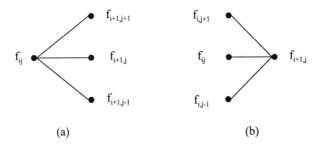

Fig. 8.2 View of explicit (a) and implicit (b) schemes to solve the Black-Scholes PDE.

looks a bit like a trinomial lattice method, which is a generalization of the binomial lattice approach we have described in section 1.4.4 (see figure 8.2a). We can make this interpretation clearer by deriving an alternative version of the explicit method. Following [1, chapter 16], we assume that the first- and second-order derivatives with respect to S at point (i, j) are equal to those at point $(i + 1, j)$:

$$\frac{\partial f}{\partial S} = \frac{f_{i+1,j+1} - f_{i+1,j-1}}{2\delta S}$$

$$\frac{\partial^2 f}{\partial S^2} = \frac{f_{i+1,j+1} + f_{i+1,j-1} - 2f_{i+1,j}}{\delta S^2}.$$

An alternative view is that we substitute the right-hand term $f_{i,j}$ in equation (8.2) by $f_{i-1,j}$. This introduces an error which is bounded and tends to zero as the grid is refined.

The finite difference equation is now

$$\frac{f_{i+1,j} - f_{i,j}}{\delta t} + rj\,\delta S\frac{f_{i+1,j+1} - f_{i+1,j-1}}{2\delta S} +$$
$$\frac{1}{2}\sigma^2 j^2\,\delta S^2\,\frac{f_{i+1,j+1} + f_{i+1,j-1} - 2f_{i+1,j}}{\delta S^2} = rf_{i,j},$$

which may be rewritten (for $j = 1, 2, \ldots, M - 1$ and $i = 0, 1, \ldots, N - 1$) as

$$f_{i,j} = \hat{a}_j\,f_{i+1,j-1} + \hat{b}_j\,f_{i+1,j} + \hat{c}_j\,f_{i+1,j+1},$$

where

$$\hat{a}_j = \frac{1}{1 + r\,\delta t}\left(-\frac{1}{2}rj\,\delta t + \frac{1}{2}\sigma^2 j^2\,\delta t\right) = \frac{1}{1 + r\,\delta t}\pi_d$$

$$\hat{b}_j = \frac{1}{1 + r\,\delta t}\left(1 - \sigma^2 j^2\,\delta t\right) = \frac{1}{1 + r\,\delta t}\pi_0$$

$$\hat{c}_j = \frac{1}{1 + r\,\delta t}\left(\frac{1}{2}rj\,\delta t + \frac{1}{2}\sigma^2 j^2\,\delta t\right) = \frac{1}{1 + r\,\delta t}\pi_u.$$

This scheme is again explicit and is subject to numerical instabilities as well. However, the coefficients $\hat{a}_j$, $\hat{b}_j$, and $\hat{c}_j$ lend themselves to a nice interpretation. Recall that in a binomial lattice we obtain an option value in a node as the discounted expected value of the values in the successor nodes, where expectation is taken with respect to a risk-neutral probability measure. In fact, the coefficients above include a $1/(1 + r\,\delta t)$ term, which may be interpreted as a discount factor over a time interval of length δt. Furthermore, we have

$$\pi_d + \pi_0 + \pi_u = 1.$$

This suggests interpreting the coefficients as probabilities, times a discount factor. Are they risk-neutral probabilities? We should first check which is the expected value of the increase in the asset price during the time interval δt:

$$\mathrm{E}[\Delta] = -\delta S \pi_d + 0\pi_0 + \delta S \pi_u = rj\,\delta S\,\delta t = rS\,\delta t,$$

which is exactly what we would expect in a risk-neutral world. As to the variance of the increment, we have

$$\mathrm{E}[\Delta^2] = (-\delta S)^2 \pi_d + 0\pi_0 + (\delta S)^2 \pi_u = \sigma^2 j^2 (\delta S)^2\,\delta t.$$

Hence, for small δt

$$\mathrm{Var}[\Delta] = \mathrm{E}[\Delta^2] - \mathrm{E}^2[\Delta] = \sigma^2 S^2\,\delta t - r^2 S^2 (\delta t)^2 \approx \sigma^2 S^2\,\delta t,$$

which is also coherent with geometric Brownian motion in a risk-neutral world. So we see that indeed the explicit method could be regarded as a trinomial lattice approach, except for a little problem. The "probabilities" π_d and π_0 may be negative.

One possibility to avoid the trouble, described in [1], is to change variables. By rewriting the Black-Scholes equation in terms of $Z = \ln S$, simple conditions for stability may be derived. However, a change of variables may not be a good idea for certain exotic options. In the next section we implement a fully implicit approach that avoids the stability issue altogether.

8.3   PRICING A VANILLA EUROPEAN OPTION BY A FULLY IMPLICIT METHOD

To overcome the stability issues of the explicit method, we may resort to an implicit method. This is obtained by using a forward difference to approximate the partial derivative with respect to time. We get the grid equations

$$\frac{f_{i+1,j} - f_{i,j}}{\delta t} + rj\,\delta S \frac{f_{i,j+1} - f_{i,j-1}}{2\,\delta S}$$
$$+ \frac{1}{2}\sigma^2 j^2\,\delta S^2 \frac{f_{i,j+1} + f_{i,j-1} - 2f_{i,j}}{\delta S^2} = r f_{i,j},$$

which we may rewrite (for $j = 1, 2, \ldots, M - 1$ and $i = 0, 1, \ldots, N - 1$) as

$$a_j f_{i,j-1} + b_j f_{i,j} + c_j f_{i,j+1} = f_{i+1,j}, \qquad (8.4)$$

where, for each j,

$$
\begin{aligned}
a_j &= \frac{1}{2} r j \, \delta t - \frac{1}{2} \sigma^2 j^2 \, \delta t \\
b_j &= 1 + \sigma^2 j^2 \, \delta t + r \, \delta t \\
c_j &= -\frac{1}{2} r j \, \delta t - \frac{1}{2} \sigma^2 j^2 \, \delta t.
\end{aligned}
$$

Here we have three unknown values linked to one known value (see figure 8.2b). First note that for each time layer we have $M - 1$ equations in $M - 1$ unknowns; the boundary conditions yield the two missing values for each time layer and the terminal conditions give the values in the last time layer. As in the explicit case, we must go backward in time, solving for $i = N - 1, \ldots, 0$ a sequence of systems of linear equations. The system of time layer i is the following:

$$
\begin{bmatrix}
b_1 & c_1 \\
a_2 & b_2 & c_2 \\
 & a_3 & b_3 & c_3 \\
 & & \ddots & \ddots & \ddots \\
 & & & a_{M-2} & b_{M-2} & c_{M-2} \\
 & & & & a_{M-1} & b_{M-1}
\end{bmatrix}
\begin{bmatrix}
f_{i1} \\
f_{i2} \\
f_{i3} \\
\vdots \\
f_{i,M-2} \\
f_{i,M-1}
\end{bmatrix}
$$

$$
=
\begin{bmatrix}
f_{i+1,1} \\
f_{i+1,2} \\
f_{i+1,3} \\
\vdots \\
f_{i+1,M-2} \\
f_{i+1,M-1}
\end{bmatrix}
-
\begin{bmatrix}
a_1 f_{i+1,0} \\
0 \\
0 \\
\vdots \\
0 \\
c_{M-1} f_{i+1,M}
\end{bmatrix}.
$$

We may note that the matrix is tridiagonal and that it is constant for each time layer i. So we may speed up the computation by resorting to a LU-factorization.[1] All of this is accomplished by the MATLAB code in figure 8.3.

```
>> [c,p] = blsprice(50,50,0.1,5/12,0.4);
>> p
```

---

[1] Due to the sparse structure of the matrix, it would be much better to write a specific code to solve the sequence of linear systems. Here we use just the ready-to-use MATLAB functionalities.

```
% EuPutImpl.m
function price = EuPutImpl(S0,X,r,T,sigma,Smax,dS,dt)
% set up grid and adjust increments if necessary
M = round(Smax/dS);
dS = Smax/M;
N = round(T/dt);
dt = T/N;
matval = zeros(M+1,N+1);
vetS = linspace(0,Smax,M+1)';
veti = 0:N;
vetj = 0:M;
% set up boundary conditions
matval(:,N+1) = max(X-vetS,0);
matval(1,:) = X*exp(-r*dt*(N-veti));
matval(M+1,:) = 0;
% set up the tridiagonal coefficients matrix
a = 0.5*(r*dt*vetj-sigma^2*dt*(vetj.^2));
b = 1+sigma^2*dt*(vetj.^2)+r*dt;
c = -0.5*(r*dt*vetj+sigma^2*dt*(vetj.^2));
coeff = diag(a(3:M),-1) + diag(b(2:M)) + diag(c(2:M-1),1);
[L,U] = lu(coeff);
% solve the sequence of linear systems
aux = zeros(M-1,1);
for i=N:-1:1
   aux(1) = - a(2) * matval(1,i);
   matval(2:M,i) = U \ (L \ (matval(2:M,i+1) + aux));
end
% find closest point to S0 on the grid and return price
% possibly with a linear interpolation
jdown = floor(S0/dS);
jup = ceil(S0/dS);
if jdown == jup
   price = matval(jdown+1,1);
else
   price = matval(jdown+1,1) + ...
      (S0 - jdown*dS)*(matval(jup+1,1) - matval(jup+1,1))/dS;
end
```

Fig. 8.3 MATLAB code to price a vanilla European option by an implicit method.

```
p =
    4.0760
>> EuPutImpl(50,50,0.1,5/12,0.4,100,0.5,5/2400)
ans =
    4.0718
```

The results are fairly accurate and may be improved by a refined grid without the risk of running into numerical instabilities. Another way to improve accuracy is to exploit the Crank-Nicolson method; we will do this in the next section for a barrier option.

8.4 PRICING A BARRIER OPTION BY THE CRANK-NICOLSON METHOD

The Crank-Nicolson method has been introduced in section 5.3.3 as a way to improve accuracy by combining the explicit and implicit methods. Applying this idea to the Black-Scholes equation leads to the following grid equation:

$$
\frac{f_{ij} - f_{i-1,j}}{\delta t} + \frac{rj\,\delta S}{2}\left(\frac{f_{i-1,j+1} - f_{i-1,j-1}}{2\,\delta S}\right) + \frac{rj\,\delta S}{2}\left(\frac{f_{i,j+1} - f_{i,j-1}}{2\,\delta S}\right)
$$
$$
+ \frac{\sigma^2 j^2 (\delta S)^2}{4}\left(\frac{f_{i-1,j+1} - 2f_{i-1,j} + f_{i-1,j-1}}{(\delta S)^2}\right)
$$
$$
+ \frac{\sigma^2 j^2 (\delta S)^2}{4}\left(\frac{f_{i,j+1} - 2f_{i,j} + f_{i,j-1}}{(\delta S)^2}\right)
$$
$$
= \frac{r}{2} f_{i-1,j} + \frac{r}{2} f_{ij}.
$$

These equations may be rewritten as

$$
-\alpha_j f_{i-1,j-1} + (1 - \beta_j) f_{i-1,j} - \gamma_j f_{i-1,j+1} = \alpha_j f_{i,j-1} + (1 + \beta_j) f_{ij} + \gamma_j f_{i,j+1},
\tag{8.5}
$$

where

$$
\alpha_j = \frac{\delta t}{4}(\sigma^2 j^2 - rj)
$$
$$
\beta_j = -\frac{\delta t}{2}(\sigma^2 j^2 + r)
$$
$$
\gamma_j = \frac{\delta t}{4}(\sigma^2 j^2 + rj).
$$

We consider here the down-and-out put option we have introduced in section 7.3.1, assuming continuous barrier monitoring. In this case we need only to consider the domain $S \geq S_b$; the boundary conditions are

$$
f(t, S_{\max}) = 0 \qquad f(t, S_b) = 0.
$$

Taking these boundary conditions into account, we may rewrite equation (8.5) in matrix form:

$$\mathbf{M}_1 \mathbf{f}_{i-1} = \mathbf{M}_2 \mathbf{f}_i, \tag{8.6}$$

where

$$\mathbf{M}_1 = \begin{bmatrix} 1-\beta_1 & -\gamma_1 & & & & & \\ -\alpha_2 & 1-\beta_2 & -\gamma_2 & & & & \\ & -\alpha_3 & 1-\beta_3 & -\gamma_3 & & & \\ & & \ddots & \ddots & \ddots & & \\ & & & -\alpha_{M-2} & 1-\beta_{M-2} & -\gamma_{M-2} \\ & & & & -\alpha_{M-1} & 1-\beta_{M-1} \end{bmatrix}$$

$$\mathbf{M}_2 = \begin{bmatrix} 1+\beta_1 & \gamma_1 & & & & & \\ \alpha_2 & 1+\beta_2 & \gamma_2 & & & & \\ & \alpha_3 & 1+\beta_3 & \gamma_3 & & & \\ & & \ddots & \ddots & \ddots & & \\ & & & \alpha_{M-2} & 1+\beta_{M-2} & \gamma_{M-2} \\ & & & & \alpha_{M-1} & 1+\beta_{M-1} \end{bmatrix}$$

$$\mathbf{f}_i = [f_{i1}, f_{i2}, \ldots, f_{i,M-1}]^T.$$

The MATLAB code is displayed in figure 8.4. The result may be compared with those obtained by the analytical pricing formula of section 7.3.1:

```
>> DOPut(50,50,0.1,5/12,0.4,40)
ans =
    0.5424
>> DOPutCK(50,50,0.1,5/12,0.4,40,100,0.5,1/1200)
ans =
    0.5414
```

Barrier options come in a variety of forms; more on the application of PDEs to barrier options may be found in [6].

8.5 DEALING WITH AMERICAN OPTIONS

While pricing a vanilla European option by finite differences is certainly instructive, it is not very practical. We may apply the idea to American options, for which exact formulas are not available (and using Monte Carlo simulation is difficult).[2]

The main difficulty in pricing an American option is the free boundary due to the possibility of early exercise. To avoid arbitrage, the option value at

---

[2]We should note that approximate formulas for American options are available and that pricing American options by simulation is an active research area (see, e.g., [2, p. 230]).

```
% DOPutCK.m
function price = DOPutCK(S0,X,r,T,sigma,Sb,Smax,dS,dt)
% set up grid and adjust increments if necessary
M = round((Smax-Sb)/dS);
dS = (Smax-Sb)/M;
N = round(T/dt);
dt = T/N;
matval = zeros(M+1,N+1);
vetS = linspace(Sb,Smax,M+1)';
veti = 0:N;
vetj = vetS / dS;
% set up boundary conditions
matval(:,N+1) = max(X-vetS,0);
matval(1,:) = 0;
matval(M+1,:) = 0;
% set up the coefficients matrix
alpha = 0.25*dt*( sigma^2*(vetj.^2) - r*vetj );
beta = -dt*0.5*( sigma^2*(vetj.^2) + r );
gamma = 0.25*dt*( sigma^2*(vetj.^2) + r*vetj );
M1 = -diag(alpha(3:M),-1) + diag(1-beta(2:M)) - diag(gamma(2:M-1),1);
[L,U] = lu(M1);
M2 = diag(alpha(3:M),-1) + diag(1+beta(2:M)) + diag(gamma(2:M-1),1);
% solve the sequence of linear systems
for i=N:-1:1
   matval(2:M,i) = U \ (L \ (M2*matval(2:M,i+1)));
end
% find closest point to S0 on the grid and return price
% possibly with a linear interpolation
jdown = floor((S0-Sb)/dS);
jup = ceil((S0-Sb)/dS);
if jdown == jup
   price = matval(jdown+1,1);
else
   price = matval(jdown+1,1) + ...
       (S0 - Sb - jdown*dS)*(matval(jup+1,1) - matval(jup+1,1))/dS;
end
```

Fig. 8.4 MATLAB code to price a down-and-out put option by the Crank-Nicolson method.

each point in the (t, S) space cannot be less than the intrinsic value (i.e., the immediate payoff if the option is exercised). For a vanilla American put this means

$$f(t, S) \geq \max\{X - S(t), 0\}.$$

From a strictly practical point of view, taking this condition into account is not very difficult, at least in an explicit scheme. We could simply apply the procedure of section 8.2 with a small modification. After computing f_{ij}, we should check for the possibility of early exercise, and set

$$f_{ij} = \max[f_{ij}, X - j\delta S].$$

Due to instability issues, we might prefer adopting an implicit scheme. In this case, there is an additional complication, as the relationship above requires knowing f_{ij} already, which is not the case in an implicit scheme. To get past this difficulty, we may resort to an iterative method to solve the linear system rather than to a direct method based on LU-factorization. In section 2.2.4 we considered the Gauss-Seidel method with overrelaxation. We recall the idea here for convenience. Given a system of linear equations such as

$$\mathbf{Ax} = \mathbf{b},$$

we should apply the following iterative scheme, starting from an initial point $\mathbf{x}^{(0)}$:

$$x_i^{(k+1)} = x_i^{(k)} + \frac{\omega}{a_{ii}} \left(b_i - \sum_{j=1}^{i-1} a_{ij} x_j^{(k+1)} - \sum_{j=i}^{N} a_{ij} x_j^{(k)} \right) \qquad i = 1, \ldots, N,$$

where k is the iteration counter and ω is the overrelaxation parameter, until a convergence criterion is met, such as

$$\| \mathbf{x}^{(k+1)} - \mathbf{x}^{(k)} \| < \epsilon,$$

where ϵ is a tolerance parameter.

Now, suppose that we want to apply the Crank-Nicolson method to price an American put option. We have to solve more or less the same system as (8.6), but here the boundary conditions are a bit different, since there is no barrier on which the option value is zero. The systems we should solve backward in time look like:

$$\mathbf{M}_1 \mathbf{f}_{i-1} = \mathbf{r}_i,$$

where the right-hand side is

$$\mathbf{r}_i = \mathbf{M}_2 \mathbf{f}_i + \alpha_1 \begin{bmatrix} f_{i-1,0} + f_{i,0} \\ 0 \\ \vdots \\ 0 \end{bmatrix}.$$

The additional term takes the customary boundary conditions for a put into account. The overrelaxation scheme should both take into account the tridiagonal nature of the matrix $\mathbf{M}_1$ and be adjusted for early exercise. Let g_j, $j = 1, \ldots, M - 1$ be the intrinsic value when $S = j \, \delta S$. For each time layer i, we have the iterative scheme

$$
f_{i1}^{(k+1)} = \max \left\{ g_1, \, f_{i1}^{(k)} \right.
$$
$$
\left. + \frac{\omega}{1 - \beta_1} \left[r_1 - (1 - \beta_1) f_{i1}^{(k)} + \gamma_1 f_{i2}^{(k)} \right] \right\}
$$

$$
f_{i2}^{(k+1)} = \max \left\{ g_2, \, f_{i2}^{(k)} \right.
$$
$$
\left. + \frac{\omega}{1 - \beta_2} \left[r_2 + \alpha_2 f_{i1}^{(k+1)} - (1 - \beta_2) f_{i2}^{(k)} + \gamma_2 f_{i3}^{(k)} \right] \right\}
$$

$$
\vdots
$$

$$
f_{i,M-1}^{(k+1)} = \max \left\{ g_{M-1}, \, f_{i,M-1}^{(k)} \right.
$$
$$
\left. + \frac{\omega}{1 - \beta_{M-1}} \left[r_{M-1} + \alpha_{M-1} f_{i,M-2}^{(k+1)} - (1 - \beta_{M-1}) f_{i,M-1}^{(k)} \right] \right\}.
$$

When passing from a time layer to the next one, it may be reasonable to initialize the iteration with a starting vector equal to the outcome of the previous time layer. The resulting code is displayed in figures 8.5 and 8.6. The code is a bit tricky because MATLAB starts indexing vectors from 1, but it should be clear enough. In this case we have not set up a matrix to contain all of the f_{ij} values, and the sparse matrix $\mathbf{M}_1$ has not been stored; the iterations above are best carried out by using the vectors $\boldsymbol{\alpha}$, $\boldsymbol{\beta}$, and $\boldsymbol{\gamma}$ directly.

The code may be compared with the `binprice` function, available in the Financial toolbox, which prices American options by a binomial lattice method (see section 1.4.4).

```
>> tic,[pr,opt] = binprice(50,50,0.1,5/12,1/1200,0.4,0);,toc
elapsed_time =
    14.1700
>> opt(1,1)
ans =
    4.2830
>> tic,AmPutCK(50,50,0.1,5/12,0.4,100,1,1/600,1.5,0.001),toc
ans =
    4.2815
elapsed_time =
    59.4800
>> tic,AmPutCK(50,50,0.1,5/12,0.4,100,1,1/600,1.8,0.001),toc
ans =
```

```
% AmPutCK.m
function price = AmPutCK(S0,X,r,T,sigma,Smax,dS,dt,omega,tol)
M = round(Smax/dS); dS = Smax/M; % set up grid
N = round(T/dt); dt = T/N;
oldval = zeros(M-1,1); % vectors for Gauss-Seidel update
newval = zeros(M-1,1);
vetS = linspace(0,Smax,M+1)';
veti = 0:N; vetj = 0:M;
% set up boundary conditions
payoff = max(X-vetS(2:M),0);
pastval = payoff; % values for the last layer
boundval = X*exp(-r*dt*(N-veti)); % boundary values
% set up the coefficients and the right hand side matrix
alpha = 0.25*dt*( sigma^2*(vetj.^2) - r*vetj );
beta = -dt*0.5*( sigma^2*(vetj.^2) + r );
gamma = 0.25*dt*( sigma^2*(vetj.^2) + r*vetj );
M2 = diag(alpha(3:M),-1) + diag(1+beta(2:M)) + diag(gamma(2:M-1),1);
% solve the sequence of linear systems by SOR method
aux = zeros(M-1,1);
for i=N:-1:1
   aux(1) = alpha(2) * (boundval(1,i) + boundval(1,i+1));
   % set up right hand side and initialize
   rhs = M2*pastval(:) + aux;
   oldval = pastval;
   error = REALMAX;
   while tol < error
      newval(1) = max ( payoff(1), ...
         oldval(1) + omega/(1-beta(2)) * (...
         rhs(1) - (1-beta(2))*oldval(1) + gamma(2)*oldval(2)));
      for k=2:M-2
         newval(k) = max ( payoff(k), ...
            oldval(k) + omega/(1-beta(k+1)) * (...
            rhs(k) + alpha(k+1)*newval(k-1) - ...
            (1-beta(k+1))*oldval(k) + gamma(k+1)*oldval(k+1)));
      end
      newval(M-1) = max( payoff(M-1),...
         oldval(M-1) + omega/(1-beta(M)) * (...
         rhs(M-1) + alpha(M)*newval(M-2) - ...
         (1-beta(M))*oldval(M-1)));
      error = norm(newval - oldval);
      oldval = newval;
   end
   pastval = newval;
end
```

Fig. 8.5   MATLAB code to price an American put option (continued in figure 8.6).

```
% find closest point to S0 on the grid and return price
% possibly with a linear interpolation
newval = [boundval(1) ; newval ; 0]; % add missing values
jdown = floor(S0/dS);
jup = ceil(S0/dS);
if jdown == jup
   price = newval(jdown+1,1);
else
   price = newval(jdown+1,1) + ...
       (S0 - jdown*dS)*(newval(jup+1,1) - newval(jup+1,1))/dS;
end
```

Fig. 8.6   MATLAB code to price an American put option (continued from figure 8.5).

```
    4.2794
elapsed_time =
  136.3300
>> tic,AmPutCK(50,50,0.1,5/12,0.4,100,1,1/600,1.2,0.001),toc
ans =
    4.2800
elapsed_time =
   26.8100
>> tic,AmPutCK(50,50,0.1,5/12,0.4,100,1,1/1200,1.2,0.001),toc
ans =
    4.2828
elapsed_time =
   55.0900
>> tic,AmPutCK(50,50,0.1,5/12,0.4,100,1,1/100,1.2,0.001),toc
ans =
    4.2778
elapsed_time =
   10.3300
```

From these examples you see that the overrelaxation parameter ω has a significant effect on the convergence of the iterative methods. In terms of computational speed, the finite difference approach may not seem competitive with the binomial lattice. However, an important point to keep in mind is that in practice you are most interested in Greeks. Having a whole grid of values rather than a binomial lattice allows us to obtain better estimates of some of the sensitivities (those involved in the Black-Scholes equation). Furthermore, the finite difference approach may be preferable to deal with complex exotic options.

For further reading

- Many examples of how the PDE approach may be exploited in financial engineering are given in [4] or [5], which include interesting chapters on finite difference methods.

- You may also find [2] useful.

- A book aimed specifically at finite differences in financial engineering is [3].

REFERENCES

1. J.C. Hull. *Options, Futures, and Other Derivatives (4th ed.).* Prentice Hall, Upper Saddle River, NJ, 2000.

2. Y.K. Kwok. *Mathematical Models of Financial Derivatives.* Springer-Verlag, Berlin, 1998.

3. D. Tavella and C. Randall. *Pricing Financial Instruments: The Finite Difference Method.* Wiley, New York, 2000.

4. P. Wilmott. *Derivatives: The Theory and Practice of Financial Engineering.* Wiley, Chichester, West Sussex, England, 1999.

5. P. Wilmott. *Quantitative Finance (vols. I and II).* Wiley, Chichester, West Sussex, England, 2000.

6. R. Zvan, K.R. Vetzal, and P.A. Forsyth. PDE Methods for Pricing Barrier Options. *Journal of Economic Dynamics and Control,* 24:1563–1590, 2000.

Part IV

Appendices

Appendix A
Introduction to
MATLAB programming

We give here a brief outline of the MATLAB basics, referring to the user manual for a full treatment. You may also type **demo** to see a demonstration of both MATLAB and the toolboxes you are interested in. Actual use of the features we describe is illustrated in the remainder of the book. A rich online documentation is available in the MATLAB environment; the reader should take advantage of this whenever a piece of code in the book is not clear.

A.1 MATLAB ENVIRONMENT

- MATLAB is an interactive computing environment. You may enter expressions and obtain an immediate evaluation.

```
>> rho = 1+sqrt(5)/2
rho =
    2.1180
```

By entering a command like this, you also define a variable `rho` which is added to the current environment and may be referred to in any other expression.

- There is a rich set of predefined functions. Try typing `help elfun`, `help elmat`, and `help ops` to get information on elementary mathematical functions, matrix manipulation, and operators, respectively. For each predefined function there is an online help.

```
>> help sqrt
 SQRT   Square root.
    SQRT(X) is the square root of the elements of X. Complex
    results are produced if X is not positive.
    See also SQRTM.
```

The `help` command should be used when you know the name of the function you are interested in, but you need additional information. Otherwise, `lookfor` may be tried.

- MATLAB is case sensitive (`Pi` and `pi` are different).

```
>> pi
ans =
    3.1416
>> Pi
??? Undefined variable or capitalized internal function Pi;
    Caps Lock may be on.
```

- MATLAB is a matrix-oriented environment. Vectors and matrices are the basic data structures, and more complex ones have been introduced in the more recent MATLAB versions. Functions and operators are available to deal with vectors and matrices directly. You may enter row and column vectors as follows:

```
>> V1=[22 5 3]
V1 =
    22     5     3

>> V2 = [33; 7; 1]
V2 =
    33
     7
     1
```

- The `who` and `whos` commands may be used to check the user defined variables in the current environment, which can be cleared by the `clear` command.

```
>> who
Your variables are:
V1              V2
>> whos
   Name         Size           Bytes   Class
   V1           1x3               24   double array
   V2           3x1               24   double array
Grand total is 6 elements using 48 bytes
>> clear
>> whos
>>
```

- You may also enter matrices (note the difference between ; and ,).

```
>> A=[1 2 3; 4 5 6]
A =
     1     2     3
     4     5     6
>> B=[V2 , V2]
B =
    33    33
     7     7
     1     1
>> C=[V2 ; V2]
C =
    33
     7
     1
    33
     7
     1
```

Also note the effects of the following commands:

```
>> M1=zeros(2,2)
M1 =
     0     0
     0     0
>> M1=rho
M1 =
    2.1180
>> M1=zeros(2,2);
>> M1(:,:)=rho
M1 =
    2.1180    2.1180
```

           2.1180   2.1180

The semicolon ; may be used at the end of a command to suppress printing of the value returned by an expression.

- The colon (:) is used to spot subranges of an index in a matrix.

```
>> M1=zeros(2,3)
M1 =
     0    0    0
     0    0    0
>> M1(2,:)=4
M1 =
     0    0    0
     4    4    4
>> M1(1,2:3)=6
M1 =
     0    6    6
     4    4    4
```

- The dots (...) may be used to write multiline commands.

```
>> M1=ones(2,
??? M1=ones(2,

Missing variable or function.
>> M1=ones(2,...
2)
M1 =
     1    1
     1    1
```

- The zeros and ones commands are useful to initialize and preallocate matrices. This is recommended for efficiency. In fact, matrices are resized automatically by MATLAB whenever you assign a value to an element beyond the current row or column size, but this may be time consuming and should be avoided when possible.

- [] is the empty vector. You may also use it to delete submatrices:

```
>> M1

M1 =
     0    6    6
     4    4    4
```

```
>> M1(:,2)=[]
M1 =
     0     6
     4     4
```

- Matrices can be transposed and multiplied easily (if dimensions fit):

```
>> M1'
ans =
     0     4
     6     4
>> M2=rand(2,3)
M2 =
     0.9501     0.6068     0.8913
     0.2311     0.4860     0.7621
>> M1*M2
ans =
     1.3868     2.9159     4.5726
     4.7251     4.3713     6.6136
>> M1+1
ans =
     1     7
     5     5
```

The **rand** command yields a matrix with random entries, uniformly distributed in the (0,1) interval.

- Note the use of the dot . to operate on matrix elements as scalars:

```
>> A=0.5*ones(2,2)
A =
     0.5000     0.5000
     0.5000     0.5000
>> M1
M1 =
     0     6
     4     4
>> M1*A
ans =
     3     3
     4     4
>> M1.*A
ans =
     0     3
     2     2
```

```
>> I=[1 2; 3 4]
I =
       1     2
       3     4
>> I^2
ans =
       7    10
      15    22
>> I.^2
ans =
       1     4
       9    16
```

- Subranges may be used to build vectors. For instance, to compute the factorial:

```
>> 1:10
ans =
    1    2    3    4    5    6    7    8    9    10
>> prod(1:10)
ans =
      3628800
>> sum(1:10)
ans =
     55
```

You may also specify an optional increment step in these expressions:

```
>> 1:0.8:4
ans =
    1.0000    1.8000    2.6000    3.4000
```

- Note the use of the special quantities Inf (infinity) and NaN (not a number):

```
>> l=1/0
Warning: Divide by zero.
l =
   Inf
>> l
l =
   Inf
>> prod(1:200)
ans =
```

```
    Inf
>> 1/0 - prod(1:200)
Warning: Divide by zero.
ans =
    NaN
```

- Useful functions to operate on matrices are: eye, diag, inv, eig, det, rank.

```
>> eye(3)
ans =
     1     0     0
     0     1     0
     0     0     1
>> K=eye(3)*[1 2 3]'
K =
     1
     2
     3
>> K=diag([1 2 3])
K =
     1     0     0
     0     2     0
     0     0     3
>> K=inv(K)
K =
     1.0000         0         0
          0    0.5000         0
          0         0    0.3333
>> eig(K)
ans =
     1.0000
     0.5000
     0.3333
>> rank(K)
ans =
     3
>> det(K)
ans =
     0.1667
```

- Some functions operate on matrices columnwise:

```
>> A = [1 3 5 ; 2 4 6 ];
>> sum(A)
```

```
ans =
     3     7     11
>> mean(A)
ans =
     1.5000     3.5000     5.5000
```

However, it is possible to specify the dimension along which they should work:

```
>> sum(A,2)
ans =
     9
     12
>> mean(A,2)
ans =
     3
     4
```

- Systems of linear equations are easily solved:

```
>> A = [3 5 -1; 9 2 4;   4 -2 -9];
>> b = (1:3)';
>> X = A\b
X =
     0.3119
    -0.0249
    -0.1892
>> A*X
ans =
     1.0000
     2.0000
     3.0000
```

- The efficiency of a function or a command may be checked by using the commands `tic` and `toc` as follows:

```
>> tic, inv(rand(500,500));, toc
elapsed_time =
     14.3400
```

A.2   MATLAB GRAPHICS

- Plotting a function of a single variable is easy. Try the following commands:

```
>> x = 0:0.01:2*pi;
>> plot(x,sin(x))
>> axis([0 2*pi -1 1])
```

The `axis` command may be used to resize plot axes at will. There is also a rich set of ways to annotate a plot.

- Different types of plots may be obtained by using optional parameters of the `plot` command. Try with

```
>> plot(0:20, rand(1,21), 'o')
>> plot(0:20, rand(1,21), 'o-')
```

- To obtain a tridimensional surface, the `surf` command may be used.

```
>> f = inline('exp(-3*(x.^2 + y.^2)).*(sin(5*pi*x)+ ...
   cos(10*pi*y))');
>> [X Y] = meshgrid(-1:0.01:1 , -1:0.01:1);
>> surf(X,Y,f(X,Y))
```

In this snapshot we have first defined an inline function using the dot operator . in such a way that it can properly handle vector and matrix arguments; if the input parameters are matrices, the output is the matrix of function values. Without the dots, input matrices would be multiplied. Then we set a grid of points up by the `meshgrid` command, and finally, plot the function.

A.3 MATLAB PROGRAMMING

- MATLAB toolboxes extend considerably the capabilities of the MATLAB core. They consist of a set of functions that are coded in the MATLAB programming language. They are contained in M-files, which are plain text files with default extension *.m. It is quite instructive to open some of these files in the MATLAB editor to see how a robust and flexible code is written.

- You may also write your own functions. You have simply to open the MATLAB editor and save the file in a directory which is on the MATLAB path.

- A simple function is displayed in figure A.1. The function consists of the function header, which specifies the input and output arguments. Note how multiple output arguments are expressed. The comments below the function heading are displayed if you ask for some `help` about the function:

```
function [xout, yout] = samplefile(x,y)
% a simple m-file to do some pointless computation
% this comment is printed by issuing the help samplefile
% command
[m,n] = size(x);
[p,q] = size(y);
z = rand(10,m)*x*rand(n,10) + rand(10,p)*y*rand(q,10);
xout = sum(z);
yout = sin(z);
```

Fig. A.1   Typical structure of a MATLAB function.

```
>> help samplefile
   a simple m-file to do some pointless computation
   this comment is printed by issuing the help samplefile
   command
```

Then the function body is given, which may contain further comment lines and arbitrarily complex control structures.

- In general, you may write a function in which some input arguments are optional and are given default parameters. To see a simple example, try typing the following commands:

```
>> help mean
```

and

```
>> type mean.m
```

Alternatively, you may open mean.m within the MATLAB editor.

- The function body includes a sequence of instructions, which in turn are built by:

    - Using control structures common to any other programming language, such as if, for, while, etc.
    - Calling other predefined functions.
    - Building expressions based on the familiar arithmetic, relational and logical operators.

```
function p = myprimes(N)
found = 0;
trynumber = 2;
p = [];
while (found < N)
   if isprime(trynumber)
      p = [p , trynumber];
      found = found + 1;
   end
   trynumber = trynumber + 1;
end
```

Fig. A.2 MATLAB function to return the first N prime numbers.

- For instance, suppose you want to write a function that returns the first N prime numbers. MATLAB provides the user with two related functions, **primes** and **isprime**. The function **primes** returns the prime numbers that are less than or equal to an input number:

```
>> primes(11)
ans =
     2    3    5    7    11
```

whereas **isprime** returns 1 if the input number is prime, 0 otherwise:

```
>> isprime([3 4 5])
ans =
     1    0    1
```

primes is not what we need, since we want the first N prime numbers. One way to accomplish our aim is illustrated in figure A.2. Note how the **if** statement treats 1 as "true" and 0 as "false."

```
>> myprimes(8)
ans =
     2    3    5    7    11    13    17    19
```

The function can and should be improved. To begin with, even numbers larger than 2 cannot be prime and should not be checked; furthermore, the vector **p** should be preallocated, rather than dynamically resized. These improvements are left as an exercise.

- When developing m-files, a most useful tool is the debugger. We refer the reader to the manual for details.

Appendix B
Refresher on probability theory

In this appendix we recall briefly some basic facts about probability theory and parameter estimation. This is not meant as a substitute for a thorough treatment, for which we refer the reader to the references. We also give information on some functions of the MATLAB Statistics toolbox.

B.1 SAMPLE SPACE, EVENTS, AND PROBABILITY

Probability is defined based on random events that take place within a sample space. A sample space S contains the possible outcomes of a random experiment or a sequence of random experiments. An event E is any subset of the sample space S. The empty set $\emptyset$ is a particular event. For any event E, we may consider its complement E^c; since the sample space S contains all the possible outcomes, we have $S^c = \emptyset$. Given any two events E_1 and E_2, we may consider their union $E_1 \cup E_2$ and their intersection $E_1 E_2$. If the intersection of two events is empty, i.e., if $E_1 E_2 = \emptyset$, we say that the two events are mutually exclusive. More generally, we may consider the union and the intersection of an arbitrary number of events.

For each event E on a sample space S, we define a probability measure $P(E)$ which must satisfy the following three conditions:

1. $0 \leq P(E) \leq 1$.

2. $P(S) = 1$.

3. For any sequence of mutually exclusive events $E_1, E_2, E_3, \ldots$ (i.e., such that $E_i E_j = \emptyset$, for $i \neq j$), we have

$$P\left(\bigcup_{i=1}^{\infty} E_i\right) = \sum_{i=1}^{\infty} P(E_i).$$

Different properties may be proven as a consequence of these conditions. For instance, it can be shown that

$$P(E) + P(E^c) = 1$$

and that

$$P(E_1 \cup E_2) = P(E_1) + P(E_2) - P(E_1 E_2).$$

Often we are interested in the probability of an event E conditional on the occurrence of another event F, denoted by $P(E \mid F)$. It is natural to define the conditional probability as[1]

$$P(E \mid F) = \frac{P(EF)}{P(F)}.$$

This follows from the observation that if we know that the event F has occurred, the new sample space is F, so that probabilities must be adjusted accordingly. Finally, we say that two events are independent if

$$P(EF) = P(E)P(F),$$

which in turn implies that

$$P(E \mid F) = P(E).$$

So, for independent events, knowing that F has occurred tells us nothing about the probability of the occurrence of E. Note that mutually exclusive events are not independent; if we know that one has occurred, we know that the other cannot.

---

[1]In the axiomatic treatment of probability theory, this may be considered a further axiom.

B.2   RANDOM VARIABLES, EXPECTATION, AND VARIANCE

When we associate numerical values of one or more variables to events, we obtain random variables. Random variables may be thought of as mappings from events to real or integer numbers. Usually, a random variable is denoted by a capital letter such as X; the value assumed by a random variable on a particular realization of the events is denoted by a lowercase letter such as x. When X takes values on a finite or countable domain, such as nonnegative integer numbers, we speak of a discrete random variable. For a discrete random variable, we define the *probability mass function* $p(\cdot)$ for each possible outcome value x_i:

$$p(x_i) = P\{X = x_i\}.$$

We have

$$\sum_{i=1}^{\infty} p(x_i) = 1.$$

We also define the *distribution function* $F(\cdot)$:

$$F(a) = P\{X \leq a\} = \sum_{x_i \leq a} p(x_i).$$

It is easy to see that the distribution function for a discrete random variable is a piecewise constant, nondecreasing function.

Example B.1 A typical example of discrete probability distribution is the Poisson random variable, with parameter λ. In this case the random variable X takes values in the set $\{0, 1, 2, 3, \ldots\}$, and its probability mass function is

$$p(i) = P\{X = i\} = e^{-\lambda}\frac{\lambda^i}{i!} \qquad i = 0, 1, 2, \ldots.$$

We may check that this is indeed a probability mass function:

$$\sum_{i=0}^{\infty} p(i) = e^{-\lambda} \sum_{i=0}^{\infty} \frac{\lambda^i}{i!} = e^{-\lambda}e^{\lambda} = 1.$$

In practice, one usually works with a parameter λt, where λ is the rate at which certain events occurs over time and t is the length of the time interval we observe. For instance, this could model the number of shocks we observe over a time interval on the price of a stock or the credit rating of a bond issuer.   ⬜

If the random variable may take values over a continuous set, such as a bounded interval on the real line, say (a, b), or the entire line $(-\infty, +\infty)$, we have a continuous random variable. In this case, we cannot define a probability mass function; since the outcome values are infinite and uncountable, the probability that X takes a specific value will be zero.[2] We must define a nonnegative probability density function $f(x)$ for $x \in (-\infty, +\infty)$ such that for a certain subset B of real numbers,

$$P\{X \in B\} = \int_B f(x)\, dx.$$

Then we have

$$P\{a \leq X \leq b\} = \int_a^b f(x)\, dx,$$

and

$$\int_{-\infty}^{+\infty} f(x)\, dx = 1.$$

To understand what the probability density means, consider the following:

$$P\{X \in (x, x + \Delta x)\} = \int_x^{x+\Delta x} f(y)\, dy \approx f(x)\, \Delta x$$

for a small Δx. We may also define the distribution function:

$$F(a) = P\{X \leq a\} = \int_{-\infty}^a f(x)\, dx,$$

from which we obtain[3]

$$\frac{dF(x)}{dx} = f(x).$$

Given a random variable, we may compute its expected value using the probability mass function or the density function. In the discrete case we have

$$E[X] = \sum_i x_i p(x_i),$$

whereas for the continuous case,

$$E[X] = \int_{-\infty}^{+\infty} x f(x)\, dx.$$

An important property of the expectation operator is

$$E[\alpha X + \beta] = \alpha\, E[X] + \beta.$$

---

[2] We are not considering mixed probability distributions, which are a hybrid between discrete and continuous distributions.

[3] The distribution function is not everywhere differentiable in the case of mixed distributions, which we do not consider.

Example B.2 Let us compute the expected value of a Poisson random variable. Applying the definition yields

$$E[X] \;=\; \sum_{i=0}^{\infty} i e^{-\lambda}\frac{\lambda^i}{i!} = \lambda e^{-\lambda}\sum_{i=1}^{\infty}\frac{\lambda^{i-1}}{(i-1)!}$$

$$=\; \lambda e^{-\lambda}\sum_{k=0}^{\infty}\frac{\lambda^k}{k!} = \lambda.$$

This may be interpreted as follows. If events occur at a rate of λ events per time unit, the expected number of events over a unit interval is actually the rate λ. By the same token, the expected number of events occurring over an interval of length t is λt. ▯

By the same token, we may compute the expected value of a function $g(X)$ of a random variable:

$$E[g(X)] = \begin{cases} \displaystyle\sum_i g(x_i)p(x_i) & \text{for the discrete case} \\[2mm] \displaystyle\int_{-\infty}^{+\infty} g(x)f(x)\,dx & \text{for the continuous case.} \end{cases}$$

A particular quantity of interest is the *variance*:

$$\mathrm{Var}(X) = E[(X - E[X])^2].$$

This quantity, or its square root, called *standard deviation*, gives a measure of the dispersion of a random variable around its expected value. A couple of properties of the variance are the following:

$$\mathrm{Var}(X) = E[X^2] - E^2[X]$$
$$\mathrm{Var}(\alpha X + \beta) = \alpha^2\,\mathrm{Var}(X).$$

We see immediately that, unlike the expectation, the variance operator is not linear. Indeed, it is *not* true in general that the variance of a sum of random variables is the sum of their variances (see later).

B.2.1   Common continuous random variables

Uniform random variable   A random variable is distributed uniformly over the interval (a, b) if its density function is

$$f(x) = \begin{cases} 1/(b-a) & \text{if } x \in (a, b) \\ 0 & \text{otherwise.} \end{cases}$$

A typical case is the uniform distribution over the interval $(0, 1)$. It is easy to see that

$$E[X] = \int_a^b \frac{x}{b-a}\,dx = \frac{b^2 - a^2}{2(b-a)} = \frac{b+a}{2}$$

and

$$
\begin{aligned}
\mathrm{Var}(X) &= \mathrm{E}[X^2] - \mathrm{E}^2[X] = \int_a^b \frac{x^2}{b-a}\, dx - \left(\frac{a+b}{2}\right)^2 \\
&= \frac{b^3 - a^3}{3(b-a)} - \frac{(b+a)^2}{4} = \frac{(b-a)^2}{12}.
\end{aligned}
$$

Exponential random variable   The exponential random variable may only assume nonnegative values, and its density is given by

$$
f(x) = \begin{cases} \lambda e^{-\lambda x} & \text{if } x \geq 0 \\ 0 & \text{if } x < 0 \end{cases}
$$

for some parameter $\lambda > 0$. The distribution function is

$$
F(a) = \int_0^a \lambda e^{-\lambda x}\, dx = 1 - e^{-\lambda a}.
$$

The expected value is

$$
\mathrm{E}[X] = \int_0^\infty x\lambda e^{-\lambda x}\, dx = \frac{1}{\lambda},
$$

and the variance is $1/\lambda^2$. It is interesting to note that if the time elapsing between events is exponentially distributed with parameter λ, the events occur at a rate λ, and the distribution of the number of events over a time interval of length t is Poisson with parameter λt.

Normal random variable   The normal random variable has an infinite support, i.e., it may take values over the whole real line, and its density function is the bell-shaped function:

$$
f(x) = \frac{1}{\sqrt{2\pi}\,\sigma} e^{-(x-\mu)^2/2\sigma^2} \qquad -\infty < x < +\infty,
$$

for given parameters μ and σ^2. The distribution function for the normal distribution is not known in closed form, but it can be computed by numerical approximations (see example 2.10). With some calculations it can be shown that the parameters μ and σ have a precise meaning:

$$
\mathrm{E}[X] = \mu \qquad \mathrm{Var}[X] = \sigma^2.
$$

Example B.3 The parameter μ influences where the maximum of the density is located, whereas the variance σ^2, or the standard deviation σ, tells how stretched the function is. We may plot the density functions for two normal distributions with $\mu = 0$, and $\sigma = 1$ and $\sigma = 3$.

```
>> x=-10:0.1:10;
```

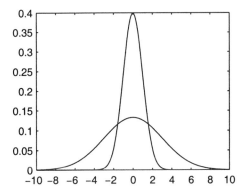

Fig. B.1 Normal density function for $\mu = 0$, and $\sigma = 1$ and $\sigma = 3$.

```
>> plot(x, normpdf(x,0,1))
>> hold on
>> plot(x, normpdf(x,0,3))
```

The result is plotted in figure B.1. The Statistics toolbox includes functions to compute the probability functions for the main probability distributions.

□

An important property of the normal distribution is that if X is normally distributed with parameters μ and σ^2, then $\alpha X + \beta$ is normally distributed with parameters $\alpha\mu + \beta$ and $\alpha^2\sigma^2$. In particular, $Z = (X - \mu)/\sigma$ is normally distributed with parameters 0 and 1 and is called the *unit* or *standard normal distribution*. By working with standard normal variables, we are actually able to deal with the more general case. For instance, to compute the distribution function for an arbitrary normal variable, it is sufficient to come up with an approximation for the standard case:

$$N(x) = \frac{1}{\sqrt{2\pi}} \int_{-\infty}^{x} e^{-z^2/2}\, dz.$$

Example B.4 The function normcdf(x,sigma,mu) yields the distribution function. To compute the probability that a standard normal variable lies in the interval $(-2, 2)$:

```
>> p = normcdf([-2 2]);
>> p(2) - p(1)
ans =
    0.9545
```

Similarly:

```
>> p = normcdf([-3 3]);
```

```
>> p(2)-p(1)
ans =
    0.9973
```

from which we see that for a normal distribution, the probability of falling outside the interval $(\mu - 3\sigma, \mu + 3\sigma)$ is quite small. In fact, the normal distribution is a debatable model for asset returns, as in practice these exhibit fat tails, i.e., the occurrence of extreme values is more likely than it should be with the normal distribution.

You may also invert the distribution function. Compare x and xnew in the following:

```
>> x=[-3:0.2:0.3];
>> xnew=norminv(normcdf(x,0,1),0,1);                                        []
```

The importance of normal variables, apart from their many properties, stems from the central limit theorem. Roughly speaking, it states that if we sum many identically distributed and independent random variables, their sum tends to have a normal distribution as the number of summed random variables goes to infinity.

Lognormal random variable   Due to the central limit theorem, a normal random variable may be thought of as the limit of a sum of random variables. The lognormal variable may be thought of as the limit of a *product* of random variables. Formally, we say that a random variable Z is lognormally distributed if $\ln Z$ is normally distributed; put another way, if X is normal, then e^X is lognormal. The density function is

$$f(z) = \frac{1}{\sqrt{2\pi}\,\sigma z} e^{-\frac{1}{2\sigma^2}(\ln z - \nu)^2}$$

for two parameters σ and ν. The following formulas illustrate the relationships between the parameters of a normal and a lognormal distribution:

$$\begin{aligned}
\mathrm{E}[Z] &= e^{\nu + \sigma^2/2} \\
\mathrm{E}[\ln Z] &= \nu \\
\mathrm{Var}(Z) &= e^{2\nu + \sigma}(e^{\sigma^2} - 1) \\
\mathrm{Var}(\ln Z) &= \sigma^2.
\end{aligned}$$

B.3 JOINTLY DISTRIBUTED RANDOM VARIABLES

When considering multiple random variables, we may follow the same route as in the scalar case. We illustrate in the bidimensional case, as the generalization is straightforward. Given two random variables X and Y, we may define the joint distribution function:

$$F(x, y) = P\{X \le x, Y \le y\}.$$

In the discrete case we also consider the probability mass function:

$$p(x, y) = P\{X = x, Y = y\},$$

whereas continuous variables are characterized by a density $f(x, y)$ such that, for a region D in the plane,

$$P\{(X, Y) \in D\} = \iint_D f(x, y) \, dy \, dx.$$

From the joint distribution we may derive the marginal distributions for the single variables. For instance

$$\begin{aligned} P\{X \in A\} &= P\{X \in A, Y \in (-\infty, +\infty)\} = \int_A \int_{-\infty}^{+\infty} f(x, y) \, dy \, dx \\ &= \int_A f_X(x) \, dx, \end{aligned}$$

where

$$f_X(x) = \int_{-\infty}^{+\infty} f(x, y) \, dy$$

is the marginal density for the random variable X; the other density $f_Y(y)$ may be defined similarly.

The computation of expected values is quite similar to the scalar case. Given a function $g(X, Y)$ of the two random variables, we have

$$E[g(X, Y)] = \begin{cases} \displaystyle\sum_i \sum_j g(x_i, y_j) p(x_i, y_j) & \text{in the discrete case} \\ \displaystyle\int_{-\infty}^{+\infty} \int_{-\infty}^{+\infty} g(x, y) f(x, y) \, dy \, dx & \text{in the continuous case.} \end{cases}$$

From the linearity of these operations it is easy to see that the expected value of a linear combination of random variables

$$Z = \sum_{i=1}^n \lambda_i X_i$$

is the same linear combination of the expected values:

$$E[Z] = \sum_{i=1}^n \lambda_i E[X_i].$$

However, a similar result does not hold, in general, for the variance. Similarly, for jointly distributed variables it is *not* true in general that

$$E[g(X)h(Y)] = E[g(X)]E[h(Y)].$$

To investigate this matter we must deal with the dependence or independence between the random variables.

B.4   INDEPENDENCE, COVARIANCE, AND CONDITIONAL EXPECTATION

Two random variables X and Y are independent if the two events $(X \leq a)$ and $(Y \leq b)$ are independent, i.e.,

$$F(a, b) = P\{X \leq a, Y \leq b\} = P\{X \leq a\}P\{Y \leq b\} = F_X(a)F_Y(b).$$

This in turn implies that

$$p(x, y) = p_X(x)p_Y(y) \qquad f(x, y) = f_X(x)f_Y(y),$$

for discrete and continuous variables, respectively. If the variables are independent, it is easy to show that

$$E[g(X)h(Y)] = E[g(X)]E[h(Y)]$$

holds.

If there is some degree of dependence between random variables, we should try to measure it somehow. One measure of mutual dependence is the covariance:

$$\mathrm{Cov}(X, Y) = E[(X - E[X])(Y - E[Y])] = E[XY] - E[X]E[Y].$$

If X and Y are independent, their covariance is zero (but the converse is not necessarily true, as the covariance is only one measure of dependence). If $\mathrm{Cov}(X, Y) > 0$, Y tends to be large when X is, and small when X is; a similar observation holds when the covariance is negative. The following properties of the covariance are useful:

- $\mathrm{Cov}(X, X) = \mathrm{Var}(X),$

- $\mathrm{Cov}(X, Y) = \mathrm{Cov}(Y, X),$

- $\mathrm{Cov}(aX, Y) = a\,\mathrm{Cov}(Y, X),$

- $\mathrm{Cov}(X, Y + Z) = \mathrm{Cov}(X, Y) + \mathrm{Cov}(X, Z).$

Using these properties (or the definitions), it can be shown that

$$\begin{aligned}
\mathrm{Var}(X + Y) &= \mathrm{Var}(X) + \mathrm{Var}(Y) + 2\,\mathrm{Cov}(X, Y), \\
\mathrm{Var}(X - Y) &= \mathrm{Var}(X) + \mathrm{Var}(Y) - 2\,\mathrm{Cov}(X, Y).
\end{aligned}$$

More generally

$$\mathrm{Var}\left(\sum_{i=1}^{n} X_i\right) = \sum_{i=1}^{n} \mathrm{Var}(X_i) + 2\sum_{i=1}^{n}\sum_{j<i} \mathrm{Cov}(X_i, X_j).$$

Thus, for mutually independent variables, the variance of a sum is the sum of the variances.

Example B.5 We often have to work with multivariate normals. Let

$$\mathbf{X} = \begin{bmatrix} X_1 \\ X_2 \\ \vdots \\ X_n \end{bmatrix}$$

be a vector of normal random variables with expected value μ and covariance matrix

$$\Sigma = \mathrm{E}[(\mathbf{X} - \mu)(\mathbf{X} - \mu)^T].$$

Then the joint density function is given by

$$f(\mathbf{x}) = \frac{1}{(2\pi)^{n/2} \mid \Sigma \mid^{1/2}} e^{-\frac{1}{2}(\mathbf{X}-\mu)^T \Sigma^{-1}(\mathbf{X}-\mu)},$$

where $\mid \Sigma \mid$ is the determinant of the covariance matrix. If the normal variables are mutually uncorrelated, then both the matrix Σ and its inverse are diagonal. This implies that the density function may be factorized into separate components, one for each X_i; hence, uncorrelated normal variables are also independent.

Another property of jointly normal variables is that they may combined linearly to yield other jointly normal variables. Given a matrix $\mathbf{T} \in \mathbb{R}^{m,n}$, $\mathbf{TX}$ is a vector of m jointly normal variables. ▯

The value of the covariance depends on the magnitude of the random variables involved. Often, a normalized measure of dependence is preferred, the coefficient of correlation:

$$\rho_{XY} = \frac{\mathrm{Cov}(X,Y)}{\sqrt{\mathrm{Var}(X)}\sqrt{\mathrm{Var}(Y)}}.$$

It can be shown that $\rho_{XY} \in [-1, 1]$.

Example B.6 The correlation is often used in finance. However, it is important to realize its limitations. Consider the following example.

```
>> x = -1:0.001:1;
>> y = sqrt(1-x.^2);
>> cov(x,y)
ans =
    0.3338    0.0000
    0.0000    0.0501
```

Here we have a random variable X which is distributed uniformly on $(-1, 1)$, and a random variable Y which is deterministically linked to X, as

$$Y = \sqrt{1 - X^2}.$$

However, the covariance and the correlation are zero, since

$$\text{Cov}(X, Y) = \text{E}[XY] - \text{E}[X]\text{E}[Y],$$

but $\text{E}[X] = 0$, and (because of symmetry)

$$\text{E}[XY] = \int_{-1}^{1} x \frac{1}{2} \sqrt{1 - x^2} \, dx = 0.$$

The key issue is that the correlation is a measure of *linear* dependence. Here the dependence is nonlinear, as the points (X, Y) lie on the upper half of the unit circle $X^2 + Y^2 = 1$. ▯

A general tool to investigate dependence is conditioning. Just as we have defined conditional probabilities for events, we may define conditional expectation. This means that we want to know how an event like $(Y = y)$ influences the distribution of X. For discrete random variables we have

$$\text{E}[X \mid Y = y_j] = \sum_i x_i P\{X = x_i \mid Y = y_j\} = \frac{\sum_i x_i P\{X = x_i, Y = y_j\}}{P\{Y = y_j\}}.$$

Similarly, for continuous variables

$$\text{E}[X \mid Y = y] = \frac{\int x f(x, y) \, dx}{\int f(x, y) \, dx}.$$

Conditioning is a useful way to solve many problems. A fundamental property is the following:

$$\text{E}[X] = \text{E}[\text{E}[X \mid Y]]. \tag{B.1}$$

In practice, this may be used when fixing the value of a random variable makes working with another one easier. Equation (B.1) may be rewritten, in concrete, as

$$\text{E}[X] = \begin{cases} \displaystyle\sum_j \text{E}[X \mid Y = y_j] P\{Y = y_j\} & \text{in the discrete case} \\[2ex] \displaystyle\int \text{E}[X \mid Y = y] f_Y(y) \, dy & \text{in the continuous case.} \end{cases}$$

We may also define a conditional variance:

$$\text{Var}(X \mid Y) = \text{E}\left[(X - \text{E}[X \mid Y])^2 \mid Y\right].$$

The following formula may be proved for the conditional variance:

$$\text{Var}(X) = \text{E}[\text{Var}(X \mid Y)] + \text{Var}(\text{E}[X \mid Y]). \tag{B.2}$$

This formula may be used to obtain a variance by conditioning, but it implies also that

$$\text{Var}(X) \leq \text{E}[\text{Var}(X \mid Y)]$$
$$\text{Var}(X) \leq \text{Var}(\text{E}[X \mid Y]),$$

since variance is a nonnegative quantity by definition. These properties may be used for variance reduction in Monte Carlo simulation (see section 4.4).

B.5   PARAMETER ESTIMATION

In many situations we are given a set of sample data and we must estimate one or more parameters to characterize their underlying probability distribution or to infer some property. The sample data might come from the real world (e.g., stock prices) or from a Monte Carlo simulation. Typical parameters we want to estimate are the mean, the variance, or the covariance matrix; furthermore, we would also like to quantify the reliability of the estimate.

Suppose that $X_1, X_2, \ldots, X_n$ are independent and identically distributed random variables. Say that the expected value of the underlying population is μ and the variance is σ^2; these parameters are unknown, and we would like to come up with a reasonable estimate of them. An intuitive way to estimate μ is to use the sample mean:

$$\bar{X}(n) = \frac{1}{n} \sum_{i=1}^{n} X_i.$$

Note that this estimator is itself a random variable. It is a reasonable estimator, in the sense that it is unbiased:

$$\text{E}[\bar{X}(n)] = \mu.$$

The more samples we get, the better, in the sense that the variance of the estimator decreases:

$$\text{Var}(\bar{X}(n)) = \frac{1}{n^2} \text{Var}\left(\sum_{i=1}^{n} X_i\right) = \frac{1}{n^2} \sum_{i=1}^{n} \text{Var}(X_i) = \sigma^2/n.$$

It is fundamental to understand that in this derivation we have assumed the independence of the samples; if the samples are not independent, reasoning this way may lead to an underestimate of the uncertainty of the estimator.[4]

----

[4]See, e.g., [1] for a clear discussion of this point.

By the same token, we may estimate σ^2 by the sample variance:

$$S^2(n) = \frac{1}{n-1} \sum_{i=1}^{n} \left[X_i - \bar{X}(n) \right]^2 .$$

Note the $1/(n-1)$ factor, which is needed to make the estimator unbiased. By a similar expression we may estimate the covariance matrix. This task is accomplished by MATLAB functions. The basic versions are available in the MATLAB core; some advanced functionalities are included only in the Statistics toolbox.

Example B.7 The function `mean` yields the sample mean. For instance, let us use the `normrnd` function to generate a set of independent normally distributed data values:[5]

```
>> x=normrnd(2,3,1000,2);
>> mean(x)
ans =
    2.0468    1.7942
```

The first two parameters of `normrnd` are the expected value and the standard deviation of the normal variable; the remaining two are optional and define the size of the matrix to generate. The matrix, which here has 1000 rows and two columns, is interpreted columnwise, as 1000 realizations of two random variables. This is why two means are estimated, one per column of the data matrix. The function `cov(x)` estimates the covariance matrix (assuming a column-oriented data matrix).

```
>> x=normrnd(10,2,10000,4);
>> cov(x)
ans =
    4.0243    0.0613    0.0030    0.0269
    0.0613    4.0405   -0.0172    0.0465
    0.0030   -0.0172    4.0539    0.0038
    0.0269    0.0465    0.0038    4.0238
```

Note that the values on the diagonal are close to the "correct" variance $\sigma^2 = 4$ for each of the four independent variables; off-diagonal elements should be zero, as the samples should be independent. Given the limited number of samples, it is not surprising that the results do not match exactly what we would expect in theory.

---

[5]The numbers you get may differ from the following ones, depending on the current state and the initial seed of the random number generator; the issue is explained in section 4.2. Furthermore, if you repeat the experiment, you will get different outcomes.

In practice, estimating parameters may be a tough problem. Consider drawing 100 samples from a normal distribution with known parameters and checking if the sample mean corresponds to the known expected value. Let us repeat ten of these experiments:

```
>> x = normrnd(0.3,2,100,10);
>> mean(x)'
ans =
    0.3181
    0.4450
    0.3554
    0.5102
    0.0590
    0.3113
    0.1077
    0.6897
    0.6433
    0.0664
```

You see that the estimated mean value may be quite different from the correct value $\mu = 0.3$. Actually, if you repeat the experiment a few times, you will even get negative sample means. This is due to the fact that the expected value is small with respect to the variance of the data; if you think of estimating stock returns over short periods, using historical data when volatility is high, you will realize that this is not a hypothetical circumstance. This phenomenon, called *mean blur*, is described, e.g., in [2, chapter 8]. Another point worth mentioning is that if you use historical data, you might question the validity of the old data; however, using only the recent ones may lead to unreliable estimates. The Financial toolbox includes a more sophisticated function (ewstats) to compute a covariance matrix by applying a "forgetting factor," reducing the weight of the old data. []

When we estimate a mean value by the sample mean, it is customary to associate a confidence interval. If we assume that the sample mean is normally distributed, which may be reasonable thanks to the central imit theorem, if the number of samples n is large, we may consider the random variable

$$Z = \frac{\bar{X}(n) - \mu}{\sqrt{S^2(n)/n}},$$

which is approximately distributed as a standard normal variable. Let $z_{1-\alpha/2}$ be the upper $1 - \alpha/2$ critical number for the standard normal distribution, i.e., a number such that

$$P\{Z \le z_{1-\alpha/2}\} = \int_{-\infty}^{z_{1-\alpha/2}} e^{-y^2/2} \, dy = 1 - \alpha/2.$$

Then we may argue that

$$P\left\{-z_{1-\alpha/2} \leq \frac{\bar{X}(n) - \mu}{\sqrt{S^2(n)/n}} \leq z_{1-\alpha/2}\right\}$$
$$= P\left\{\bar{X}(n) - z_{1-\alpha/2}\sqrt{S^2(n)/n} \leq \mu \leq \bar{X}(n) + z_{1-\alpha/2}\sqrt{S^2(n)/n}\right\}$$
$$\approx 1 - \alpha.$$

So we may build an approximate confidence interval at level $(1 - \alpha)$:

$$\bar{X}(n) \pm z_{1-\alpha/2}\sqrt{S^2(n)/n}.$$

The idea is that if we repeat the sampling and estimation procedure over and over, the percentage of cases in which the "true" value falls within the interval should be $100 \times (1 - \alpha)$. Typical values of α are 0.05 and 0.01.

Example B.8 Calling the function [muhat, sigmahat, muci, sigmaci] = normfit(x) yields, under a normality assumption, an estimate of the expected value and the standard deviation and the respective 95% confidence intervals.

```
>> x=normrnd(1,2,100,1);
>> [mu,s,mci,sci] = normfit(x)
mu =
    0.8469
s =
    1.9838
mci =
    0.4533
    1.2406
sci =
    1.7418
    2.3046
```

It is possible to specify a different confidence level by calling the function with an optional parameter: normfit(x,alpha).   ▯

We stress that the confidence interval above is only an approximation. In fact, for a small number n of samples, we should use the critical numbers from Student's t distribution $(t_{n-1,1-\alpha/2})$, which basically yields a larger confidence interval. Since this distribution is close to the normal distribution for a large number of samples, we will just use the normal. Furthermore, the confidence interval is more or less reliable, depending on the skewness of the underlying distribution of the samples X_i. The skewness may be measured by the coefficient $\nu = E[(X - \mu)^3]/(\sigma^2)^{3/2}$, which is zero for a symmetric distribution such as the normal. The point is that the central limit theorem is a limit theorem, and the number of samples we should sum in order to obtain a sample mean which is approximately normally distributed may be large if the samples themselves have a skewed distribution. Finally, we stress again that the way we have computed the confidence interval requires the independence of the sample points.

For further reading

There are many excellent books on probability theory, ranging from the elementary to the very sophisticated. An introductory book characterized by a remarkable clarity, plenty of insightful examples, and a wide range of topics is [3], which does not rely on measure-theoretic concepts. If you are interested in a more advanced treatment, based on rigorous axiomatic foundations, see, e.g., [4]. Apart from good statistics books, a quick and readable introduction to parameter estimation may be found in simulation books such as [1].

REFERENCES

1. A.M. Law and W.D. Kelton. *Simulation Modeling and Analysis (2nd ed.)*. McGraw-Hill, New York, 1991.

2. D.G. Luenberger. *Investment Science*. Oxford University Press, New York, 1998.

3. S. Ross. *Introduction to Probability Models (6th ed.)*. Academic Press, San Diego, CA, 1997.

4. A.N. Shiryaev. *Probability (2nd ed.)*. Springer-Verlag, New York, 1996.

Index

WILEY SERIES IN PROBABILITY AND STATISTICS
ESTABLISHED BY WALTER A. SHEWHART AND SAMUEL S. WILKS

Editors
Peter Bloomfield, Noel A. C. Cressie, Nicholas I. Fisher, Iain M. Johnstone,
J. B. Kadane, Louise M. Ryan, David W. Scott, Bernard W. Silverman,
Adrian F. M. Smith, Jozef L. Teugels
Editors Emeriti: *Vic Barnett, Ralph A. Bradley, J. Stuart Hunter,*
David G. Kendall

Wiley Series in Probability and Statistics is well established and authoritative. It covers many topics of current research interest in both pure and applied statistics and probability theory. Written by leading statisticians and institutions, the titles span both state-of-the-art developments in the field and classical methods.

Reflecting the wide range of current research in statistics, the series encompasses applied, methodological and theoretical statistics, ranging from applications and new techniques made possible by advances in computerized practice to rigorous treatment of theoretical approaches.

This series provides essential and invaluable reading for all statisticians, whether in academia, industry, government, or research.

ABRAHAM and LEDOLTER · Statistical Methods for Forecasting
AGRESTI · Analysis of Ordinal Categorical Data
AGRESTI · An Introduction to Categorical Data Analysis
AGRESTI · Categorical Data Analysis
ANDĚL · Mathematics of Chance
ANDERSON · An Introduction to Multivariate Statistical Analysis, *Second Edition*
*ANDERSON · The Statistical Analysis of Time Series
ANDERSON, AUQUIER, HAUCK, OAKES, VANDAELE, and WEISBERG ·
    Statistical Methods for Comparative Studies
ANDERSON and LOYNES · The Teaching of Practical Statistics
ARMITAGE and COLTON · Encyclopedia of Biostatistics: Volumes 1 to 6 with Index
ARMITAGE and DAVID (editors) · Advances in Biometry
ARNOLD, BALAKRISHNAN, and NAGARAJA · A First Course in Order Statistics
ARNOLD, BALAKRISHNAN, and NAGARAJA · Records
*ARTHANARI and DODGE · Mathematical Programming in Statistics
ASMUSSEN · Applied Probability and Queues
BACCELLI, COHEN, OLSDER, and QUADRAT · Synchronization and Linearity:
    An Algebra for Discrete Event Systems
*BAILEY · The Elements of Stochastic Processes with Applications to the Natural
    Sciences
BARNETT · Comparative Statistical Inference, *Third Edition*
BARNETT and LEWIS · Outliers in Statistical Data, *Third Edition*
BARTHOLOMEW, FORBES, and McLEAN · Statistical Techniques for Manpower
    Planning, *Second Edition*
BARTOSZYNSKI and NIEWIADOMSKA-BUGAJ · Probability and Statistical Inference
BASILEVSKY · Statistical Factor Analysis and Related Methods: Theory and
    Applications
BASU and RIGDON · Statistical Methods for the Reliability of Repairable Systems
BATES and WATTS · Nonlinear Regression Analysis and Its Applications
BECHHOFER, SANTNER, and GOLDSMAN · Design and Analysis of Experiments for
    Statistical Selection, Screening, and Multiple Comparisons

*Now available in a lower priced paperback edition in the Wiley Classics Library.

BELSLEY · Conditioning Diagnostics: Collinearity and Weak Data in Regression
BELSLEY, KUH, and WELSCH · Regression Diagnostics: Identifying Influential Data and Sources of Collinearity
BENDAT and PIERSOL · Random Data: Analysis and Measurement Procedures, *Third Edition*
BERRY, CHALONER, and GEWEKE · Bayesian Analysis in Statistics and Econometrics: Essays in Honor of Arnold Zellner
BERNARDO and SMITH · Bayesian Statistical Concepts and Theory
BHAT · Elements of Applied Stochastic Processes, *Second Edition*
BHATTACHARYA and JOHNSON · Statistical Concepts and Methods
BHATTACHARYA and WAYMIRE · Stochastic Processes with Applications
BILLINGSLEY · Convergence of Probability Measures, *Second Edition*
BILLINGSLEY · Probability and Measure, *Second Edition*
BIRKES and DODGE · Alternative Methods of Regression
BLISCHKE AND MURTHY · Reliability: Modeling, Prediction, and Optimization
BLOOMFIELD · Fourier Analysis of Time Series: An Introduction, *Second Edition*
BOLLEN · Structural Equations with Latent Variables
BOROVKOV · Asymptotic Methods in Queuing Theory
BOROVKOV · Ergodicity and Stability of Stochastic Processes
BOULEAU · Numerical Methods for Stochastic Processes
BOX · Bayesian Inference in Statistical Analysis
BOX · R. A. Fisher, the Life of a Scientist
BOX and DRAPER · Empirical Model-Building and Response Surfaces
*BOX and DRAPER · Evolutionary Operation: A Statistical Method for Process Improvement
BOX, HUNTER, and HUNTER · Statistics for Experimenters: An Introduction to Design, Data Analysis, and Model Building
BOX and LUCEÑO · Statistical Control by Monitoring and Feedback Adjustment
BRANDIMARTE · Numerical Methods in Finance: A MATLAB-Based Introduction
BRANDT, FRANKEN, and LISEK · Stationary Stochastic Models
BROWN and HOLLANDER · Statistics: A Biomedical Introduction
BUCKLEW · Large Deviation Techniques in Decision, Simulation, and Estimation
BUNKE and BUNKE · Nonlinear Regression, Functional Relations and Robust Methods: Statistical Methods of Model Building
CAINES · Linear Stochastic Systems
CAIROLI and DALANG · Sequential Stochastic Optimization
CHATTERJEE and HADI · Sensitivity Analysis in Linear Regression
CHATTERJEE and PRICE · Regression Analysis by Example, *Third Edition*
CHERNICK · Bootstrap Methods: A Practitioner's Guide
CHILÈS and DELFINER · Geostatistics: Modeling Spatial Uncertainty
CHOW and LIU · Design and Analysis of Clinical Trials: Concepts and Methodologies
CLARKE and DISNEY · Probability and Random Processes: A First Course with Applications, *Second Edition*
*COCHRAN and COX · Experimental Designs, *Second Edition*
CONOVER · Practical Nonparametric Statistics, *Second Edition*
CONSTANTINE · Combinatorial Theory and Statistical Design
COOK · Regression Graphics
COOK and WEISBERG · Applied Regression Including Computing and Graphics
COOK and WEISBERG · An Introduction to Regression Graphics
CORNELL · Experiments with Mixtures, Designs, Models, and the Analysis of Mixture Data, *Second Edition*
COVER and THOMAS · Elements of Information Theory
COX · A Handbook of Introductory Statistical Methods

*COX · Planning of Experiments
CRESSIE · Statistics for Spatial Data, *Revised Edition*
CSÖRGŐ and HORVÁTH · Weighted Approximations in Probability Statistics
CSÖRGŐ and HORVÁTH · Limit Theorems in Change Point Analysis
DANIEL · Applications of Statistics to Industrial Experimentation
DANIEL · Biostatistics: A Foundation for Analysis in the Health Sciences, *Sixth Edition*
*DANIEL · Fitting Equations to Data: Computer Analysis of Multifactor Data,
    Second Edition
DAVID · Order Statistics, *Second Edition*
*DEGROOT, FIENBERG, and KADANE · Statistics and the Law
DETTE and STUDDEN · The Theory of Canonical Moments with Applications in
    Statistics, Probability, and Analysis
DEY and MUKERJEE · Fractional Factorial Plans
DILLON and GOLDSTEIN · Multivariate Analysis: Methods and Applications
DODGE · Alternative Methods of Regression
*DODGE and ROMIG · Sampling Inspection Tables, *Second Edition*
*DOOB · Stochastic Processes
DOWDY and WEARDEN · Statistics for Research, *Second Edition*
DRAPER and SMITH · Applied Regression Analysis, *Third Edition*
DRYDEN and MARDIA · Statistical Shape Analysis
DUDEWICZ and MISHRA · Modern Mathematical Statistics
DUNN and CLARK · Applied Statistics: Analysis of Variance and Regression, *Second
    Edition*
DUNN and CLARK · Basic Statistics: A Primer for the Biomedical Sciences,
    Third Edition
DUPUIS and ELLIS · A Weak Convergence Approach to the Theory of Large Deviations
*ELANDT-JOHNSON and JOHNSON · Survival Models and Data Analysis
ETHIER and KURTZ · Markov Processes: Characterization and Convergence
EVANS, HASTINGS, and PEACOCK · Statistical Distributions, *Third Edition*
FELLER · An Introduction to Probability Theory and Its Applications, Volume I,
    Third Edition, Revised; Volume II, *Second Edition*
FISHER and VAN BELLE · Biostatistics: A Methodology for the Health Sciences
*FLEISS · The Design and Analysis of Clinical Experiments
FLEISS · Statistical Methods for Rates and Proportions, *Second Edition*
FLEMING and HARRINGTON · Counting Processes and Survival Analysis
FREEMAN and SMITH · Aspects of Uncertainty: A Tribute to D. V. Lindley
FULLER · Introduction to Statistical Time Series, *Second Edition*
FULLER · Measurement Error Models
GALLANT · Nonlinear Statistical Models
GHOSH, MUKHOPADHYAY, and SEN · Sequential Estimation
GIFI · Nonlinear Multivariate Analysis
GLASSERMAN and YAO · Monotone Structure in Discrete-Event Systems
GNANADESIKAN · Methods for Statistical Data Analysis of Multivariate Observations,
    Second Edition
GOLDSTEIN and LEWIS · Assessment: Problems, Development, and Statistical Issues
GREENWOOD and NIKULIN · A Guide to Chi-Squared Testing
GROSS and HARRIS · Fundamentals of Queueing Theory, *Third Edition*
GUTTORP · Statistical Inference for Branching Processes
*HAHN · Statistical Models in Engineering
HAHN and MEEKER · Statistical Intervals: A Guide for Practitioners
HALD · A History of Probability and Statistics and their Applications Before 1750
HALD · A History of Mathematical Statistics from 1750 to 1930
HALL · Introduction to the Theory of Coverage Processes
HAMPEL · Robust Statistics: The Approach Based on Influence Functions

*Now available in a lower priced paperback edition in the Wiley Classics Library.

*Now available in a lower priced paperback edition in the Wiley Classics Library.

*Now available in a lower priced paperback edition in the Wiley Classics Library.

*MILLER · Survival Analysis, *Second Edition*

MONTGOMERY, PECK, and VINING · Introduction to Linear Regression Analysis, *Third Edition*

MORGENTHALER and TUKEY · Configural Polysampling: A Route to Practical Robustness

MUIRHEAD · Aspects of Multivariate Statistical Theory

MURRAY · X-STAT 2.0 Statistical Experimentation, Design Data Analysis, and Nonlinear Optimization

MYERS and MONTGOMERY · Response Surface Methodology: Process and Product in Optimization Using Designed Experiments

NELSON · Accelerated Testing, Statistical Models, Test Plans, and Data Analyses

NELSON · Applied Life Data Analysis

NEWMAN · Biostatistical Methods in Epidemiology

OCHI · Applied Probability and Stochastic Processes in Engineering and Physical Sciences

OKABE, BOOTS, and SUGIHARA · Spatial Tesselations: Concepts and Applications of Voronoi Diagrams

OLIVER and SMITH · Influence Diagrams, Belief Nets and Decision Analysis

PANKRATZ · Forecasting with Dynamic Regression Models

PANKRATZ · Forecasting with Univariate Box-Jenkins Models: Concepts and Cases

*PARZEN · Modern Probability Theory and Its Applications

PEÑA, TIAO, and TSAY · A Course in Time Series Analysis

PIANTADOSI · Clinical Trials: A Methodologic Perspective

PORT · Theoretical Probability for Applications

POURAHMADI · Foundations of Time Series Analysis and Prediction Theory

PRESS · Bayesian Statistics: Principles, Models, and Applications

PRESS and TANUR · The Subjectivity of Scientists and the Bayesian Approach

PUKELSHEIM · Optimal Experimental Design

PURI, VILAPLANA, and WERTZ · New Perspectives in Theoretical and Applied Statistics

PUTERMAN · Markov Decision Processes: Discrete Stochastic Dynamic Programming

RACHEV · Probability Metrics and the Stability of Stochastic Models

RAO · Asymptotic Theory of Statistical Inference

RAO · Linear Statistical Inference and Its Applications, *Second Edition*

RAO and SHANBHAG · Choquet-Deny Type Functional Equations with Applications to Stochastic Models

RENCHER · Linear Models in Statistics

RENCHER · Methods of Multivariate Analysis

RENCHER · Multivariate Statistical Inference with Applications

RÉNYI · A Diary on Information Theory

RIPLEY · Spatial Statistics

RIPLEY · Stochastic Simulation

ROBERTSON, WRIGHT, and DYKSTRA · Order Restricted Statistical Inference

ROGERS and WILLIAMS · Diffusions, Markov Processes, and Martingales, Volume I: Foundations, *Second Edition;* Volume II: Îto Calculus

ROHATGI and SALEH · An Introduction to Probability and Statistics, *Second Edition*

ROLSKI, SCHMIDLI, SCHMIDT, and TEUGELS · Stochastic Processes for Insurance and Finance

ROSS · Introduction to Probability and Statistics for Engineers and Scientists

ROUSSEEUW and LEROY · Robust Regression and Outlier Detection

RUBIN · Multiple Imputation for Nonresponse in Surveys

RUBINSTEIN · Simulation and the Monte Carlo Method

RUBINSTEIN and MELAMED · Modern Simulation and Modeling

RUBINSTEIN and SHAPIRO · Discrete Event Systems: Sensitivity Analysis and Stochastic Optimization by the Score Function Method

*Now available in a lower priced paperback edition in the Wiley Classics Library.

RUZSA and SZEKELY · Algebraic Probability Theory
RYAN · Modern Regression Methods
RYAN · Statistical Methods for Quality Improvement, *Second Edition*
SCHEFFE · The Analysis of Variance
SCHIMEK · Smoothing and Regression: Approaches, Computation, and Application
SCHOTT · Matrix Analysis for Statistics
SCHUSS · Theory and Applications of Stochastic Differential Equations
SCOTT · Multivariate Density Estimation: Theory, Practice, and Visualization
*SEARLE · Linear Models
SEARLE · Linear Models for Unbalanced Data
SEARLE · Matrix Algebra Useful for Statistics
SEARLE, CASELLA, and McCULLOCH · Variance Components
SEARLE and WILLETT · Matrix Algebra for Applied Economics
SEBER · Linear Regression Analysis
SEBER · Multivariate Observations
SEBER and WILD · Nonlinear Regression
SENNOTT · Stochastic Dynamic Programming and the Control of Queueing Systems
SERFLING · Approximation Theorems of Mathematical Statistics
SHAFER and VOVK · Probability and Finance: It's Only a Game!
SHORACK and WELLNER · Empirical Processes with Applications to Statistics
SMALL and McLEISH · Hilbert Space Methods in Probability and Statistical Inference
STAPLETON · Linear Statistical Models
STAUDTE and SHEATHER · Robust Estimation and Testing
STOYAN, KENDALL, and MECKE · Stochastic Geometry and Its Applications, *Second Edition*
STOYAN and STOYAN · Fractals, Random Shapes and Point Fields: Methods of Geometrical Statistics
STOYANOV · Counterexamples in Probability
STYAN · The Collected Papers of T. W. Anderson: 1943–1985
TANAKA · Time Series Analysis: Nonstationary and Noninvertible Distribution Theory
THOMPSON · Empirical Model Building
THOMPSON · Sampling
THOMPSON · Simulation: A Modeler's Approach
THOMPSON and SEBER · Adaptive Sampling
TIAO, BISGAARD, HILL, PEÑA, and STIGLER (editors) · Box on Quality and Discovery: with Design, Control, and Robustness
TIERNEY · LISP-STAT: An Object-Oriented Environment for Statistical Computing and Dynamic Graphics
TIJMS · Stochastic Modeling and Analysis: A Computational Approach
TIJMS · Stochastic Models: An Algorithmic Approach
TITTERINGTON, SMITH, and MAKOV · Statistical Analysis of Finite Mixture Distributions
UPTON and FINGLETON · Spatial Data Analysis by Example, Volume 1: Point Pattern and Quantitative Data
UPTON and FINGLETON · Spatial Data Analysis by Example, Volume II: Categorical and Directional Data
VAN RIJCKEVORSEL and DE LEEUW · Component and Correspondence Analysis
VIDAKOVIC · Statistical Modeling by Wavelets
WEISBERG · Applied Linear Regression, *Second Edition*
WELSH · Aspects of Statistical Inference
WESTFALL and YOUNG · Resampling-Based Multiple Testing: Examples and Methods for *p*-Value Adjustment
WHITTAKER · Graphical Models in Applied Multivariate Statistics
WHITTLE · Systems in Stochastic Equilibrium

*Now available in a lower priced paperback edition in the Wiley Classics Library.

WONNACOTT and WONNACOTT · Econometrics, *Second Edition*
WOODING · Planning Pharmaceutical Clinical Trials: Basic Statistical Principles
WOOLSON · Statistical Methods for the Analysis of Biomedical Data
WU and HAMADA · Experiments: Planning, Analysis, and Parameter Design
    Optimization
YANG · The Construction Theory of Denumerable Markov Processes
*ZELLNER · An Introduction to Bayesian Inference in Econometrics